Schriftenreihe aus dem Institut für Strömungsmechanik

Herausgeber
J. Fröhlich, S. Odenbach, K. Vogeler

Institut für Strömungsmechanik
Technische Universität Dresden
D-01062 Dresden

Band 8

Stephan Schwarz

An immersed boundary method for particles and bubbles in magnetohydrodynamic flows

TUD*press*
2014

Die vorliegende Arbeit wurde am 30. September 2013 an der Fakultät Maschinenwesen der Technischen Universität Dresden als Dissertation eingereicht und am 23. Januar 2014 erfolgreich verteidigt.

This work was submitted as a PhD thesis to the Faculty of Mechanical Science and Engineering of TU Dresden on 30 September 2013 and successfully defended on 23 January 2014.

Gutachter | Reviewers
Prof. Dr.-Ing. habil. Jochen Fröhlich, TU Dresden
PD Dr. rer. nat. habil. Thomas Boeck, TU Ilmenau

Bibliografische Information der Deutschen Nationalbibliothek
Die Deutsche Nationalbibliothek verzeichnet diese Publikation in der Deutschen Nationalbibliografie; detaillierte bibliografische Daten sind im Internet über http://dnb.d-nb.de abrufbar.

Bibliographic information published by the Deutsche Nationalbibliothek
The Deutsche Nationalbibliothek lists this publication in the Deutsche Nationalbibliografie; detailed bibliographic data are available in the Internet at http://dnb.d-nb.de.

ISBN 978-3-944331-58-4

Verlag der Wissenschaften GmbH
Bergstr. 70 | D-01069 Dresden
Tel.: 0351/47 96 97 20 | Fax: 0351/47 96 08 19
http://www.tudpress.de

Gesetzt vom Autor. | Typeset by the author.
Printed in Germany.

Technische Universität Dresden
Fakultät Maschinenwesen

An immersed boundary method for particles and bubbles in magnetohydrodynamic flows

Dissertation

zur Erlangung des akademischen Grades

Doktor-Ingenieur (Dr.-Ing.)

von

Dipl.-Ing. Stephan Schwarz

Gutachter:	1.	Prof. Dr.-Ing. habil. Jochen Fröhlich
	2.	PD Dr. rer. nat. habil. Thomas Boeck
Tag der Einreichung:		30.09.2013
Tag der Verteidigung:		23.01.2014
Gesamtnote:		*summa cum laude*

Abstract

This thesis presents a numerical method for the phase-resolving simulation of rigid particles and deformable bubbles in viscous, magnetohydrodynamic flows. The presented approach features solid robustness and high numerical efficiency. The implementation is three-dimensional and fully parallel suiting the needs of modern high-performance computing.
In addition to the steps towards magnetohydrodynamics, the thesis covers method development with respect to the immersed boundary method which can be summarized in simple words by "From rigid spherical particles to deformable bubbles". The development comprises the extension of an existing immersed boundary method to non-spherical particles and very low particle-to-fluid density ratios. A detailed study is dedicated to the complex interaction of particle shape, wake and particle dynamics.
Furthermore, the representation of deformable bubble shapes, i.e. the coupling of the bubble shape to the fluid loads, is accounted for. The topic of bubble interaction is surveyed including bubble collision and coalescence and a new coalescence model is introduced.
The thesis contains applications of the method to simulations of the rise of a single bubble and a bubble chain in liquid metal with and without magnetic field highlighting the major effects of the field on the bubble dynamics and the flow field. The effect of bubble coalescence is quantified for two closely adjacent bubble chains.
A framework for large-scale simulations with many bubbles is provided to study complex multiphase phenomena like bubble-turbulence interaction in an efficient manner.

Zusammenfassung

Die vorliegende Dissertation behandelt eine numerische Methode für die phasenauflösende Simulation von festen Partikeln und deformierbaren Blasen in viskosen, magnetohydrodynamischen Strömungen. Der vorgestellte Ansatz zeichnet sich durch solide Robustheit und hohe numerische Effizienz aus. Die Implementierung ist dreidimensional, vollständig parallel und entspricht den Ansprüchen des modernen Hochleistungsrechnens.
Neben den notwendigen Erweiterungen mit Bezug auf die Magnetohydrodynamik widmet sich die Dissertation der Methodenentwicklung bezüglich der Immersed-Boundary-Methode, welche mit den einfachen folgenden Worten zusammengefasst werden kann: "Von festen, sphärischen Partikeln zu deformierbaren Blasen". Die Entwicklungsschritte umfassen die Erweiterung einer existierenden Immersed-Boundary-Methode auf nicht-sphärische Partikel sowie auf sehr niedrige Verhältnisse von Partikel- zu Fluiddichte. Eine umfassende Studie untersucht die komplexe Wechselwirkung zwischen Partikelform, dem Nachlauf und der Partikeldynamik.
Des Weiteren wird der Abbildung deformierbarer Blasenformen Rechnung getragen, d.h. die Kopplung von Blasenform und Fluidlasten wird behandelt. Die Thematik der Blaseninteraktion, welche Blasenkollision und -koaleszenz beinhaltet, wird im Sinne eines Überblicks diskutiert und in der Folge wird ein neues Koaleszenzmodell entwickelt und eingeführt.
Die Dissertation enthält Anwendungen der Methode bezüglich der Simulation des Aufstiegs einer Einzelblase und einer Blasenkette in Flüssigmetall, in denen die Auswirkung eines Magnetfeldes auf die Blasendynamik und das Strömungsfeld beleuchtet wird. Ferner wird der Einfluss der Behandlung der Blasenkoaleszenz am Beispiel zweier nah beieinander aufsteigender Blasenketten quantifiziert.
Die vorliegende Arbeit stellt das Rüstzeug für umfassende Simulationen von Strömungen mit vielen Blasen im großen Maßstab. Komplexe Mehrphasenphänomene wie Blasen-Turbulenz-Interaktion können nun in effizienter Weise untersucht werden.

Contents

1 Introduction and Research Goals

1.1 The simulation of multiphase flows

Multiphase flows with rigid particles, drops and bubbles are of significant relevance in many engineering applications reaching from microfluidic devices as lab on a chip gear to environmental flows like sediment erosion processes in river beds. In metallurgical applications, bubbles are injected into the liquid metal to stir, to refine the melt, and to homogenize the physical and chemical properties of the alloy [74]. In these gas-liquid metal flows, external magnetic fields are used for contactless, electromagnetic flow control [87]. Due to the opacity of the liquid metal, the impact of the magnetic field on the bubbles is still unclear [319].
The topic multiphase flows in itself is of great interest in fundamental research. However, it is also vital for the understanding and optimization of complex industrial applications. Here, the subject multiphase flows is examined with special focus on magnetohydrodynamic, gas-liquid metal flows.
A variety of numerical methods has been developed to study the wide range of multiphase flows [215, 307]. Turbulent dispersed multiphase flows in dilute systems are reviewed in [8] with a specific focus on point particle methods. To clarify the terminology used here, the term particle denotes an element of the dispersed phase, be it a bubble, drop or a rigid body [37]. In point particle simulations, the characteristic length of the particle is smaller than the grid spacing and therefore the fluid-particle interaction needs to be modeled. In contrast, the geometry of the particle, as well as the fluid-particle interaction are directly resolved in phase-resolving simulations. For bubbles, the Navier-Stokes equations are often solved in a one-fluid formulation and the interface between the gaseous and liquid phase is dealt with for example by a level-set [253] or a front-tracking algorithm [275]. In the one-fluid formulation, gas and liquid are treated as a single fluid with spatially varying physical properties while the surface tension force is introduced at the gas-liquid interface. With this approach, the deformation of the bubble is directly accounted for. Bubbles in liquid metals are characterized by high Reynolds numbers, large density ratios and high surface tension and are therefore difficult to handle with classical methods [74, 319]. The large density ratio between the phases yields a high pressure jump at the interface and the surface tension force becomes problematic [275]. Both aspects can lead to numerical issues and reduced computational efficiency.
For rigid particles, the geometry-resolving simulation by means of an immersed boundary method (IBM) has become increasingly popular in recent years [179, 282, 136]. The IBM is an Euler-Lagrange approach and the motion of the solid particle is determined by solving an equation of motion for the rigid body. The particle-fluid-coupling is achieved, e.g., by introducing additional forcing terms to the Navier-Stokes equation to account for the no-slip condition at the particle surface. High numerical efficiency can be obtained by the usage

of an IBM for the simulation of many particles in turbulent flow [297, 283]. It is therefore tempting to use an IBM for the representation of bubbles in contaminated systems, such as liquid metal systems, where a no-slip condition is appropriate at the interface [168, 2].
The development of such an IBM, as a numerical method for the simulation of bubbles in metallurgical, magnetohydrodynamic flows, is the main goal of this thesis.
If research in fluid mechanics is briefly classified into the three phases, 1) method development, 2) gathering of data and physical interpretation, and 3) derivation of lower order models, closures and correlations, this thesis is mainly assigned to the first segment. Throughout the method development, limitations of the approach, possible improvements, and future challenges towards large scale applications are addressed.
The remainder of the introductory chapter is structured as follows. At first, the class of multiphase flows, which is of interest in this thesis, is approached from a physical point of view. The physical parameters describing the multiphase flows that will be considered are introduced in Section 1.1.1 below. The capabilities of the original IBM [136, 134], forming the base of this thesis and chosen to study these flows, are stated briefly and necessary improvements are motivated. In Section 1.2, the governing equations and the original numerical method for their solution are stated. Finally, the detailed goals of the thesis are formulated in Section 1.3.

1.1.1 Physical parameters

This paragraph outlines the physical parameters which characterize the multiphase flows to be studied. Starting from the parameters for a single particle in an unbounded fluid, further non-dimensional numbers are introduced to address multiple particles, bounded geometries and the impact of a magnetic field.

Single particles
The flow of an incompressible fluid is characterized by the Reynolds number, Re, describing the ratio of inertial to viscous forces. A typical example is the flow around a fixed sphere, where Re is based on the free-stream velocity and the diameter of the sphere. The problem of a single, homogeneous particle rising or falling under the effect of buoyancy in a Newtonian fluid, which is at rest at infinity, is described by the Galilei number, G, the particle-to-fluid density ratio, π_ρ and at least one particle-shape parameter, X, [63]. The Galilei number is a Reynolds number based on the gravitational velocity scale as discussed below, and it determines the mean vertical velocity of the body to be used for the definition of Re. Hence, three parameters adequately characterize the single particle problems,

$$Re = \frac{u_{ref}\, L_{ref}}{\nu} , \qquad \pi_\rho = \frac{\rho_p}{\rho_f} , \qquad X = f(\text{shape}) , \tag{1.1}$$

with u_{ref} and L_{ref} being a reference velocity and length scale, respectively. The particle density is denoted as ρ_p, the fluid density as ρ_f, and ν is the kinematic viscosity of the carrier fluid. The viscosity ratio between the phases appears as another parameter for drops and bubbles in clean systems, but it is dispensable for rigid particles or bubbles in contaminated systems [2], which are of interest here. Hence, the Reynolds number characterizes the wake behind the body, the density ratio describes the relation of particle to fluid inertia, and the shape parameter marks the level of geometric anisotropy [63].
The interaction of the fluid flow with the particle motion and shape is one of the key aspects of this thesis. Figure 1.1 was inspired by [63] and provides an explanatory view of

how the physical parameters of (1.1) are used to investigate this interaction. The sketch composes the three-dimensional parameter space spanned by Re, π_ρ and X. A large amount of work has been invested towards the understanding of the dynamics of single particles [168, 63, 113, 37]. Figure 1.1 is used to summarize some of this research and to point to the chapters of this thesis where it is detailed.

The inset Figure 1.1a) shows a regime map of wake forms for fixed spheroids and disks as a function of Re and X modified from [36] and including two wake visualizations from present simulations. Fixed bodies can be interpreted as particles with $\pi_\rho \to \infty$, i.e. they have a very large inertia and do not react to excitations by the flow. In Section 3.4, the wake regimes are discussed in detail and the interplay between the wake and the particle dynamics is investigated by moving towards smaller π_ρ and successively releasing specific degrees of freedom with respect to the particle motion.

Figure 1.1b) displays trajectory regimes of rising and falling spheres as a function of Re and π_ρ modified from [113]. The trajectories comprise vertical, oblique and intermittent paths, as well as a zig-zag regime. Further, the plot indicates the corresponding wake forms where R, $2R$ and $4R$ denotes the number of vortex rings per cycle.

Bubbles can be seen as very light particles with $\pi_\rho \ll 1$. Their shape is governed by Re and the ratio of deforming fluid loads to stabilizing surface tension. The latter is expressed by the Eötvös number, Eo, introduced below. Figure 1.1c) illustrates a regime map of bubble shapes modified from [37, 165]. The aspect ratio of a bubble rarely exceeds a value of four [165]. In the presence of vortex shedding for high Re, the shape of the bubble can become time-dependent. The representation of bubble shapes is studied in Chapter 4.

For illustration, Figure 1.1d) shows the dynamics of falling disks with $X > 10$ modified from [69, 172]. The ordinate is formed by the dimensionless moment of inertia $I^* = I_{disk}/(\rho_f d^5) = \pi/64\, X^{-1}\pi_\rho$, where $X = d/t$ with d the diameter and t the thickness of the disk. For low Re, a steady fall is observed for all I^*. Periodic oscillations without flips are found for low I^* and the path is a roughly vertical. The disk tumbles for high I^*. It continuously drifts sideways while turning end-over-end [69]. In between, a chaotic regime is observed where the disk glides sideways and flips with irregular time lags. Particles of high aspect ratio are not considered in this thesis, but some similarities are also apparent for moderately anisotropic ellipsoids.

The blue-shaded area in Figure1.1 designates the region of applicability of the original IBM [136] for spheres, $X = 1$, with $\pi_\rho \gtrsim 0.4$. The zig-zag regime of spheres is thus is not accessible by the original IBM. The original method is discussed in Section 1.2 below. The limitation in terms of Re is indistinct and it is basically only given by the available computational resources and the chosen parallelization of the code addressed briefly in Section 3.1.3.

For spheroidal particles, the length scale is given by the sphere-volume equivalent diameter $L_{ref} = d_{eq} = \sqrt[3]{6V_p/\pi}$, where V_p is the particle volume. The gravitational velocity scale, u_g, is an appropriate choice for the reference velocity, u_{ref}, of a single spheroidal particle freely ascending or settling under gravity g. This reference velocity is given by $u_{ref} = u_g = \sqrt{|\pi_\rho - 1|\, g\, d_{eq}}$. The Reynolds number based on u_g and d_{eq} is denoted as Galilei number (Archimedes number) and is defined as

$$G = \frac{\sqrt{|\pi_\rho - 1|\, g\, d_{eq}^3}}{\nu} = \frac{u_g\, d_{eq}}{\nu} = Re_g\,. \tag{1.2}$$

It characterizes the ratio of buoyancy to viscous forces and is known a priori in experiments

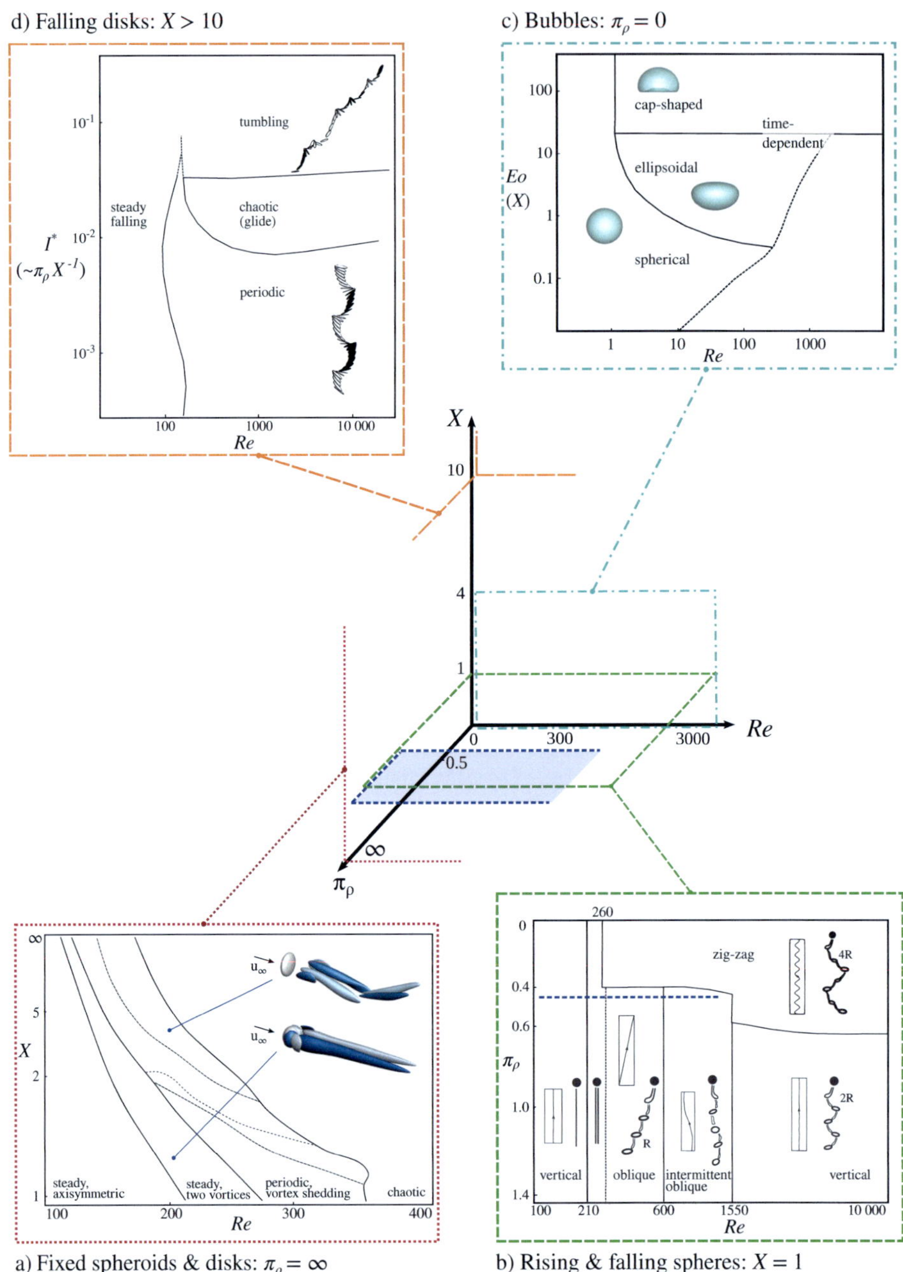

Figure 1.1 Parameter space for single particles spanned by Reynolds number, Re, density ratio, π_ρ, and shape parameter, X, [63]. a) Fixed spheroids and disks. b) Rising and falling spheres. c) Bubble shapes. d) Falling disks. All figures have been modified from the literature. In particular, a) the regime borders relate to [36], b) the sketches and regimes are adopted from [113], c) the curves are digitized from [37], and sub-figure d) is a concentrate of [69, 172]. The blue-shaded area shows the region of applicability of the original IBM [136] for spheres with $\pi_\rho \gtrsim 0.4$.

on free settling or ascent. For these kinds of problems, the Reynolds number based on the average vertical velocity is an outcome of the experiment.
For bubbles, the shape parameter X can be correlated to the Eötvös number, Eo, as the ratio of buoyancy force to surface tension force,

$$Eo = \frac{|\rho_f - \rho_p|\, g\, d_{eq}^2}{\sigma}, \tag{1.3}$$

where σ denotes the surface tension.
The governing non-dimensional numbers of any specific problem can be derived employing the Buckingham theorem [28]. Additional and alternative parameters can be obtained by the combination of these non-dimensional numbers and the usage of, e.g., length ratios. It is not of interest at this stage to go into the details of other dimensionless numbers, such as the Weber, Ohnesorg, or Stokes number. These are defined later on where used.

Multiple particles
The method development documented in this thesis strives towards an approach capable of dealing efficiently with many rigid particles or bubbles. The fractional volume occupied by the disperse phase governs the degree of particle interaction, such as the rate of inter-particle collisions or the occurrence of bubble coalescence. For the rather dilute and anisotropic systems studied in Chapter 6 and 7, the number of bubbles is parametrized by a local gas volume flux expressed as a non-dimensionalized bubble detachment frequency.

Bounded geometries
With the method used throughout the thesis, the outer geometry of the computational domain is cuboidal. The lengths are grouped in a list and are expressed as a multitude of the characteristic length, $\mathbf{L} = (L_x,\, L_y,\, L_z)\, L_{ref}$. In the presence of outer walls, the above statement provides the confinement ratio(s). Other relevant lengths, like for example the position of the bubble injection nozzle in Chapter 6, are also expressed in terms of the reference length scale.

Magnetic field
When a static, external magnetic field is applied and the magnitude of this field dominates the flow-induced field in the electrically conducting fluid, a single further parameter describes the relative strength of the electromagnetic forces. Further details are provided in Chapter 2. The Hartmann number, Ha, is traditionally used for channel flows,

$$Ha = B\, L_{ref} \sqrt{\frac{\sigma_e}{\mu_f}}, \qquad N = \frac{\sigma_e\, B^2\, L_{ref}}{\rho_f\, u_{ref}} = \frac{Ha^2}{Re}, \tag{1.4}$$

where Ha represents the ratio of electromagnetic forces to viscous forces [43, 139]. Alternatively, the magnetic interaction parameter (also known as Stuart number), N, can be employed, which is often done when dealing with immersed objects. A combination of Ha and Re yields N according to (1.4), and N is the ratio of electromagnetic forces to inertial forces [139, 319]. The electric conductivity is denoted by σ_e and the dynamic viscosity by μ_f. Immersed objects are assumed to be perfectly insulating. In the studies of this thesis, homogeneous magnetic fields $\mathbf{B}$ are applied with constant magnitude B.

1.2 Immersed boundary method and multiphase code PRIME

This section is dedicated to the statement of the original method that forms the basis of this thesis. The method represents an Euler-Lagrange approach for the phase-resolving simulation of particulate flows. The Navier-Stokes equations of an incompressible fluid are solved and an immersed boundary method (IBM) is employed for spherical, rigid particles. The method was reported by Uhlmann in [282], and was successively improved by Kempe and Fröhlich [136, 135, 134]. First, the governing equations are given in continuous form. Then the numerical implementation into the code PRIME (Phase-Resolving SIMulation Environment) is stated for future reference.

1.2.1 Continuous phase

The unsteady three-dimensional Navier-Stokes equations are solved for a Newtonian fluid of constant density, which represents the continuous or carrier phase in the multiphase context. The continuity equation and the momentum equations read

$$\nabla \cdot \mathbf{u} = 0\,, \tag{1.5}$$

$$\frac{\partial \mathbf{u}}{\partial t} + (\mathbf{u} \cdot \nabla)\,\mathbf{u} = -\frac{1}{\rho_f}\nabla p + \nu \nabla^2 \mathbf{u} + \mathbf{f}\,, \tag{1.6}$$

with the usual nomenclature $\mathbf{u} = (u,\, v,\, w)^T$ being the velocity of the continuous phase along the Cartesian coordinates x, y, z. Furthermore, p is the pressure and $\mathbf{f}$ a volumetric force. The latter is of specific interest as it carries the coupling information to the disperse phase. The coupling between the embedded particle and the fluid is realized by introducing additional volume forces $\mathbf{f}$ in the right hand side of (1.6). These volume forces are localized at the particle surface and impose a no-slip condition between the phases [282, 136, 179]. Moreover, $\mathbf{f}$ will also be extended to assimilate the Lorentz force in magnetohydrodynamics as discussed in Chapter 2.

1.2.2 Disperse phase

The motion of a rigid particle (index p) can be described by a translation of its center of mass $\mathbf{x}_p$ and the rotation of the solid-fluid interface around $\mathbf{x}_p$. For a single spherical particle, the motion is obtained by solving the Newtonian linear and angular momentum equations

$$m_p \frac{d\mathbf{u}_p}{dt} = \rho_f \oint_S \boldsymbol{\tau} \cdot \mathbf{n}_S \, dS + V_p(\rho_p - \rho_f)\,\mathbf{g} + \mathrm{F}_{col}\,, \tag{1.7}$$

$$I_p \frac{d\boldsymbol{\omega}_p}{dt} = \rho_f \oint_S \mathbf{r} \times (\boldsymbol{\tau} \cdot \mathbf{n}_S)\, dS + \mathrm{M}_{col}\,. \tag{1.8}$$

where $\boldsymbol{\tau} = -\mathbf{I}p/\rho_f + \nu\left(\nabla\mathbf{u} + \nabla\mathbf{u}^T\right)$ is the hydrodynamic stress tensor, with $\mathbf{I}$ being the identity matrix, and p the pressure without its hydrostatic part. The surface integrals thus give the force and torque acting from the fluid on the particle. Here, $\mathbf{u}_p$ and $\boldsymbol{\omega}_p$ designate the linear and angular velocity of the spherical particle, while m_p denotes its mass and $I_p = 2/5\, m_p r_p^2$ its moment of inertia. The vector $\mathbf{n}_S$ denotes the outward-pointing normal vector of the interface S, and $\mathbf{r}$ identifies a point on S with respect to the position of the center of mass of the particle, $\mathbf{x}_p$. Note that the hydrostatic component in the continuous pressure field p has been eliminated in (1.6). Therefore, the buoyancy force appears as a source term proportional to the density difference on the right hand side of (1.7). The forces and moments due to inter-particle and particle-wall collision are denoted as $\mathbf{F}_{col}$ and $\mathbf{M}_{col}$ and their modeling is introduced separately in [135, 134] and Section 7.4. For simplicity, the terms are excluded from the discussion of the fluid-particle interaction until then.

1.2.3 Original method

The original method is outlined in detail in [136, 134]. The original IBM was implemented in the code PRIME and is re-called here in the present notation for completeness and later reference. The equations for the continuous phase are solved on a Cartesian grid with staggered grid arrangement employing a second-order finite volume method. More details on the discrete variable arrangement and the staggered control volumes are provided in Section 2.1. The grid is structured with coordinates $\mathbf{x}_{i,j,k}$ and has a constant grid spacing h in all three directions. Details on the spatial discretization are given in [134] and not reproduced here. Within the IBM, the surface of each individual particle is described using a set of Lagrangian forcing points (index fp). Figure 1.2 illustrates the method and also summarizes the present nomenclature. The transfer between the Cartesian grid and the Lagrangian points is accomplished by regularized delta functions δ_h addressed in Section 3.3.4.

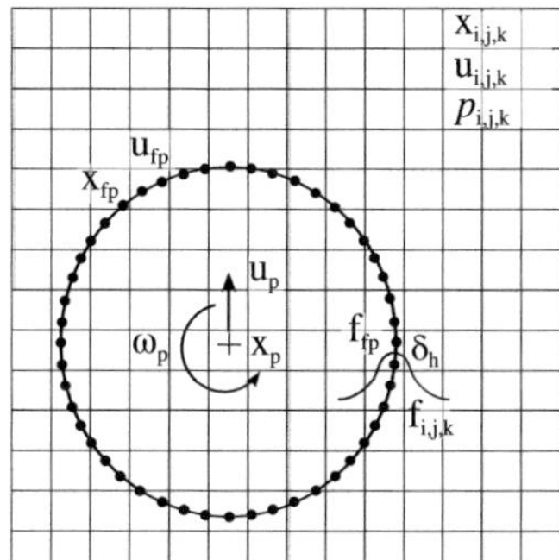

Figure 1.2 Schematic graph of IBM.

A low-storage Runge-Kutta three-step method with implicit treatment of the viscous terms by a Crank-Nicolson scheme is used for time integration. The full algorithm of the original IBM for Runge-Kutta step k reads:

$$\frac{\tilde{\mathbf{u}} - \mathbf{u}^{k-1}}{\Delta t} = 2\alpha^k\,\nu\,\nabla^2\mathbf{u}^{\,k-1} - 2\,\alpha^k\,\nabla\left(\frac{p^{\,k-1}}{\rho_f}\right) - \gamma^k\,\nabla\cdot(\mathbf{uu})^{\,k-1} - \zeta^k\,\nabla\cdot(\mathbf{uu})^{\,k-2}\,, \tag{1.9a}$$

$$\mathbf{u}^i(\mathbf{x}_{fp}^{(l)}) = \sum_{i=1}^{N_x}\sum_{j=1}^{N_y}\sum_{k=1}^{N_z} \tilde{\mathbf{u}}(\mathbf{x}_{i,j,k})\,\delta_h(\mathbf{x}_{i,j,k} - \mathbf{x}_{fp}^{(l)})\,h^3\,, \tag{1.9b}$$

$$\mathbf{f}_{fp}^{(l)} = \frac{\mathbf{u}^d(\mathbf{x}_{fp}^{(l)}) - \mathbf{u}^i(\mathbf{x}_{fp}^{(l)})}{\Delta t}\,, \tag{1.9c}$$

$$\mathbf{f}(\mathbf{x}_{i,j,k}) = \sum_{p=1}^{N_p} \sum_{l=1}^{N_L} \mathbf{f}_{fp}^{(l)} \, \delta_h \left(\mathbf{x}_{i,j,k} - \mathbf{x}_{fp}^{(l)} \right) \Delta V_L \,, \tag{1.9d}$$

$$\nabla^2 \mathbf{u}^* - \frac{\mathbf{u}^*}{\alpha^k \, \nu \, \Delta t} = -\frac{1}{\alpha^k \, \nu} \left(\frac{\tilde{\mathbf{u}}}{\Delta t} + \mathbf{f} \right) + \nabla^2 \mathbf{u}^{k-1} \,, \tag{1.9e}$$

$$\nabla^2 (\delta p)^k = \frac{\nabla \cdot \mathbf{u}^*}{2 \alpha^k \Delta t} \,, \tag{1.9f}$$

$$p^k = p^{k-1} + (\delta p)^k - \alpha^k \Delta t \, \nu \nabla^2 (\delta p)^k \,, \tag{1.9g}$$

$$\mathbf{u}^k = \mathbf{u}^* - 2 \alpha^k \Delta t \, \nabla (\delta p)^k \,, \tag{1.9h}$$

$$\mathbf{u}_p^k = \mathbf{u}_p^{k-1} - 2 \, \alpha^k \, \Delta t \, \frac{\rho_p \, \rho_f}{m_p \, (\rho_p - \rho_f)} \left(\sum_{l=1}^{N_L} \mathbf{f}_{fp}^{(l)} \, \Delta V_L + \mathbf{g} \right) \,, \tag{1.9i}$$

$$\boldsymbol{\omega}_p^k = \boldsymbol{\omega}_p^{k-1} - 2 \, \alpha^k \, \Delta t \, \frac{\rho_p \, \rho_f}{I_p \, (\rho_p - \rho_f)} \sum_{l=1}^{N_L} (\mathbf{x}_{fp}^{(l)} - \mathbf{x}_p^k) \times \mathbf{f}_{fp}^{(l)} \, \Delta V_L \,, \tag{1.9j}$$

$$\mathbf{x}_p^k = \mathbf{x}_p^{k-1} + \alpha^k \, \Delta t \left(\mathbf{u}_p^k + \mathbf{u}_p^{k-1} \right) \,, \tag{1.9k}$$

$$\mathbf{u}^d(\mathbf{x}_{fp}^{(l)}) = \mathbf{u}_{fp}^{(l)} = \mathbf{u}_p^k + \boldsymbol{\omega}_p^k \times \left(\mathbf{x}_{fp}^{(l)} - \mathbf{x}_p^k \right) \,. \tag{1.9l}$$

An intermediate velocity field $\tilde{\mathbf{u}}$ is computed explicitly with (1.9a), which is used for the IBM forcing procedure. This velocity field is interpolated to the Lagrangian forcing points by the regularized delta function (1.9b). A Lagrangian force $\mathbf{f}_{fp}$ is obtained at each forcing point by the direct forcing method (1.9c) from the difference of the desired velocity at this forcing point $\mathbf{u}^d(\mathbf{x}_{fp})$ and the interpolated velocity $\mathbf{u}^i(\mathbf{x}_{fp})$ divided by the time step Δt. Then, $\mathbf{f}_{fp}$ is transferred to the Cartesian grid yielding $\mathbf{f}(\mathbf{x}_{i,j,k})$, where the spreading (1.9d) is realized again by the regularized delta function. The Helmholtz equation (1.9e) is obtained by including the Eulerian force, reformulating (1.9a) with implicit viscous terms and subtracting the resulting equation from (1.9a) [136]. Note that, the resulting velocity field $\mathbf{u}^*$ does not fulfill the solenoidal constraint. A pressure projection step is conducted by solving the Poisson equation (1.9f), correcting the pressure field (1.9g) and computing the divergence-free velocity field $\mathbf{u}^k$ (1.9h). The particle momentum equations (1.7) and (1.8) are discretized and advanced in time employing the assumption of rigid body motion to the particle interior [282]. This yields (1.9i) and (1.9j) which are solved for the new particle velocities $\mathbf{u}_p^k$ and $\boldsymbol{\omega}_p^k$. The particle position is updated (1.9k), while the change of orientation is irrelevant for spherical particles. Finally, the velocity at the particle surface is determined by (1.9l) to be employed in the forcing procedure in the next Runge-Kutta step. The Runge-Kutta coefficients used here are [258, 136]:

$$\begin{aligned} \alpha^{k=1} &= 4/15\,, & \alpha^{k=2} &= 1/15\,, & \alpha^{k=3} &= 1/6\,, \\ \gamma^{k=1} &= 8/15\,, & \gamma^{k=2} &= 5/12\,, & \gamma^{k=3} &= 3/4\,, \\ \zeta^{k=1} &= 0\,, & \zeta^{k=2} &= -17/60\,, & \zeta^{k=3} &= -5/12\,. \end{aligned} \tag{1.10}$$

The overall scheme is very efficient and robust. Since the fluid density is spatially constant throughout the entire Cartesian grid, a constant coefficient Poisson equation is obtained without a marked pressure jump. The forcing procedure is only conducted at the particle surface and an artificial fluid motion develops inside the particle. Even though not stated explicitly here, the method is fully parallelized to suit the needs of modern high performance

computing [281, 134].
Several modifications to the original method from [282] were applied, and three of them are explained briefly here with more detail provided in [136, 135]. The first modification improves the accuracy of the boundary condition at the particle surface. Additional forcing loops are introduced in an efficient way excluding multiple solutions of the Poisson equation as discussed in [136]. The above scheme was modified as follows:

0) Compute once (1.9a) to (1.9e) and set $\mathbf{u}^{(m=0)} = \mathbf{u}^*$.

`Do` $m = 1, n_f$

1) Perform interpolation (1.9b), forcing (1.9c) and spreading (1.9d).
2) Conduct a velocity correction $\mathbf{u}^{(m)} = \mathbf{u}^{(m-1)} + \Delta t\ \mathbf{f}^m(\mathbf{x}_{i,j,k})$.

`end do`

3) Set $\mathbf{u}^* = \mathbf{u}^{(n_f)}$, solve the Poisson equation for the pressure correction (1.9f) and continue the algorithm (1.9a).
4) In the particle momentum equations (1.9i) and (1.9j), the force and torque acting on the particle are determined using the information from 0) and all additional loops, i.e. employing $\sum_0^{n_f} \mathbf{f}_{fp}^{(l)}$.

With the additional forcing, the error in the no-slip boundary condition is reduced. The excellent computational performance is retained as the costly solution of the Poisson equation is excluded from the iterative procedure.
The second improvement documented in [136] involves a modification of particle momentum equations (1.9i) and (1.9j). The assumption of a rigid body rotation for the 'artificial' fluid inside the particle is removed enabling particle-to-fluid density ratios smaller than unity ($\pi_\rho \gtrsim 0.4$). This is revisited in more detail in Section 3.2.3. Finally, a very elaborate collision model for spherical particles was developed in [135, 134]. It allows for the numerically efficient treatment of oblique collisions of spherical particles in viscous flow and is discussed in Section 7.4. This IBM provides a very effective framework for the simulation of many spherical particles in bounded geometries.

1.3 Research objectives

The original IBM stated in Section 1.2 is specifically tailored for the efficient, phase-resolving simulation of multiphase flows with many rigid particles. It allows for particle-to-fluid density ratios slightly smaller than unity. The particle shape is restricted to a spherical geometry. The ultimate goal of this thesis is the application of a variant of this IBM towards bubbles in liquid metals and the study of the impact of magnetic fields. To achieve this, numerous improvements of the IBM are necessary. Apart from the changes in methodology, the underlying physics shall also be studied. The scenarios dealt with range from a single rigid particle to a flow with many interacting bubbles. The thesis has been structured as follows:

- Chapter 2 is dedicated to the extension towards magnetohydrodynamics (MHD). The Navier-Stokes equations are altered by the Lorentz force extension and coupled to the electromagnetic equations. A rigorous validation of the MHD implementation is conducted and the main effects of a magnetic field on laminar and turbulent channel flows are characterized. Furthermore, immersed insulating objects are accounted for.

- Chapter 3 deals with the evolution of the IBM towards a wider range of applicability. The method is modified to account for the orientation and motion of non-spherical particles and for very light particles. The flow around spheroidal particles is studied with and without magnetic field, and the physical mechanisms leading to path oscillations, like the zig-zag motion of a bubble, are examined.

- In Chapter 4, an approach is introduced to represent deformable bubbles accurately and efficiently. The bubble shape is described analytically and it is determined from the unsteady fluid loads.

- Chapter 5 presents an extensive study on the rise of a single bubble in an aligned magnetic field in liquid metal. The interplay between the bubble wake, the bubble dynamics and the field is analyzed.

- A bubble chain exposed to an aligned magnetic field is examined in Chapter 6. The disperse phase statistics and the induced global flow field in the mold are compared with and without magnetic field. This gives implications towards electromagnetic flow control.

- Chapter 7 provides a view on bubble interaction comprising collision and coalescence. The computation of the inter-particle distance is addressed for complex particle shapes. Also the current status towards bubble collision modeling is surveyed. A new coalescence model is developed and applied to quantify the impact of bubble coalescence.

A much more versatile numerical method is proposed which retains the numerical efficiency of the original IBM. Chapter 8 summarizes the modifications and findings and further research directions are suggested.

2 Magnetohydrodynamics

2.1 Physical and numerical model for MHD

The present work was partly conducted in the collaborative research center 609 (*Sonderforschungsbereich, SFB 609*) entitled "Electromagnetic Flow Control in Metallurgy, Crystal Growth and Electrochemistry". This framework inter-connected more than twenty research projects and existed over a period of twelve years. All of these projects had the common ground of magnetohydrodynamics (MHD) studied by both experiments and numerical simulations and reaching from fundamental research to industrial applications. A large deal of the research in this field is reviewed and gathered in a special topics issue of the European Physical Journal [87].

The mathematical backbone of MHD is built upon the Navier-Stokes Equations covering the hydrodynamics and the Maxwell Equations describing electromagnetism. The coupling of both fields of physics is realized via the Lorentz force. A general access to MHD is provided for instance by the textbooks [43, 182], and the historical evolution of MHD and future trends are reviewed in [229].

A categorization of the vast number of phenomena related to MHD can by achieved by means of the magnetic Reynolds number, $R_m = \mu_0 \sigma_e L\, u$, where μ_0 is the magnetic permeability of free space, σ_e is the electric conductivity and L and u are a characteristic length scale and velocity scale, respectively. High magnetic Reynolds numbers, $R_m \gg 1$, relate to solar and astrophysical processes, as for instance the generation of Alfvén waves [3]. Another example are problems associated with the earth's magnetic field addressed, e.g., by experiments and numerical studies of the geodynamo [88, 261].

Basically all terrestrial MHD applications fall in the range of low magnetic Reynolds number, i.e. $R_m \ll 1$. Here, only a few examples from this extensive field of MHD applications shall be provided for illustration. These references originate from the research gathered in the SFB 609. Stirring and mixing of liquid metals can be achieved effectively and contactless by magnetic fields and was studied numerically and experimentally in [262, 263, 218]. Further contents within the electromagnetic processing of materials are casting and solidification which were, e.g., studied in [271, 177, 255]. Liquid metal two-phase flows [74, 244, 320] will be dealt with in more detail in Chapter 5 and Chapter 6. Electromagnetic flow control is a key aspect to influence the crystal growth process [105, 76] which furnishes the materials for the semiconductor industry. Besides the liquid metal applications also the flow in low-conducting fluids can be altered by magnetic fields. One interesting field is electrochemistry, where magnetic fields show a substantial impact on the evolution of hydrogen bubbles during electrolysis [117, 301] or on the electrodeposition process [192, 191].

For low magnetic Reynolds numbers, the induced magnetic field is negligible compared to the applied, external magnetic field. The mathematical formulation simplifies substantially compared to the full MHD equations. The resulting equations and their numerical treatment are outlined in the next paragraph.

2.1.1 Lorentz force

For an electrically conducting fluid moving in a magnetic field, the Lorentz force appears as an additional volumetric force on the right hand side of the Navier-Stokes equation (1.6). The Lorentz force is determined by

$$\mathbf{f}_L = \frac{1}{\rho_f} \left(\mathbf{j} \times \mathbf{B} \right), \tag{2.1}$$

with $\mathbf{B}$ being the magnetic induction or magnetic flux density, which is often just referred to as magnetic field. In the present work, only temporally constant magnetic fields are addressed. The current density, $\mathbf{j}$, is governed by Ohm's law which reads

$$\mathbf{j} = \sigma_e \left(-\nabla\Phi + \mathbf{u} \times \mathbf{B} \right), \tag{2.2}$$

where the scalar Φ denotes the electric potential. We now discuss the case of constant electric conductivity $\sigma_e = const$. From charge conservation,

$$\nabla \cdot \mathbf{j} = 0, \tag{2.3}$$

follows a Poisson equation for the electric potential

$$\nabla^2 \Phi = \nabla \cdot \left(\mathbf{u} \times \mathbf{B} \right). \tag{2.4}$$

A closed system of equations is given by (2.4) for the electric potential Φ, (2.2) yielding the current density, $\mathbf{j}$, and finally (2.1) provides the Lorentz force $\mathbf{f}_L$ to be included in the Navier-Stokes equation. The numerical treatment of this system and the coupling to the fluid momentum equation are addressed next.

2.1.2 Coupling to Navier-Stokes solver

The Lorentz force extension to the existing flow solver, outlined in Section 1.2 and [134, 136], is developed here. A similar strategy is prosecuted as shown in [145, 94, 152]. There are several ways of coupling the electrodynamic equations to the hydrodynamic equations. Since the introduction of a second Poisson equation is computationally costly, multi-time stepping might be beneficial for some applications. Here, the electrodynamic equations are solved in every sub-step of the Runge-Kutta time-integration scheme. For completeness and comparison to the original method (1.9), the full MHD algorithm is given for one Runge-Kutta sub-step. For the k-th Runge-Kutta sub-step, the following procedure is run through. A preliminary current is computed,

$$\mathbf{j}^* = \sigma_e \left(-\nabla\Phi^{k-1} + \mathbf{u}^{k-1} \times \mathbf{B} \right), \tag{2.5}$$

which does not fulfill the solenoidal constraint (2.3). Thus, then a Poisson equation for the correction of the electric potential $\delta\Phi$ is solved,

$$\nabla^2 \delta\Phi = \nabla \cdot \mathbf{j}^*, \tag{2.6}$$

to obtain a divergence free current,

$$\mathbf{j}^k = \mathbf{j}^* - \nabla \delta \Phi, \tag{2.7}$$

and the electric potential is corrected accordingly

$$\Phi^k = \Phi^{k-1} + \delta \Phi. \tag{2.8}$$

The solution of the Poisson equation is obtained using the same solvers as for the pressure correction (2.12), possibly with a different convergence criterion. These highly efficient solvers are provided by the library *hypre* [65] and incorporate pre-conditioning and multigrid techniques. Finally, the Lorentz force is computed from

$$\mathbf{f}_L^k = \frac{1}{\rho_f} \left(\mathbf{j}^k \times \mathbf{B} \right). \tag{2.9}$$

The Lorentz force is added to the volume force term $\tilde{\mathbf{f}}^k = \mathbf{f}_b^k + \mathbf{f}_L^k$, with $\mathbf{f}_b$ being a given volumetric force distribution, e.g. to drive a channel flow.
A predictor step, also including the pressure gradient, is conducted to compute the intermediate velocity field $\tilde{\mathbf{u}}$,

$$\tilde{\mathbf{u}} = \mathbf{u}^{k-1} + \Delta t \left(2\alpha^k \left[\nu \nabla^2 \mathbf{u} - \frac{1}{\rho_f} \nabla p \right]^{k-1} + 2\alpha^k \tilde{\mathbf{f}}^k - \gamma^k \left[(\mathbf{u} \cdot \nabla) \mathbf{u} \right]^{k-1} - \zeta^k \left[(\mathbf{u} \cdot \nabla) \mathbf{u} \right]^{k-2} \right). \tag{2.10}$$

At this stage, the IBM volume force, $\mathbf{f}_{IBM}^k$, obtained from (1.9d), is incorporated by $\mathbf{f}^k = \tilde{\mathbf{f}}^k + \mathbf{f}_{IBM}^k$ to account for immersed boundaries. A Helmholtz equation, required for the implicit treatment of viscous terms, then yields the non-divergence free velocity field $\mathbf{u}^*$,

$$\nabla^2 \mathbf{u}^* - \frac{\mathbf{u}^*}{2\alpha^k \nu \Delta t} = -\frac{1}{\alpha^k \nu} \left(\frac{\tilde{\mathbf{u}}}{\Delta t} + \mathbf{f}^k \right) . + \nabla^2 \mathbf{u}^{k-1} \tag{2.11}$$

The Poisson equation for the correction of the pseudo-pressure is solved,

$$\nabla^2 \delta p = \frac{\nabla \cdot \mathbf{u}^*}{2\alpha^k \Delta t}, \tag{2.12}$$

The pressure correction is conducted and the divergence-free velocity field is obtained.

$$p^k = p^{k-1} + \delta p - \alpha^k \Delta t \, \nu \nabla^2 \delta p \tag{2.13}$$

$$\mathbf{u}^k = \mathbf{u}^* - 2\alpha^k \Delta t \nabla \delta p \tag{2.14}$$

While the three-step Runge-Kutta scheme is of third order accuracy, the implicit Crank-Nicolson scheme for the viscous terms is only of second order so that overall second order convergence is obtained for the time integration scheme [219, 134, 282]. This order of convergence is retained when considering the MHD extension as shown below. The Lorentz force, and the electrodynamic quantities are computed based on the divergence free velocity field. Using e.g. $\mathbf{u}^*$ leads to significant errors as discussed in the student thesis of Koschichow [142]. In contrast, the usage of $\mathbf{u}^*$ as the basis for the forcing procedure in the IBM also yields low errors in the inner boundary conditions for $\mathbf{u}$ after the pressure correction.
The spatial discretization of the electrodynamic equations is realized by second-order finite

volumes on a staggered Cartesian grid analogous to the hydrodynamic equations [97, 134]. The discretized equations are given in Appendix A. A comparison of collocated versus staggered MHD variable arrangement is provided in [94] where the authors also favor the staggered variant for reasons of numerical stability. Figure 2.1 shows a schematic drawing of a single control volume. The scalar quantities - pressure and electric potential - are saved cell-centered, whereas vector quantities - velocity, current density, volume forces - are stored on the faces of the cell. Only homogeneous magnetic fields are considered in this work. The implementation in PRIME, however, was done in a more general way to allow for spatially varying magnetic fields, as for instance used in [104]. Then the components of the magnetic field are also saved on the faces of the cell. The staggered arrangement can also be interpreted as a shift of the control volumes for, e.g., the velocity components by half the width of the cell in the respective coordinate direction. The pressure component in the momentum balance is then already determined on the faces of this shifted control volume without the need for additional interpolation.

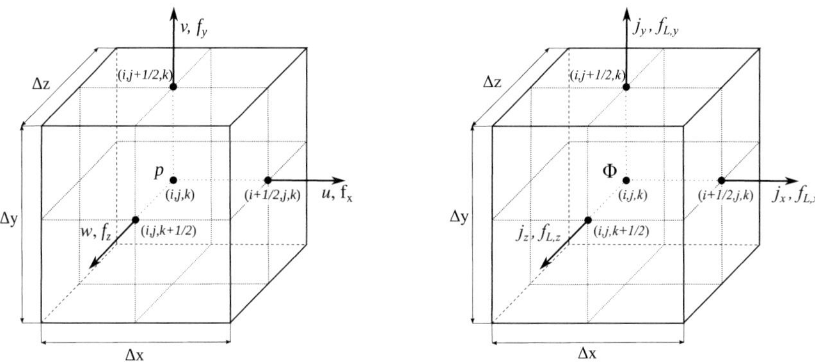

Figure 2.1 Staggered arrangement for hydrodynamic and electrodynamic variables.

The evaluation of the cross product terms to determine the current density (2.2) and the Lorentz force (2.1) requires the interpolation of the respective values to the correct face positions on the staggered grid [94]. In the case of no externally applied current, the net current in the computational domain has to vanish, i.e. $\int \mathbf{j}\, dV = 0$. A non-conservative interpolation leads to an error in this integral formulation and to a parasitic Lorentz force [145]. The used interpolation scheme is based on the procedure developed in [195] and is outlined in Appendix A. For the uniform and mildly stretched grids studied in this work, the present variable arrangement and interpolation strategy provided good conservation properties. Very strong clustering of grid points was not considered here, since also the implemented *hypre*-solvers for the Poisson equation show degraded performance in this case (see [134]-Appendix B). In the case of a cell-centered variable arrangement, a fully conservative scheme for the current density on rectangular collocated grids is provided in [196], and the extension to arbitrary collocated grids is given in [197]. Conservation properties of finite difference schemes on non-uniform meshes are also discussed in detail in [184].

2.1.3 Boundary conditions

Appropriate boundary conditions have to be stated for the solution of the MHD equations. The treatment of the hydrodynamic boundary conditions and their implementation into the code PRIME is discussed in [134] and not repeated in detail here. In PRIME, only cuboidal computational domains are considered. Periodic boundary conditions, inlet and convective outlet conditions, as well as no-slip and free-slip walls are considered for the Navier-Stokes equations.
For the simulations with inflow and outflow conditions presented, e.g., in Section 3.4, an additional correction of the outflow velocity profile was implemented to improve global mass conservation. First, an already very good estimate for the outlet velocity, $\mathbf{u}^0_{out}$, is obtained from the time integration of

$$\frac{\partial \mathbf{u}_{out}}{\partial t} + U_{conv}\frac{\partial \mathbf{u}_{out}}{\partial n} = 0\,, \tag{2.15}$$

where n denotes the direction normal to the boundary pointing outwards. For the convection velocity, U_{conv}, usually a value similar to the reference velocity is chosen. Then an additional mass flow correction is performed by scaling the outlet velocity profile,

$$\mathbf{u}_{out} = \frac{\dot{m}_{in}}{\dot{m}_{out}}\mathbf{u}^0_{out}\,, \tag{2.16}$$

which is then prescribed as the Dirichlet boundary condition $\mathbf{u}|_{out} = \mathbf{u}_{out}$. Additional MPI communication is necessary to obtain the global inlet and outlet mass fluxes, e.g. in x-direction $\dot{m}_{out} = \int_{out} \rho_f u^0_{out} dA$.

The electrodynamic boundary conditions are realized in an analogous fashion. Periodic conditions, insulating walls, as well as a prescribed current density are considered.

Periodic: To account for plane periodicity, the values for $\mathbf{j}$ and Φ are copied from the last cell to the first cell in the respective direction. Periodic conditions are stated in the libraries *PETSC* and *hypre* for the parallel data structure and the solvers.

Insulating walls: The boundary conditions for a non-conducting wall [152] are given by

$$j_n = 0, \qquad \frac{\partial \Phi}{\partial n} = 0. \tag{2.17}$$

To allow for an accurate evaluation of the face-to-face interpolation discussed above at the boundaries, an extrapolation of the non-normal current components to the ghost cells is performed by Lagrangian polynomials [145] (see also Section 3.3).

Dirichlet condition: In order to study the impact of a prescribed external current, a Dirichlet condition for the current can be set on the boundaries. Attention should then be given to global charge conservation. The combined effect of a magnetic field and a prescribed current creating a defined average Lorentz force was studied for instance in [104, 103], where this Lorentz force was used to compensate gravity.

Zero gradient: As a special case of a Neumann condition, the zero gradient constraint was implemented for the current density, $\frac{\partial j_n}{\partial n} = 0$, i.e. the gradient normal to the boundary vanishes.

2.1.4 Parallelization

The parallelization of the MHD equations is based on domain decomposition using one ghost cell in each direction with no additional corner ghost cells. The approach towards large scale computations with the code PRIME is outlined in more detail in [134]. Parallel solvers are incorporated from the library *hypre* [65]. The parallel data structure is provided by the library *PETSc* [9] which is MPI based. A division of the cuboidal domain in again cuboidal subdomains is conducted taking into account proper load balancing by PETSc. In addition to the already fully parallel hydrodynamic solver in PRIME, two further MPI-vectors are introduced storing the current density and electric potential fields. The correction of the electric potential $\delta\Phi$ and correction of the pseudo-pressure δp use the same MPI-vector to save memory.

Only the gradient of the electric potential enters the equations above. Thus the level of Φ itself is not fixed. For reasons of boundedness and to reduce numerical errors, a reference value Φ_{ref} is needed. In the present studies, Φ is initialized in a reasonable way (e.g. all zeros) and during the course of the simulation Φ_{ref} is chosen as the value in the origin of the laboratory coordinate system. In the rare case of expected substantial Φ-oscillations in the origin, Φ_{ref} is set to be the spatial average of Φ over the entire domain computed by global MPI-routines.

Extensive speed-up and scale-up test were conducted for the PRIME code without magnetic field. Excellent performance was measured for up to 16384 cores and problem sizes up to $1.4 \cdot 10^9$ cells on the JUQUEEN machine in Jülich [296].

Table 2.1 provides information on the parallel performance in the case that the MHD equations need to be solved. Two different test cases are considered: The first one is the turbulent MHD channel flow addressed in Section 2.4. When the same problem with $N_{cells} = 16.8 \cdot 10^6$ is computed on four times the number of processes, the computation time per time step is reduced by a factor of 3.88 corresponding to 0.97 of the ideal speedup. The second test case addresses the ascent of a single bubble in an aligned magnetic field of Chapter 5. Using eight times the processes for eight times the computational cells yields a constant computation time per iteration and process. When the number of processes is doubled for the large problem size, the wall-clock time per time step is reduced to half the time (0.98 of the ideal speedup).

Table 2.1 Parallel performance of the MHD solver on two different machines at ZIH Dresden. The problem size is denoted as N_{cells}, the number of processes as N_{proc}, and the computational cost is given as the average wall-clock time per time step.

case	N_{cells}	N_{proc}	wall-clock time
turbulent MHD channel flow	$16.8 \cdot 10^6$	32	8.85 s
(IBM iDataPlex dx360M2)	$16.8 \cdot 10^6$	128	2.28 s
single bubble ascent	$16.8 \cdot 10^6$	16	15.8 s
(SGI Altix)	$134 \cdot 10^6$	128	15.4 s
	$134 \cdot 10^6$	256	7.85 s

In summary, the very good parallel performance of the code is retained when considering a magnetic field. The additional *CPU*-time with magnetic field compared to the case without magnetic field depends very much on the nature of the test case. For the two test cases examined here, the additional cost was only about 20%. It increased to about 60% for other cases where the solution of the Poisson equation for the electric potential required more iterations than that for the pressure.

2.2 Laminar wall-bounded MHD flows

2.2.1 MHD channel flow

In order to validate the implementation of the MHD equations outlined above, several rather simple laminar test cases are conducted. The first case is the so-called Hartmann flow named after the Danish physicist Julius Hartmann (1881-1951), who was the first to conduct experiments with liquid metals in channels in 1937 [99, 100]. The configuration of the simulation features the flow between two infinite parallel plates under the impact of a uniform magnetic field. The simulations of the MHD channel flow are conducted for Hartmann numbers of $Ha = 0$, 10 and 100, where the Hartmann number $Ha = B\,L_y\sqrt{\sigma_e/\mu_f}$ quantifies the influence of the electromagnetic forces compared to the viscous forces. A constant mass flux is applied by adjusting the driving volume force $\mathbf{f}_b$ in time, and the bulk Reynolds number is chosen as $Re_b = 100 = u_b\,2H/\nu$, with H being the channel half-width. Periodic boundary conditions are applied in streamwise direction, no-slip, insulating walls bound the flow in y-direction. A free-slip condition is imposed for the velocity and a zero-gradient condition for the current in spanwise direction.
The resulting one-dimensional problem is discretized in a domain of dimension $L_y = 2H$ with $N_y = 128$ cells, where an equidistant grid as well as a non-uniform grid are considered. In the latter case, the grid is stretched according to a geometric series with a factor of 1.025 away from the walls to obtain a better resolution of the Hartmann layers in case of a wall-normal magnetic field. The Hartmann layers are very thin boundary layers whose thickness scales as $Ha^{-1}L_y$ [189]. The time-step size is adapted to yield $CFL = 0.5$.

First, the case $B = B_y$ is considered, i.e. a wall-normal orientation of the uniform magnetic field. This test serves primarily as a validation of the calculation of the Lorentz force and the boundary conditions. The current has only one non-zero component which is a function of the wall normal coordinate, i.e. $j_z(y)$. The current streamlines run in the positive z-direction in the channel center, close at infinity and return in opposite direction in the Hartmann layers close to the wall [145]. A mean spanwise electric field exists for this case, which is independent of y. This electric field follows from $\int \mathbf{j}\,dV = 0$ and is given by $e_z = -1/2\,B_y \int u\,dy$ [152]. Charge conservation is automatically fulfilled for this case and no Poisson equation for the electric potential needs to be solved.
An analytical solution can be derived [43],

$$u(y) = u_0 \left[1 - \frac{\cosh(Ha\,(y-H)/H)}{\cosh(Ha)}\right], \tag{2.18}$$

where the centerline velocity u_0 and the bulk velocity u_b are related by $u_b \approx u_0\,(1 - 1/Ha)$ for $Ha > 5$ [15, 21].
Figure 2.2 shows the simulation results of the velocity profile $u(y)$ for the different Hartmann numbers. With increasing Ha, the velocity profile becomes flatter and the boundary layers grow thinner. The right graph of Figure 2.2 displays a zoom into the near-wall region and a

comparison to the analytical solution. The agreement is excellent.

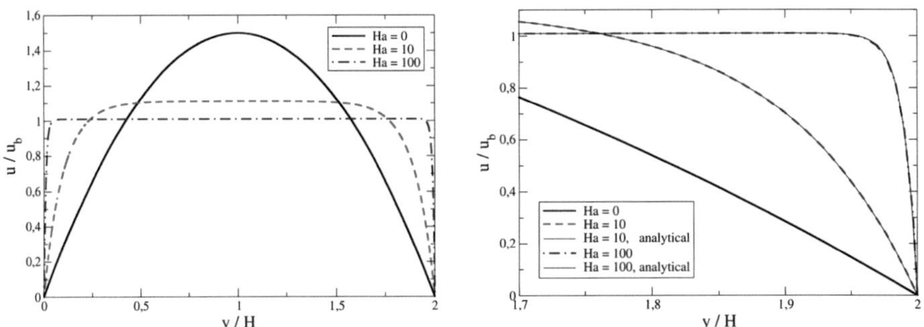

Figure 2.2 Hartmann flow with $B = B_y$. Velocity profiles $u(y)$ for $Ha = 0$, 10 and 100, as well as comparison against analytical solution [43] for closeup into the near wall region (right).

Second, the case $B = B_z$ is considered, i.e. a spanwise orientation of the uniform magnetic field. This test serves as a validation of the solution of the Poisson equation for the electric potential.
The velocity profile remains unchanged by the spanwise magnetic field compared to the purely hydrodynamic parabolic profile. Using the hydrodynamic profile with $dp/dx < 0$ [260],

$$u(y) = -\frac{L_y^2}{2\mu}\frac{dp}{dx}\left(1 - \frac{y}{L_y}\right)\frac{y}{L_y}, \tag{2.19}$$

and $v = w = 0$, the analytical solution for $\Phi(y)$ is derived from (2.2) and (2.3). The analytical solution reads in the present case

$$\Phi(y) = B_z\frac{L_y^2}{2\mu}\frac{dp}{dx}\left(\frac{y}{2} - \frac{y^2}{3L_y}\right)\frac{y}{L_y} + C_\Phi, \tag{2.20}$$

with $C_\Phi = const$. A potential difference builds up between the two plates. The comparison of the present simulation results with the analytical solution is shown in Figure 2.3a) and very good agreement is found.

2.2.2 MHD duct flow

The MHD flow in an infinitely long duct with quadratic cross section is studied. In this case, the gradient of the electric potential and the cross product of the velocity and the magnetic field contribute to the current density field simultaneously. Two simulations of the laminar flow are conducted, considering $Ha = 10$ and $Ha = 100$ for a fixed bulk Reynolds number of $Re_b = 100$. Information on the turbulent counterpart can, e.g., be found in [146]. The general effects of the wall-normal, uniform magnetic field, $B = B_y$, are sketched in Figure 2.3b). The sketch is adopted from [189] and adapted to the present nomenclature. The purely hydrodynamic profile consist of a symmetric parabola. It is altered by the Lorentz force leading to a flatter core and differing shapes of the velocity profiles in y- and z-direction. A current in the positive z-direction is induced in the channel center. Due to the presence of the insulating walls, the current stream lines have to close through the thin Hartmann layers near the y-walls in negative z-direction. A potential difference builds up in the z-direction and so-called side layers form at these wall. The thickness of the side layer scales with $Ha^{-1/2}L_z$ whereas the Hartmann layer scales with $Ha^{-1}L_y$ [189].

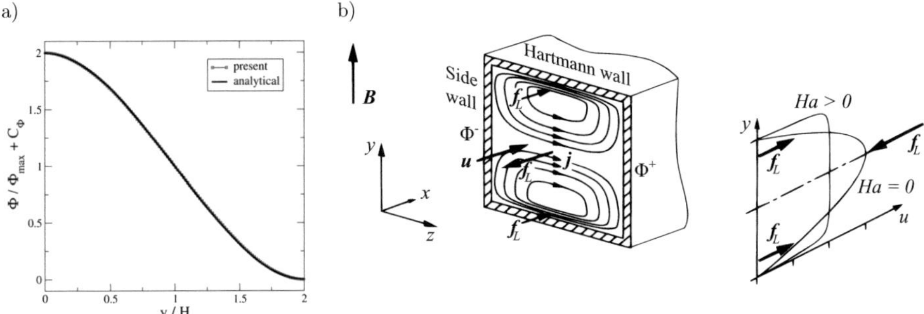

Figure 2.3 a) MHD channel flow with spanwise magnetic field. Comparison of magnetic potential against analytical solution from (2.20). b) MHD duct flow. Configuration and principal effects adopted from [189] and adapted to the present nomenclature.

For the simulation of the two-dimensional problem, a domain of extents $L_y = L_z = 2H$ is discretized with $N_y = N_z = 128$ cells. The numerical grid is stretched according to a geometric series with a factor of 1.025 away from the insulating walls. All other parameters are unchanged compared to the one-dimensional problems studied above.

The simulation results are gathered in Figure 2.4. The electromagnetic quantities are shown by streamlines of the electric current density, $\mathbf{j}$, and a contour of the magnetic potential, Φ, for $Ha = 100$ in Figure 2.4a). An analytical solution for this problem in form of a series expansion is available from [254]. Figure 2.4b) provides a comparison of the present simulation against this analytical solution for both Hartmann numbers studied. The velocity profiles $u(y)$ at $z = H$ and $u(z)$ at $y = H$ agree very well with the analytical data. An illustration of the influence of the magnetic field on the two-dimensional velocity field $u(y, z)$ is shown in sub-figures c) and d) for $Ha = 100$ and $Ha = 10$, respectively. The plots give an impression of the MHD-specific boundary layers and the asymmetry generated by the homogeneous magnetic field.

2.3 Insulating disperse phase

In order to represent non-conducting embedded objects, the electric conductivity is made phase-dependent. In case of an insulating disperse phase, the current stream lines are supposed to circumvent the immersed non-conducting particles and the current cannot penetrate the phase boundary. A phase indicator function α is introduced to distinguish the electric conductivities of the different phases in a very similar fashion as, e.g., in standard front capturing methods [307] like the volume of fluid method (VOF). Ohm's law then reads

$$\mathbf{j} = \sigma_e(\alpha)\left(-\nabla\Phi + \mathbf{u}\times\mathbf{B}\right), \tag{2.21}$$

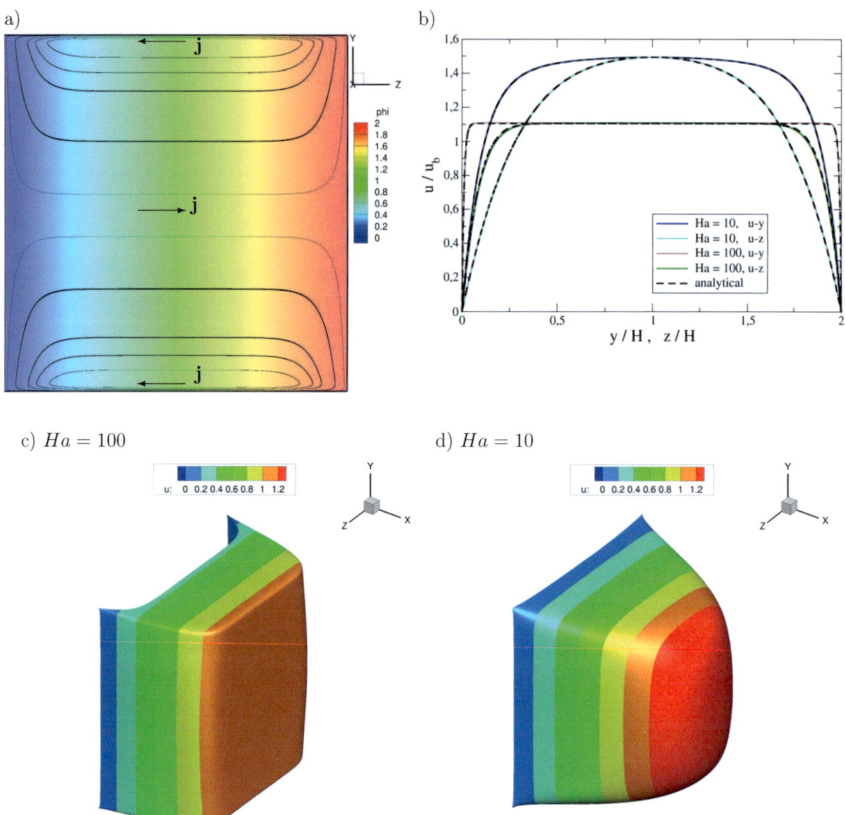

Figure 2.4 MHD duct flow. a) Streamlines of the electric current density, $\mathbf{j}$, and contour of magnetic potential, Φ, for $Ha = 100$. b) Velocity profiles $u(y)$ at $z = H$ and $u(z)$ at $y = H$ for $Ha = 10$ and $Ha = 100$ with comparison to analytical solution [254]. c) and d) Velocity field $u(y, z)$ for $Ha = 100$ and $Ha = 10$, respectively.

with

$$\alpha = \begin{cases} 0, & \text{continuous phase, e.g. liquid metal} \\ 1, & \text{disperse phase, e.g. gas bubble.} \end{cases} \tag{2.22}$$

The phase indicator $\alpha \in [0, 1]$ is obtained from a second order accurate approach for the calculation of the cut cell volumes using a signed-distance level-set representation of the phase interface [136], which is also discussed in Section 3.2.3. In the present work, the disperse phase is assumed to be perfectly insulating, i.e. its electric conductivity vanishes. The employment of electric boundary conditions on the bubble surface can be realized in the context of a magnetohydrodynamic immersed boundary method [94]. With the latter approach, an IBM correction is introduced to Ohm's law in the vicinity of the non-conducting immersed surface, S, to ensure $\mathbf{j}_S = 0$ [94]. In a very similar way, we here impose zero electric current inside the entire bubble employing a phase-dependent conductivity, $\sigma_e(\alpha)$, with $\alpha = 1$ inside

the bubble and $\sigma_e(\alpha = 1) = 0$ [102]. It follows directly from (2.21) that the current density inside the insulating disperse phase is zero. An extension to the more general case of different electric conductivities of the disperse and the continuous phase is not straight forward as the artificial flow field inside the immersed object would also lead to a non-physical current field. In the discrete representation of (2.21), the jump in the electric conductivity is smeared out over one grid cell by the phase indicator function α. The implementation was stable for all simulations presented below as well as in the computations presented in [101, 102, 104]. Only for very fine grids ($d_p/\Delta x \gtrsim 100$) and substantial magnetic interaction $N \gtrsim 1$, local fluctuations occurred directly at the interface. This is a known phenomenon in VOF-like approaches known as spurious or parasitic currents [275]. Smoothing of the interface over more cells and therefore relaxing the gradients at the phase boundary did again yield stable convergence. This was realized by a single application of a smoothing filter with a stencil width of three in all three directions, where the weight of the center cell was chosen as 0.4 and the six neighbor cells of the equidistant grid were weighted with 0.1 similar to the filter in [240].
It should be noted that the representation of immersed insulators by $\sigma_e(\alpha)$ yields a certain discrepancy concerning the wall-tangential current components in comparison to the formulation of the boundary conditions for an insulating wall coinciding with a grid line of the Cartesian grid (Section 2.1.3). Due to the isotropic electric conductivity all current components are zero inside the immersed body and thus reduced in the cut cell volumes. Also the face-to-face interpolation of the current components for the evaluation of the Lorentz force makes use of values from inside the immersed body yielding some error. Nevertheless, a phase-dependent electric conductivity based on a continuous phase indicator is a standard approach when dealing with MHD multiphase flows resulting in good agreement with experimental data as shown for instance by the VOF simulations in [107, 178]. Quantitative access to the error in the electric current near the phase boundary in comparison to an analytical solution is provided in the following sections.
The next paragraphs present several test cases for purpose of validation. They are summarized in Table 2.2.

Table 2.2 Overview of test cases for the insulating disperse phase. The abbreviation BC stands for boundary condition.

I	Current flux around insulating sphere, comparison to potential flow theory
IIa	MHD duct, single phase, $Ha = 10$, equidistant grid
IIb	MHD duct, immersed insulating sphere with only electric BC
IIc	MHD duct, moving, insulating sphere with electric and hydrodynamic BC

2.3.1 Electric current flux around an insulating sphere

The first test (case I in Table 2.2) is a purely electrodynamic one. The current flux around an insulating sphere is studied and the results are compared to the analytical solution obtained from potential flow theory. The primary aim is an estimate of the accuracy, stability and order of convergence in the absence of a flow field and in the case of electrodynamic immersed boundaries.

An insulating sphere of diameter d_p is placed in the center of a cubical domain of size $\mathbf{L} = (12.8,\ 12.8,\ 12.8)\ d_p$. Two equidistant grids are considered with a resolution of $d_p/\Delta x = 10$ and $d_p/\Delta x = 20$, respectively. Dirichlet boundary conditions (BC) are applied for the electric current with the far field assumption $\mathbf{j} = (j_0,\ 0,\ 0)$ on all outer boundaries. The Neumann condition $\partial\Phi/\partial n = 0$ is set for the electric potential on all outer boundaries. The flow solver is switched off, i.e. only the MHD equations are considered to determine the resulting current field and the magnetic potential.
An analytical solution can be derived from potential flow theory by the superposition of a three-dimensional dipole and a constant current j_0 in x-direction [153],

$$\mathbf{j} = j_0\mathbf{e}_x + \frac{j_0 d_p^3}{16}\left(\frac{1}{r^3}\mathbf{e}_x - 3\frac{\mathbf{e}_x\cdot\mathbf{r}}{r^5}\mathbf{r}\right)\ , \tag{2.23}$$

where the vector pointing from the center of the sphere $\mathbf{x}_p$ to the point considered is denoted by $\mathbf{r}$ and its absolute value is r. Note that the analytical solution is unbounded in the center of the sphere. The comparison and the error analysis are therefore restricted to points with $r \geq d_p/2$. We receive the following one-dimensional profiles of the x-component of the electric current for straights intersecting the center of the sphere. Along x, equation (2.23) yields $j_x(x,\ y_p,\ z_p) = j_0\left[1 - d_p^3/\left(8|x - x_p|^3\right)\right]$, and along y one gets $j_x(x_p,\ y,\ z_p) = j_0\left[1 + d_p^3/\left(16|y - y_p|^3\right)\right]$.
The comparison of the simulation results against the analytical solution is shown in Figure 2.5 for the two spatial resolutions studied, and very good agreement is found.

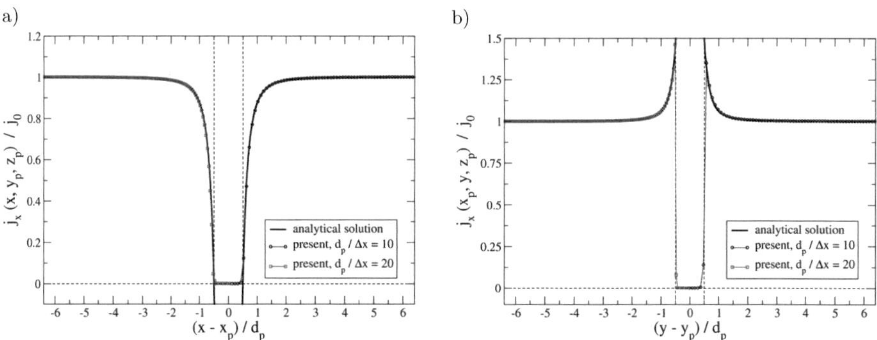

Figure 2.5 Insulating sphere in cross current. Profiles of $j_x(x)$ (b) and $j_x(y)$ (b) plotted along lines through the center of the sphere for the two spatial resolutions $d_p/\Delta x = 10$ and $d_p/\Delta x = 20$, and comparison against the analytical solution (2.23) [153].

The maximum current obtained from (2.23) is given by $j_{x,\max} = j_x(x_p,\ y_p \pm d_p/2,\ z_p) = 3/2 j_0$, in contrast to the flow around cylinder where it is $j_{x,\max} = 2j_0$. The two spatial resolutions considered yield maximum currents near the sphere of $j_{x,\max} = 1.351 j_0$ for $d_p/\Delta x = 10$ and $j_{x,\max} = 1.463 j_0$ for $d_p/\Delta x = 20$. Note that the staggered grid points do not coincide necessarily with the sphere surface. For this test case, the maximum errors occur in the direct vicinity of the sphere. This is to be expected because cut cells of mixed electric conductivity exist at the phase boundary. The current within the sphere is zero. Overall low absolute errors in the current field are observed and a good resolution of the steep gradients in the current field is obtained. Also the tangential current components agree very well with the analytical reference even very close to the sphere. Table 2.6 summarizes the error analysis. Second order convergence in the RMS absolute error compared to the analytical solution is obtained, where the errors were computed from the profile $j_x(x,\ y_p,\ z_p)$.

a)

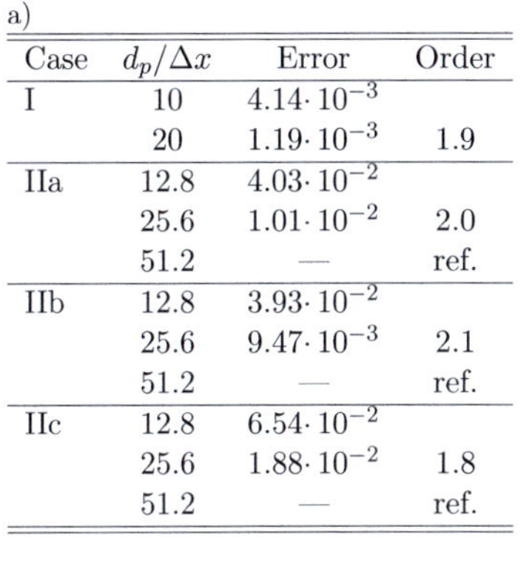

Case	$d_p/\Delta x$	Error	Order
I	10	$4.14 \cdot 10^{-3}$	
	20	$1.19 \cdot 10^{-3}$	1.9
IIa	12.8	$4.03 \cdot 10^{-2}$	
	25.6	$1.01 \cdot 10^{-2}$	2.0
	51.2	—	ref.
IIb	12.8	$3.93 \cdot 10^{-2}$	
	25.6	$9.47 \cdot 10^{-3}$	2.1
	51.2	—	ref.
IIc	12.8	$6.54 \cdot 10^{-2}$	
	25.6	$1.88 \cdot 10^{-2}$	1.8
	51.2	—	ref.

b)

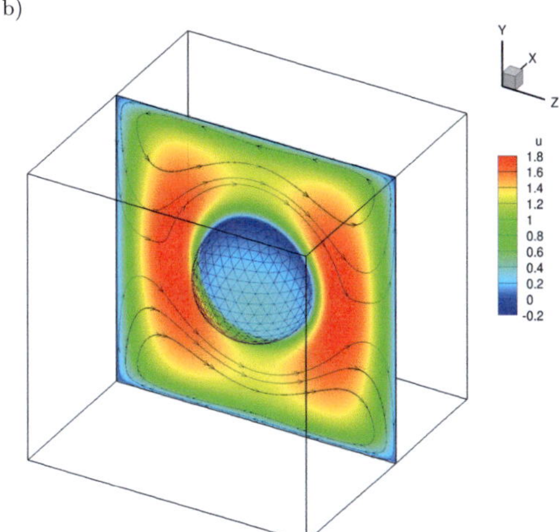

Figure 2.6 Insulating disperse phase. a) Error analysis and convergence study. b) Case IIc - MHD duct flow with immersed, moving, insulating sphere. Displayed are an instantaneous contour of the velocity u, streamlines of the electric current and a surface mesh on the sphere showing a subset of the Lagrangian discretization.

2.3.2 MHD duct flow with an immersed insulating sphere

The second test (case II in Table 2.2) deals with the previously introduced MHD duct flow, but with an additional immersed insulating sphere. The goal is to test the accuracy and the convergence of the method with electric and hydrodynamic immersed boundary conditions under the influence of a wall normal magnetic field. The non-dimensional numbers governing the single phase test case are $Re_b = 100$ and $Ha = 10$ as also studied above. Again, the domain is bounded by insulating walls in y- and z-direction and the uniform magnetic field acts in y-direction. The test case is subdivided in three variants a, b, c and three grids are considered as summarized in Table 2.6. The respective time-step size is chosen to yield $CFL = 0.5$.

Variant IIa is the standard single phase MHD duct flow as above, but on an equidistant grid and with a longer extent of the domain in x-direction. The case is repeated to control the setup and to double-check the order of convergence for the single phase MHD. Additionally, different error formulations are tested using both the analytical solution and the solution on the finest grid, as well as different error measures which will be discussed below.

For variants IIb and IIc, an insulating sphere of $d_p/L = 0.4$ is placed in the center of the cubical domain of extent $L = 1$. In the case IIb, only the electrodynamic immersed boundary conditions are considered. The sphere is embedded with $\sigma_e = 0$, but no hydrodynamic BC

are considered at the phase boundary, i.e. there is no forcing of a no-slip condition at the interface.
In the case IIc, both electric BC and a no-slip condition at the interface are considered. Furthermore, the sphere is constantly rotating and linearly oscillating with $\omega_{p,z} = 1$ and $u_p = 0.2 u_b \sin(2\pi t)$ where $u_b = 1$. A qualitative impression of the test case is given in Figure 2.6. The surface mesh on the sphere shows a subset of the Lagrangian discretization. The plot shows an instantaneous contour of the streamwise velocity u which is forced to a no-slip condition on the moving sphere. The streamlines of the electric current, which are parallel to the z-axis in the single phase case in the channel center (see Figure 2.4a), are distorted to go around the insulator.
When refining the spatial resolution of the staggered grid, the positions of the gridpoints do not coincide. Therefore, the calculation of a numerical error is not trivial and contains an additional interpolation step. Different interpolation strategies and their influence on the determination of the order of convergence were considered in the student thesis of Noack [199]. In the present case, the interpolation of the coarse grid solution to the points on the reference grid is performed accurately using either the regularized delta function (Section 3.3.4), or the Moving Least Squares (MLS) framework with quadratic basis functions (Appendix K). Both interpolations give basically the same results and both are accurate enough to not distort the results of the convergence study. For case IIa and IIb, the RMS absolute error of the velocity component u was obtained in a plane $x = const.$ which intersects the sphere very close to the center and which coincides with a x-position of the Cartesian reference grid. For case IIc, this plane was kept constant in time while the sphere moved and an additional time-averaging was performed over 50 periods of the oscillation.
Initially, a simulation with a spatial resolution of $d_p/\Delta x = 102.4$ had been planned as the reference solution. However, the very steep gradient at the interface led to poor convergence due to local fluctuations in the current and the magnetic potential, the reason for which was addressed above. The fluctuations are damped if the smoothing filter is applied. The solution is then not longer valid as a reference and the results on the $d_p/\Delta x = 51.2$ were used instead. The results for the numerical error and the order of convergence are summarized in Table 2.6. Second order convergence is obtained for test cases IIa and IIb. The original convergence behavior of the flow solver is thus retained when the MHD equations with phase-dependent electric conductivity are considered. A minor reduction of the convergence order is observed when applying the hydrodynamic IBM [136], where a value of 1.8 is computed for the present configuration.

2.3.3 Pseudo compressibility concept

In this paragraph, an alternative way of solving numerically the governing equations for the electric current and the electric potential is described, which can be especially beneficial when dealing with an insulating disperse phase.
A major difficulty for the solution of the incompressible Navier-Stokes equations is the absence of an independent equation for the pressure. Only the pressure gradient contributes to the momentum equation and the pressure is not present in the continuity equation. Usually, a combination of the latter equations, leading to a Poisson equation for the pressure (or a similar quantity), is solved to build up a pressure field and to satisfy the continuity constraint for the velocity at the same time. A different approach is the coupling of velocity and pressure by means of a pseudo or 'artificial compressibility' [35, 201].

This strategy is adopted here for the coupling of the electric current and the electric potential. A weak compressibility, i.e. a temporal change in charge density, is introduced to the charge conservation equation 2.3. The coupled system then reads

$$\begin{aligned} 0 &= \nabla \cdot \mathbf{j} + \frac{1}{\beta}\frac{\partial \Phi}{\partial \tau} \\ \mathbf{j} &= \sigma_e(\alpha)\left(-\nabla\Phi + \mathbf{u} \times \mathbf{B}\right). \end{aligned} \tag{2.24}$$

An iterative solution of (2.24) can now be obtained using a standard time integration scheme in pseudo time τ for each Runge-Kutta sub-step of physical time t.
The values for β and τ have to be configured. They determine the convergence speed and the stability of the method. The corresponding values for the hydrodynamic artificial compressibility are related to the speed of sound [201]. For the electrodynamic case, the expression $\beta\tau\sigma_e/\Delta x^2 \lesssim 1$ might be used as a first orientation.
The approach was tested for all problems presented in this chapter. Good convergence and identical results compared to the approach employing a Poisson solver could be obtained for all steady state test cases presented above, i.e. MHD channel flow, MHD duct flow, stationary insulating sphere cases. Also, good performance concerning the residual divergence of the current density per time step, as well as the computing time per time step could be achieved for smooth temporal and spatial changes in the current and electric potential. For example, the motion and ascent of insulating, spheroidal particles at moderate particle Reynolds number was re-simulated using the 'artificial compressibility' solver and showed very good agreement with the results presented in [245]. For some test cases, even less computing time per time step could be achieved than with the Poisson solver at a similar residual divergence of the current density.
To access the accuracy and the computational performance of the method, the simulation of an insulating sphere in MHD duct flow is presented, corresponding basically to the case IIc above, but for a fixed sphere. All parameters are kept unchanged to the simulations IIc discussed above, except that the sphere is not moving. A spatial resolution of $d_p/\Delta x = 51.2$ and and a time step corresponding to $CFL = 0.5$ are chosen.
The simulations were run in parallel which is a prerequisite in scientific computing. Therefore also the 'artificial compressibility' solver was parallelized based on domain decomposition and MPI. The 'artificial compressibility' solver is here denoted as Φ-Loop and was run with $\beta\Delta\tau/\Delta t = 2\cdot 10^{-2}$. The number of iterations in the pseudo-time loop is indicated with n_τ. For the solution of the Poisson equation, a BiCGStab solver (Biconjugate Gradients stabilized Method) with a PFMG pre-conditioner (Parallel Semicoarsening Multigrid Solver) was implemented from the *hypre* library [65] (additional solver parameters are a tolerance of $5\cdot 10^{-2}$ and a maximum number of iterations of 5).
The simulation is conducted for five flow through times of the very small domain starting from an initialization of the flow field with $\mathbf{u} = (u_b,\ 0,\ 0)$ and all electromagnetic quantities being zero. During the course of the simulation towards a steady state solution, the RMS value of the residual divergence of the electric current density is monitored over time. The results are shown in Figure 2.7. Four different runs are considered: A solution obtained with the Poisson solver, two simulations with the 'artificial compressibility' solver for different numbers of iterations, n_τ, and a combination of Φ-Loops and Poisson solver. The computational costs are given as the average CPU-time per time step in the table of Figure 2.7. All solvers reduce the divergence of the current density over several orders of magnitude from the initial state towards the final solution. Interestingly, the Poisson solver leaves a

larger divergence during the very initial period of time. Afterwards, the Poisson solver and the Φ-Loop with $n_\tau = 100$ yield about the same negative slope in $\text{div}(\mathbf{j})_{RMS}(t)$, whereas the Φ-Loop is substantially more computationally expensive. A lower number of $n_\tau = 20$, resulting in similar computation time, gives a flatter slope during the initial transient at higher $\text{div}(\mathbf{j})_{RMS}(t)$. The slope then becomes steeper at later stages when $\text{div}(\mathbf{j})_{RMS}(t)$ is low. A combination of Φ-Loop and Poisson solver results in the lowest residual divergence at all times with only a moderate increase in computation time. Less Poisson iterations are necessary in the combined mode.

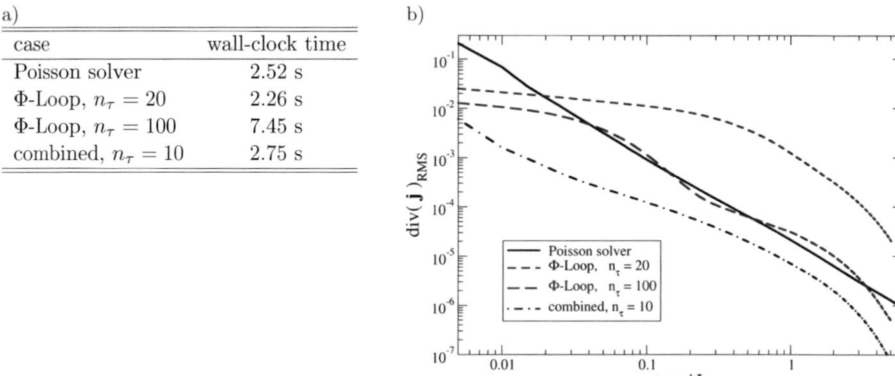

a)

case	wall-clock time
Poisson solver	2.52 s
Φ-Loop, $n_\tau = 20$	2.26 s
Φ-Loop, $n_\tau = 100$	7.45 s
combined, $n_\tau = 10$	2.75 s

Figure 2.7 Comparison of the performance of the Poisson solver and the 'artificial compressibility' solver here denoted as Φ-Loop. a) Computational costs given as average wall-clock time per time step. b) Temporal evolution of RMS residual divergence of the current density field.

Even though the first results for the 'artificial compressibility' solver look quite promising and the method has a few advantages, e.g. clear implementation (no additional libraries), the approach in the present implementation shows inferior performance for turbulent MHD channel flows (Section 2.4). The results do not converge against the reference solution at a maintainable computational effort. This could probably be improved by investing more work into the implementation. For instance, multi-grid approaches could be considered, leading to a faster reduction of the error for long wave lengths and speeding up convergence. Currently, the 'artificial compressibility' is used as an optional pre-conditioner for the MHD-Poisson solver applying only a few Φ-loops. Another important feature is that the pseudo-time loop might be an attractive gain when working on an improved treatment of immersed MHD boundary conditions. In such an MHD-IBM, iterative controller-like correction terms could be introduced similar to the IBM for the velocity.

2.4 Turbulent MHD channel flow

2.4.1 Introduction and problem definition

The objectives of this section are the validation of the solver for the magnetohydrodynamic (MHD) equations and further the study of fundamental effects of a magnetic field on a wall-bounded turbulent flow. These comprise Joule damping, turbulence anisotropy and MHD specific boundary layers [139] and are of significant importance for many metallurgical applications [42], such as continuous casting of steel and aluminum or the stirring of melts.
Turbulent MHD channel flow between infinite, insulating walls is governed by two parameters. The first is the bulk Reynolds number, $Re_b = u_b H/\nu$, where H is half of the channel height, L_y. The bulk or average velocity in streamwise direction is denoted by u_b and ν is the kinematic viscosity. The second parameter characterizes the relative strength of the magnetic field and is either the Hartmann number, $Ha = BH\sqrt{\sigma_e/(\rho_f \nu)} = \sqrt{N\,Re_b}$, or the magnetic interaction parameter, N. Herein, B is the applied magnetic field, σ_e the electric conductivity and ρ_f the fluid density.
MHD effects are studied within the quasi-static approximation [15, 143, 147], i.e. an applied, stationary magnetic field is considered and the induced magnetic fields are negligible at low magnetic Reynolds number. A validation against the full MHD equations is given in [152] for turbulent MHD channel flow with streamwise, spanwise and wall-normal magnetic field, respectively. In general, the quasi-static approximation is valid for basically all technological and laboratory MHD flows. In contrast, high magnetic Reynolds numbers occur in astrophysics, plasma research or geodynamo experiments.
Substantial effort has been put in the simulation of MHD turbulent channel flows, especially by the fluid dynamics group from TU Ilmenau, Germany [15, 52, 147]. Depending on the direction of the magnetic field different MHD effects are encountered.
A *wall-normal* field leads to the formation of Hartmann boundary layers and to possibly quasi-two-dimensional turbulence for increasing field strengths [139]. The Hartmann flow is an eminently interesting configuration concerning the concept of an electromagnetic brake, as the bulk flow is directly affected by the Lorentz force. The stability of Hartmann layers and the transition to turbulence was studied in [144]. A systematic, numerical study of the turbulent Hartmann flow for $Re_b \in (4050,\ 21600)$ and $Ha \in (10,\ 30)$ was conducted in [15] by means of direct numerical simulation with a spectral method. The mean velocity profile develops a characteristic three-layer structure at high Ha comprising a viscous near wall region, a logarithmic layer, and a plateau-like region in the center of the channel. Furthermore with increasing Ha, the wall shear stress increases, the Hartmann layers become thinner, turbulent structures of shrinking size are confined near the walls and the turbulent fluctuations are reduced [15, 145]. The damping effect is more pronounced in the Hartmann configuration than for other orientations of the homogeneous magnetic field [145, 152]. Results obtained with the code PRIME for the Hartmann flow, not shown here, agree well with the results from [15] at moderate Ha. The observations from these simulations back the conclusions by Krasnov et al. [145]: The simulation of turbulent Hartmann flow is computationally very challenging due to very thin Hartmann layers at the wall which require substantial local grid refinement and highly accurate numerical methods. Otherwise, underprediction of especially wall-normal fluctuations and earlier relaminarization are to be expected [239, 145, 140].
Growth and suppression of turbulence affected by a *streamwise* magnetic field was studied in [52] with special emphasis on relaminarization.

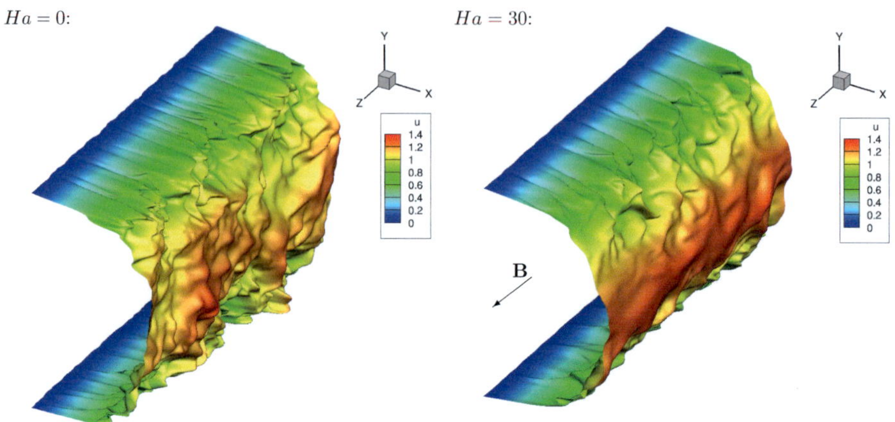

Figure 2.8 Instantaneous flow field u in a plane x = const without magnetic field ($Ha = 0$) and with spanwise magnetic field ($Ha = 30$).

The present study focuses on turbulent channel flow with a homogeneous, *spanwise* magnetic field. For the laminar flow, the parabolic velocity profile remains unchanged under the impact of the spanwise magnetic field as no Lorentz force is generated by the mean flow. In the turbulent case, the modification of the flow is therefore realized only via the influence of the magnetic field on the velocity fluctuations. The transition to turbulence was studied for channel flow with spanwise magnetic field in [143]. The turbulent flow was investigated in detail for $Re_b = 6667$ and 13333 with $Ha \in (0,\ 40)$ in [147], where also the application of LES methods was analyzed. Furthermore, the influence of the numerical approach on the predictions in MHD channel flow was studied in [145] comparing the results obtained with a highly accurate spectral method to two finite difference based methods. One case from this study serves as a reference here to validate the present MHD implementation. The turbulent channel flow with a bulk Reynolds number of $Re_b = 6667$ is investigated. Results for the purely hydrodynamic case, $Ha = 0$, are compared to those from a simulation with an additional uniform spanwise magnetic field with $Ha = 30$. The snapshots of Figure 2.8 give an impression of the flow by means of the instantaneous flow field u in a plane x = const.
The computational domain of extent $\mathbf{L} = (2\pi,\ 2,\ \pi)\, H$ is discretized with a resolution of $\mathbf{N} = (256,\ 256,\ 256)$, where the grid is stretched in the wall-normal direction (y) according to a tangens-hyperbolicus clustering as described in [145]. Equidistant grid spacing is used in streamwise (x) and spanwise direction (z), respectively. Box size and discretization are thus equivalent to those in [145]. The numerical time step is adapted to give $CFL = 1.0$. Periodic boundary conditions are employed in streamwise and spanwise direction and no-slip, non-conducting walls bound the flow in the y-direction. The flow is driven by a spatially constant volume force $f_{b,x}$ which is adjusted in time to yield the prescribed bulk velocity.

2.4.2 Results for a spanwise magnetic field

The applied spanwise magnetic field leads to a suppression of turbulent fluctuations and thus to reduced wall-normal momentum transfer. Consequently, a decrease in wall shear is observed. The velocity in the channel center increases to yield the specified bulk velocity. The turbulence appears to be confined near the walls, and the velocity profile reassumes a laminar-like shape while remaining in a fully turbulent state (Figure 2.8). Table 2.3 lists the friction Reynolds number, Re_τ, as an integral measure of wall shear and the average streamwise velocity at $y = H$. An average will be indicated by angled brackets, $\langle . \rangle$, throughout this thesis. The present results are compared to respective reference data from Krasnov et al. [143, 145], obtained with a spectral method and additionally for the spanwise case by a finite difference method with upwind-biased high-order discretization of the non-linear terms. Excellent agreement is found with and without magnetic field.

$Ha = 0$	Re_τ	$\langle u(y = H)\rangle / u_b$
present	379.45	1.1527
spectral method [143]	381.34	1.1469
$Ha = 30$		
present	305.34	1.2586
spectral method [145]	306.69	1.2542
finite difference - upw [145]	304.43	1.2630

Table 2.3 Comparison of simulation data for $Ha = 0$ and $Ha = 30$ at $Re_b = 6667$ by friction Reynolds number, Re_τ, and average streamwise velocity on the channel center line, $\langle u(y = H)\rangle / u_b$, against data from Krasnov et al. [147, 145]. The reference data with spanwise field was obtained by two different numerical approaches, i.e. a spectral method and upw indicating finite differences with upwind-biased high-order discretization of the non-linear terms.

Figure 2.9 provides the mean velocity profile and statistics for selected turbulent fluctuations in inner and outer scaling. The plots in outer scaling also present the reference data from [145] for the case with spanwise magnetic field and excellent agreement is found for all quantities. The semi-logarithmic plot of the mean streamwise velocity $u^+(y^+)$ shows the typical structure of a turbulent boundary layer for the hydrodynamic case, consisting of the viscous sublayer, the buffer layer, the logarithmic layer and the wake region. The numerical grid resolves the near wall region very well with $y_1^+ \approx 0.2$. In the presence of the spanwise field, the viscous sublayer and the lower buffer layer remain basically unaltered. Viscous stresses dominate in this region over turbulent stresses and the spanwise field only modifies the turbulent fluctuations. The velocity in the channel center is substantially higher than in the hydrodynamic case. In between, there is a log-layer-like region which, however, does not have a fully constant slope [147]. The effect of the magnetic field on the turbulent fluctuations is best judged from the plots in outer scaling as u_b is the same for both cases. The spanwise magnetic field leads to substantial suppression of turbulent fluctuations over the entire height of the channel, where the level of suppression is largest for the wall-normal fluctuations. The general shape of the profiles remains unaltered under the influence of the magnetic field. Maxima exist near the wall and the fluctuations decrease towards the channel center. The location of the maxima does not change significantly when a magnetic field is present. However, simple linear scaling does not yield a collapse of the profiles for $Ha = 0$ and $Ha = 30$. As for the hydrodynamic case, the streamwise fluctuations have the largest magnitude followed by their spanwise counterpart, and the wall-normal fluctuations.

For the hydrodynamic case, the plots in inner scaling show a comparison at similar Re_τ to the spectral DNS of [185] and results obtained by a second-order finite volume code [132, 1]. Very good agreement is found also for the purely hydrodynamic case. Instantaneous

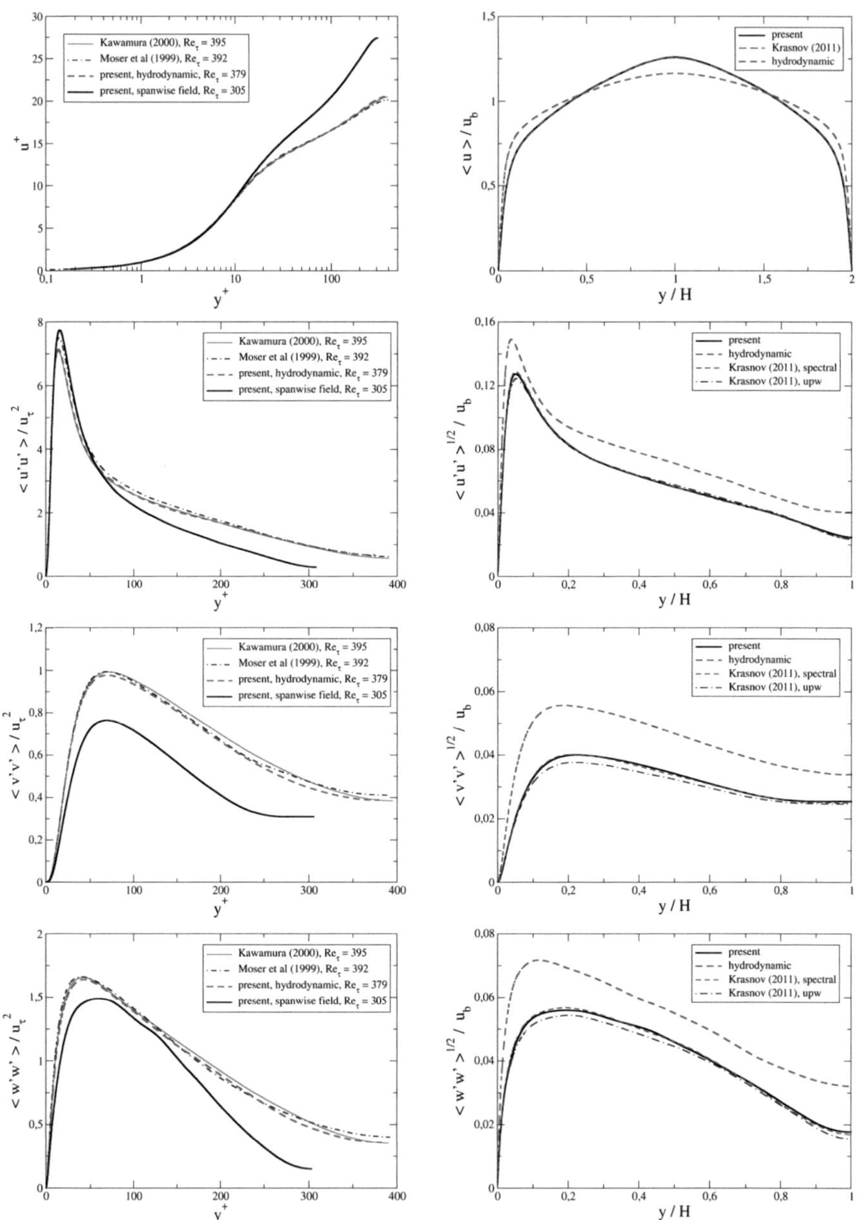

Figure 2.9 Mean velocity profile and turbulent fluctuations for turbulent MHD channel flow. Comparison of the simulation with spanwise magnetic field to the purely hydrodynamic case, and validation against numerical data of [185, 132, 1, 145]. The plots in the left column are in inner scaling, whereas those in the right column are in outer scaling.

contour plots of the normalized turbulent kinetic energy (TKE), $1/2\left(u_i' u_i'\right)/u_b^2$, are shown in Figure 2.10 in the plane $y = 0.05H$, which represents approximately the location of the maxima of the streamwise velocity fluctuations being the dominating contribution to the TKE. Streamwise streaks are observed for both cases, with and without magnetic field. These coherent structures grow substantially in size under the impact of the field and have a lower magnitude. Further, the streaks are less perturbed in spanwise direction and the spacing between them increases [147, 152].

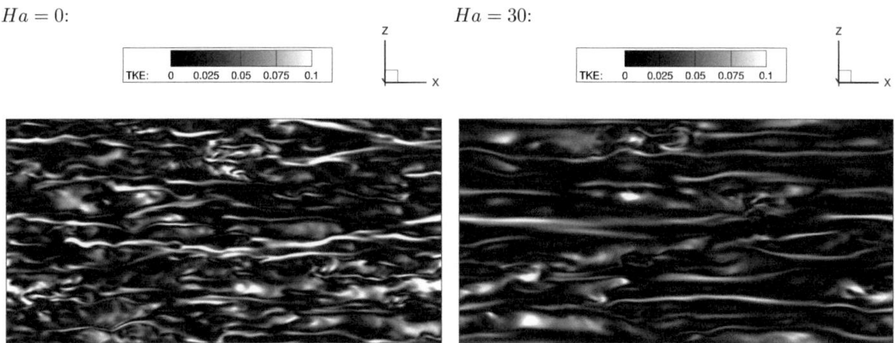

Figure 2.10 Instantaneous flow field by contour of turbulent kinetic energy, $1/2\left(u_i' u_i'\right)/u_b^2$, in the plane $y = 0.05H$.

Energy spectra were obtained for the specified plane, $y = 0.05$, to further assess the influence of the spanwise magnetic field. One- and two-dimensional spectra of the streamwise velocity component, u, will be reported. The averaging procedure for these spectra was conducted for 500 convective time units also utilizing the symmetric data at $y = 1.95H$.

Figure 2.11 provides one-dimensional spectra, E_{uu}, as functions of wave number k_x and k_z with and without magnetic field. The spectra span over at least four orders of magnitude and do not show any notable cut-off effects at the highest wave-numbers and smallest scales, respectively. In streamwise direction, E_{uu} decreases monotonously for all wavenumbers, whereas E_{uu} increases slightly at small wavenumbers, $k_z \lesssim 10$, in spanwise direction and then decays. The maxima in E_{uu} are very similar with and without magnetic field for both directions. The spanwise magnetic field introduces an additional dissipation related to the Lorentz force [29] yielding a steeper slope in the falling part of the spectrum. Consequently, less energy is contained in the smallest scales when a magnetic field is present. The spread in E_{uu} between the hydrodynamic case and the one with magnetic field starts at higher wavenumbers for k_z compared to k_x. The damping effect of a uniform magnetic field is non-isotropic and therefore leads to anisotropy of turbulence [139]. When studying anisotropy one has to distinguish between anisotropy of velocity components and anisotropy of dimensionality, i.e. imparity of characteristic length scales with direction [147]. One- and multi-dimensional spectra provide a useful tool to study anisotropy caused by Lorentz force effects, for instance in homogeneous turbulence [29, 293]. Energy spectra from experiments of MHD duct flow are provided in [57]

Here, anisotropy of dimensionality is addressed by two-dimensional spectra of the streamwise velocity, $E_{uu}^{2D} = \langle q_u\left(k_x, k_z, y\right) q_u^*\left(k_x, k_z, y\right)\rangle$. Herein, q_u denotes the Fourier coefficient of u in

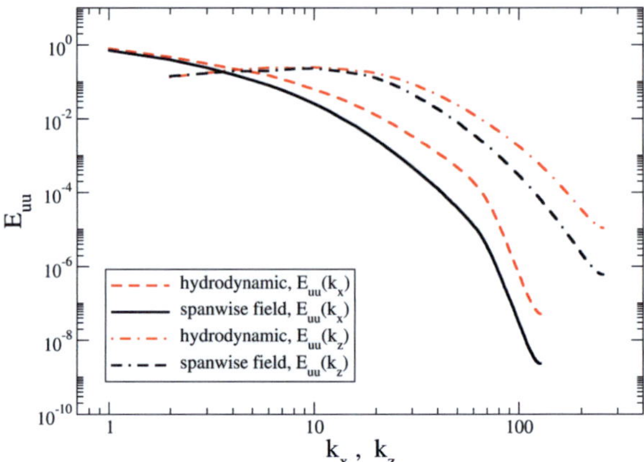

Figure 2.11 One-dimensional spectra, E_{uu}, as functions of wavenumber k_x and k_z with and without magnetic field in the plane $y = 0.05H$.

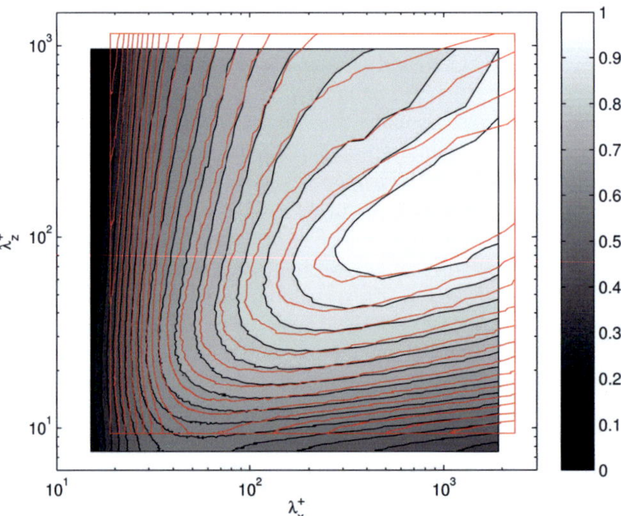

Figure 2.12 Pre-multiplied two-dimensional spectra, $k_x k_z E_{uu}^{2D}$, normalized with respective maximum as functions of wave lengths λ_x^+ and λ_z^+ at $y = 0.05H$, with magnetic field (filled contours in grayscale) and without magnetic field (red lines).

a two-dimensional Fast Fourier Transformation and q_u^* its conjugate complex. Pre-multiplied two-dimensional spectra, $k_x k_z E_{uu}^{2D}$, are displayed in Figure 2.12 at $y = 0.05H$ for the two cases with and without magnetic field. The energy spectra are normalized with the respective maximum and plotted as a function of the two wave lengths, $\lambda_x = 2\pi/k_x$ and $\lambda_z = 2\pi/k_z$, in inner scaling, i.e. normalized with the respective viscous length scale. The procedure of obtaining and evaluating E_{uu}^{2D} is also described in [47]. In this reference, anisotropic scales are compared for purely hydrodynamic turbulent channel flow at $Re_\tau = 180$ and $Re_\tau = 550$. The study was later extended up to $Re_\tau = 2003$ [48, 49, 115]. Mean shear in turbulent channel flow causes anisotropy and wall-normal non-uniformity. Here, we only address spectra obtained near the wall at $y = 0.05H$. First, conclusions for purely hydrodynamic channel flow are summarized. With increasing Re_τ, smaller and more isotropic scales are found, i.e. the iso-lines in the 2D-spectrum become rounder and the logarithmic slope of the u-spectrum becomes steeper and shifts towards the line of two-dimensional isotropy with slope one, $\lambda_x^+ \sim (\lambda_z^+)^1$. Furthermore, the maximum in the two-dimensional plot shifts towards smaller wave lengths. Or in other words, a reduction in Re_τ yields larger, streaky structures, stretched in x indicated by energy maxima at higher wave lengths and lower logarithmic slope.
Two different effects have to be distinguished in the presence of a magnetic field in turbulent channel flow. On the one hand, the spanwise magnetic field yields a decrease of the mean shear which will affect anisotropy in a similar way as discussed above. On the other hand, Joule damping due to the magnetic field acts with a preferential direction also leading to anisotropy as in homogeneous turbulence, i.e. without mean shear. Near the wall, the effect of mean shear dominates as derived in [147]. With a spanwise magnetic field, the reduction of Re_τ yields larger structures characterized by a shift of of the maximum towards larger wave lengths in the two-dimensional u-spectrum of Figure 2.12. However, a steeper logarithmic slope is found for most wave lengths compared to $Ha = 0$ indicating a relative stretching of structures in z, i.e. a stretching along the direction of the magnetic field. An exception are structures with large λ_x^+ and small to moderate λ_z^+. In the purely hydrodynamic case, the sole decrease of Re_τ leads to a decrease in slope [47]. Note that the spectra are cut off at large wave lengths, especially for λ_x^+. This means that very large structures cannot be fully represented by the present size of the computational domain [47]. Here, the box size was chosen according to the reference of [145] to obtain a one-to-one comparison with this data as the primary objective of this study.

2.4.3 Summary for turbulent MHD channel flow

Magnetohydrodynamic turbulent channel flow was studied comparing the purely hydrodynamic case to the flow with an additional uniform, spanwise magnetic field. Excellent agreement was found for the mean flow and the turbulent fluctuations with the respective reference data for both cases. Turbulent statistics, visualizations of the flow, as well as one- and two-dimensional spectra were utilized to illustrate the effects of the magnetic field. These comprise a substantial reduction of turbulent fluctuations yielding reduced wall-normal momentum transfer and therefore reduced wall shear. The velocity in the center of the channel rises compared to the hydrodynamic case. Coherent structures grow in size in both horizontal directions predominantly due to the decrease in Re_τ. The two-dimensional spectra indicate a stretching of structures in the direction of the magnetic field at most wave lengths. Future research should also put the focus on the wall-normal non-uniformity and large scale

turbulence in the channel center affected the most by the Lorentz force. The size of the computational domain of the present study is to small for this purpose. The search for analogies between the effects of a magnetic field and other mechanisms that produce a similar impact should be intensified as started in [147]. For instance, the suppression of turbulent fluctuations in turbulent channel flow with stable density stratification has been studied at various Re_τ in very large computational domains with respect to relaminarization [80].
A further key aspect is the development and improvement of subgrid models for LES, as well as RANS closures [147, 139, 291] taking into account the effects of MHD turbulence.

2.5 Concluding remarks for MHD

The field of magnetohydrodynamics is of growing interest in a large variety of applications covering traditional industries as metallurgy and reaching to modern energy concepts like nuclear fusion reactors. The PRIME code was extended by a very efficient and fully parallel solver to obtain the electric potential and the current density in the low magnetic Reynolds number approximation. A Lorentz force term is introduced into the Navier-Stokes equation representing the influence of the magnetic field on the flow. The implementation was validated for laminar MHD channel and duct flows showing excellent agreement with the analytical solutions. Also the simulation results for the turbulent MHD channel flow with a spanwise magnetic field are in very good agreement with the reference data for both, the mean profile and the turbulent statistics. One- and two-dimensional spectra were used to illustrate the influence of the magnetic field on the flow structures compared to the purely hydrodynamic case, accessing the magnetic damping and the introduction of additional anisotropy.
An insulating disperse phase is represented based on a phase-indicator function and a respective phase-dependent electric conductivity. Sound validation was achieved by comparison with the analytical solution for the electric current around an insulating sphere. For the parallel, multiphase MHD implementation, the computational performance and the order of convergence are retained compared to the purely hydrodynamic method. An alternative solution of the coupled electrodynamic equations based on a pseudo compressibility concept for the electric potential was briefly sketched showing good results for all laminar test cases, but inferior accuracy for the turbulent MHD channel flow. The concept might be beneficial in the future for an improved MHD-IBM, but for the moment it is only kept as an optional pre-conditioner providing some surplus in performance for the solution of the MHD equations.
Summarizing, a proper basis was provided for the study of magnetohydrodynamic, single phase flows, as well as MHD flows with an insulating disperse phase such as bubbles.

3 Evolution of the Immersed Boundary Method

This chapter discusses the evolution of the IBM aiming for a wider range of applicability. The necessary extension with respect to the representation of the immersed object, the IBM forcing procedure and the parallel data structure are outlined in Section 3.1, followed by applications of the modified method to the flow around a complex immersed body and the flow over periodic hills. The method is further extended to account for the orientation and motion of non-spherical particles and for very light particles in Section 3.2 and Section 3.3, respectively. The flow around spheroidal particles is examined in Section 3.4 and the physical mechanisms are discussed that lead to path oscillations of these particles. At last the influence of a magnetic field on the flow around the spheroidal, insulating objects is studied.

3.1 Distribution of forcing points for more general geometries

3.1.1 Surface grid

The original IBM [136] requires an equidistant Eulerian grid of grid spacing, h, as well as an equidistant distribution of Lagrangian markers on the surface of the sphere. Furthermore, the number of Lagrangian forcing points, N_L is pre-specified by the requirement that each forcing point controls an associated volume, ΔV_L, which is equivalent to a finite volume of the Eulerian grid, ΔV_E, and hence $\Delta V_L \approx \Delta V_E = h^3$ [282]. Already the even distribution of points on the rather simple geometry of a sphere is a non-trivial task. The usage of spherical coordinates leads to the agglomeration of points near the poles and a less dense distribution near the equator. The EQ sphere partitioning toolbox [157] employs a recursive algorithm to create zones of equal area on the unit sphere. With this tool, a database of forcing point distributions for the sphere with various N_L was built for the usage within the IBM in PRIME.

For particles with a more complex shape, e.g. ellipsoids, and generic immersed bodies, a procedure had to be established first which provides a suitable forcing point distribution. In most cases, a triangular surface mesh is created employing the commercial grid generator GAMBIT from the ANSYS FLUENT package. Either an advancing-front algorithm or Delaunay triangulation with subsequent mesh adaptation towards an even distribution are used [284]. Figure 3.1a) shows an example for a triangular surface mesh and the respective forcing point distribution obtained for a prolate ellipsoid. The Lagrangian markers are located at the corners of the triangles.

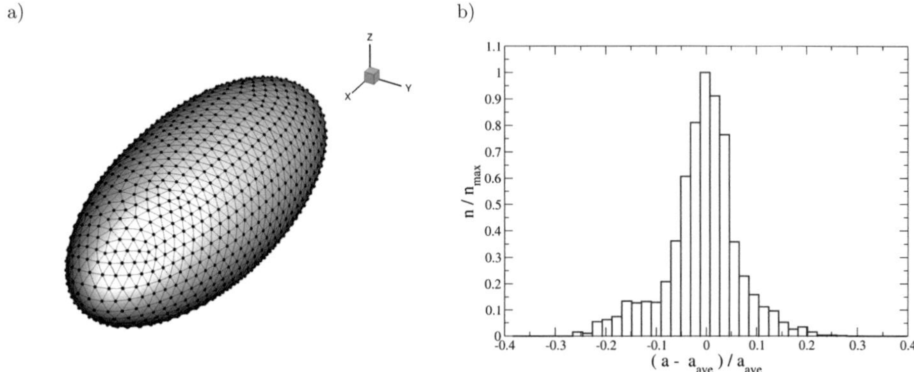

Figure 3.1 Forcing point distribution on the surface for particles of general shape. a) Surface mesh and $N_L = 1109$ forcing points on a prolate ellipsoid. b) Histogram of relative error in distance, a, between forcing points with respect to an equidistant distribution approximated by the average distance, a_{ave}.

The surface mesh is imported into PRIME utilizing the GAMBIT neutral file format. The mesh file contains in the present case, i.a., the number of forcing points, N_L, the number of triangular surface elements, N_{tri}, as header information, and further a list of the unscaled forcing point coordinates with respect to the particle center, $\mathbf{x}_{fp} - \mathbf{x}_p$, as well as a list of the three forcing points, $(fp1, fp2, fp3)$, forming each individual surface element, i.e. the connectivity. A schematic mesh file is given in Table 3.1.

Table 3.1 Schematic surface mesh file for forcing point distribution.

Header			
$N_L = \ldots$	$N_{tri} = \ldots$		
$l = 1$	$x_{fp}^{(l)} - x_p$	$y_{fp}^{(l)} - y_p$	$z_{fp}^{(l)} - z_p$
$\vdots$		$\ldots$	
N_L		$\ldots$	
$n = 1$	$l_{fp1}^{(n)}$	$l_{fp2}^{(n)}$	$l_{fp3}^{(n)}$
$\vdots$		$\ldots$	
N_{tri}		$\ldots$	

Again, a large database of mesh files is created containing mostly oblate and prolate ellipsoids of different aspect ratios for various N_L. Here, $\mathbf{x}_p$ corresponds to the center of mass and is equal to zero in the mesh file, while the small semi-axis is oriented in the y-direction and is chosen as unity. Other geometries comprise cubes and bricks, pipes or closed, finite-height cylinders, conical nozzles, a car model etc.

3.1.2 Forcing point volume

The distribution of forcing points on the surface is not equidistant. Figure 3.1b) gives an estimate of the deviation from an even distribution for a prolate ellipsoid.
Consequently, the spreading of forces in the original IBM needs to be revised to account for the uneven distribution of forcing points and to conserve momentum in the transfer between Lagrangian and Eulerian grid. In the spreading step, the force $\mathbf{f}_{fp}^{(l)}$, which is computed at the Lagrangian point $\mathbf{x}_{fp}^{(l)}$, is distributed to the underlying Eulerian grid locations, $\mathbf{x}^{(i,j,k)}$, yielding the volume force $\mathbf{f}^{(i,j,k)}$. The regularized delta function, δ_h, which is given in Section 3.3.4, provides a smooth transfer and has a stencil width of three Eulerian grid points in each direction [225]. For a single particle, the spreading reads

$$\mathbf{f}^{(i,j,k)} = \sum_{l=1}^{N_L} \mathbf{f}_{fp}^{(l)} \, \delta_h(\mathbf{x}^{(i,j,k)} - \mathbf{x}_{fp}^{(l)}) \, \Delta V_L^{(l)} \,. \tag{3.1}$$

The revision is done by actually computing the associated volume, $\Delta V_L^{(l)}$, for each Lagrangian forcing point, l, from the given point distribution instead of using the identity $\Delta V_L^{(l)} = h^3$. Equality of the total force (and analogously torque) is then again achieved in the Lagrangian and Eulerian reference frame, i.e. for a single particle:

$$\sum_{l=1}^{N_L} \mathbf{f}_{fp}^{(l)} \, \Delta V_L^{(l)} = \sum_{i,j,k} \mathbf{f}^{(i,j,k)} \, h^3 \,. \tag{3.2}$$

The union of all Lagrangian volumes, $\Delta V_L^{(l)}$, forms a thin shell of thickness h around the particle surface [282]. A single forcing point volume shall thus be computed by

$$\Delta V_L^{(l)} = A_{fp}^{(l)} \, h, \tag{3.3}$$

where $A_{fp}^{(l)}$ denotes the area associated with forcing point l. The area of a single, planar triangle n of the surface mesh is obtained by Heron's formula,

$$A_{tri}^{(n)} = \sqrt{s(s - a_\triangle)(s - b_\triangle)(s - c_\triangle)}, \text{ with } s = \frac{1}{2} \left(a_\triangle + b_\triangle + c_\triangle \right), \tag{3.4}$$

where $a_\triangle$, $b_\triangle$, $c_\triangle$ are the side lengths of the triangle, $\triangle ABC$, formed by the three forcing points according to Figure 3.2a).

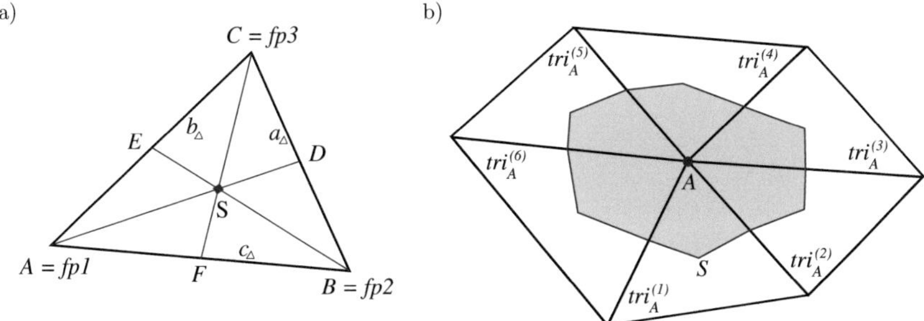

Figure 3.2 a) Single surface triangle. b) Area associated with a forcing point.

The median lines of a scalene triangle intersect each other in the center of mass S of the triangle, as illustrated in Figure 3.2a). Each median line is divided by S in a ratio of $2:1$, whereas the longer part is is located towards the corner point of the triangle. It can then be shown that S partitions the triangle in three equal areas, i.e. $A_{\square EAFS} = A_{\square FBDS} = A_{\square DCES} = \frac{1}{3} A_{\triangle ABC}$.
The area associated with a single forcing point l is then obtained by gathering one third of the area of each triangle sharing this corner point,

$$A_{fp}^{(l)} = \frac{1}{3} \sum_{i=1}^{N_{nt}^{(l)}} A_{tri}^{(i),(l)} , \tag{3.5}$$

where $N_{nt}^{(l)}$ is the number of neighbor triangles associated with this forcing point as depicted in Figure 3.2b). Note that $N_{nt}^{(l)}$ is not a constant for a non-equidistant surface mesh. In the case of a perfectly even distribution, all triangles are equilateral and have interior angles of sixty degrees, leading to $N_{nt}^{(l)} = 6 = const.$ The routines are validated using spherical and ellipsoidal particles and ensuring

$$\sum_{l=1}^{N_L} A_{fp}^{(l)} = \sum_{n=1}^{N_{tri}} A_{tri}^{(n)} \quad \text{and} \quad \lim_{N_{tri} \to \infty} \sum_{l=1}^{N_{tri}} A_{tri}^{(l)} = A_{S,analyt} , \tag{3.6}$$

where

$$A_{S,analyt.} = \begin{cases} A_{sphere} = 4\pi r_p^2 \\ A_{oblate} = 2\pi a^2 + \pi \dfrac{b^2}{\varepsilon_e} \ln \left(\dfrac{1+\varepsilon_e}{1-\varepsilon_e} \right) , & a = c > b \\ A_{prolate} = 2\pi b^2 + 2\pi \dfrac{ab}{\varepsilon_e} \arcsin(\varepsilon_e), & a > b = c \end{cases} \tag{3.7}$$

with $\varepsilon_e = \sqrt{a^2 - b^2}/a$ being the ellipticity [302]. The numerical surface area is always smaller than the analytical solution since the curved surface is approximated by planar triangles.

With the present IBM that accounts for the actual forcing point volume, the prior restriction, $\Delta V_L \approx \Delta V_E = h^3$, is relaxed to $\Delta V_L \lesssim \Delta V_E$. Appendix B provides the formulae relating the required minimum number of forcing points, N_L, to the grid spacing, h, of the Eulerian grid for ellipsoids and the sphere. Hence, multiple forcing points can share a single Eulerian volume. A rigorous validation was performed for the motion of spheroids in quiescent fluid under the action of gravity employing substantially different forcing point numbers N_L. It has been successfully checked that the obtained solutions indeed coincide independent of N_L (not shown). However, if $\Delta V_L \gtrsim \Delta V_E$, equation (3.2) is still fulfilled, but the forcing point distribution on the particle surface is insufficiently dense. The delta function, δ_h, involved in the spreading of the volume force (3.1), has a limited range of three grid points in each direction and hence fails to enforce the boundary condition on the entire surface in the latter case. Therefore, a sufficiently close forcing point arrangement has to be ensured at all times. For more complex geometries, an appropriate surface triangulation has to be created which fulfills $\max(a_\triangle, b_\triangle, c_\triangle) \lesssim h$.

3.1.3 Implementation and parallelization

Implementation

The original implementation of the IBM in the PRIME code [136, 134] was intended for the simulation of turbulent flows with many immersed heavy, spherical particles in cuboidal domains. All extensions to the code are designed as add-ons and the original implementation is recovered by undefining the respective pre-compiler flags. This is also true for all other improvements implemented throughout this thesis. The strategy ensures reproducibility and a certain safety in the code development, but also poses some limitations.
The schematic representation of the data structure for the immersed particles is given in Figure 3.3b). The particle constitutes the highest level in the data hierarchy. It is characterized by its center of mass $\mathbf{x}_p$, the translational velocity $\mathbf{u}_p$, angular velocity $\boldsymbol{\omega}_p$, material properties as the particle density ρ_p, as well as geometry information like the particle radius r_p, etc. The particle surface is represented by the set of forcing points (index fp). Each point holds information about its location $\mathbf{x}_{fp}$ and velocity $\mathbf{u}_{fp}$ in the laboratory system, etc. Each surface marker now also carries its associated area, A_{fp}, to compute the adequate weights in the spreading step as discussed above, as well as further data when necessary.
With respect to more general geometries, the triangular surface element object is added as an entity to each particle. It is included on the same level of the hierarchy as the forcing point since it is not used when dealing with spherical particles and it then can be easily switched off. The tri-elements carry as entities the connectivity ($fp1$, $fp2$, $fp3$), the surface normal vector $\mathbf{n}_{tri}$, etc.

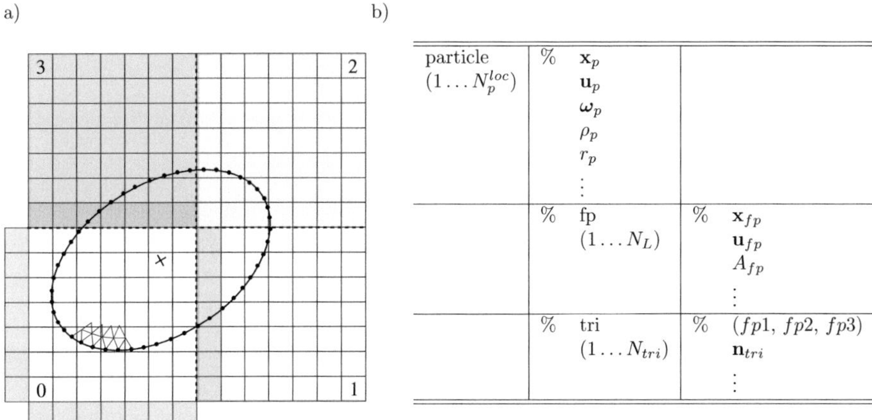

particle $(1 \dots N_p^{loc})$	% $\mathbf{x}_p$ $\mathbf{u}_p$ $\boldsymbol{\omega}_p$ ρ_p r_p ⋮	
	% fp $(1 \dots N_L)$	% $\mathbf{x}_{fp}$ $\mathbf{u}_{fp}$ A_{fp} ⋮
	% tri $(1 \dots N_{tri})$	% $(fp1, fp2, fp3)$ $\mathbf{n}_{tri}$ ⋮

Figure 3.3 a) Sketch of parallelization with domain decomposition for Eulerian grid and master-slave strategy for particles. b) Schematic of used data structure. The %-symbol denotes entities..

If the particle shape is supposed to change in time, as e.g. for deformable bubbles, a stretching of the forcing point distribution is performed in each time-step and the Lagrangian volumes, $\Delta V_L^{(l)}$ are re-calculated. To avoid costly search iterations on the unstructured surface mesh, also an extended connectivity is stored, as displayed in Figure 3.2b), e.g. for forcing point A in the form $\left(N_{nt}^{(A)}, n(tri_A^{(1)}), \dots, n(tri_A^{(N_{nt})})\right)$. The Appendix C lists the formulae for the calculation of the surface normals and curvature based on the triangulation. The surface normal vector is evidently important, e.g., for the collision modeling or the formulation of wall functions, and the curvature is needed to compute the surface tension force for deformable bubbles.

Parallelization

This paragraph and Figure 3.3a) illustrate very briefly the main concept of the parallelization. *Decomposition* of the cuboidal domain is performed for the Eulerian grid. In the example shown, the domain is split in four subdomains carried by the processes 0 to 4. One ghost cell in each direction is used to complete the stencils of the discretization. In the figure, the ghost cells of process 0 are highlighted. The MPI-communication of the ghost cell data is mostly handled by the library PETSc [9].

With respect to the particle-parallelization, a *master-slave* strategy is employed [134, 281]. The process, which contains the particle center $\mathbf{x}_p$, is the master process for this particle. In Figure 3.3a), process 0 is thus the master of the displayed ellipsoidal particle. The master deals with all particle-specific problems, like the particle momentum equations yielding $\mathbf{u}_p$ and $\boldsymbol{\omega}_p$, for a total of N_p^{loc} particles located in its domain. If a particle leaves the local domain, a new master is assigned, the particle lists containing N_p^{loc} are updated, and the particle information is transferred by master-master communication using MPI. For the extended implementation, the concept remained unchanged, but additional communication of e.g. the surface triangulation needed to be included.

If the particle surface extends to a neighbor subdomain, the neighbor process is employed as a slave process. In Figure 3.3a), processes 1, 2 and 3 carry fractions of the particle and are slaves to the master process 0. The master communicates only the necessary data with the slaves, e.g. geometry information or the velocity of the forcing points, to enable the slaves to perform certain operations in their subdomains. These operations comprise, e.g., the velocity interpolation or the spreading of the IBM forces. Further master-slave communication is related to the collision modeling. For details on the actual technical implementation of the master-slave strategy, it is referred to [134, 281]. With respect to the present extension, the detection of the slave processes is modified and the amount of data, that is exchanged with the slave processes, increases to account for the more complex geometry and the adapted forcing procedure. However, the parallel performance is not significantly altered compared to the implementation for the sphere and the high computational efficiency is retained.

With the current particle parallelization, the master can only employ its direct neighbor processes as slaves, which can pose a certain limitation when dealing with very large particles or a very high resolution per particle. For the future, one might therefore consider a parallelization based on a full decomposition of the particle surface mesh. In the case of an anisotropic distribution of particles, one might also consider an equi-distribution of the particles to the processes for an improved load-balance.

3.1.4 Flow around a complex immersed geometry

As a feasibility study, the flow around a car model was simulated employing the immersed boundary method with the present extension to more generic shapes. Figure 3.4a) shows a triangular surface mesh of the car where the grid displayed has $N_{tri} = 41112$ surface elements which is somewhat coarser then the mesh used in the simulation. The entire car is modeled as a single particle at rest. Note that it is straightforward to assemble multiple particles, where each particle can, e.g., rotate or be of time-dependent shape. The computational domain of extent $\mathbf{L} = (10,\, 5,\, 5)\ H_{car}$ is discretized equidistantly with $\mathbf{N} = (256,\, 128,\, 128)$, where H_{car} denotes the height of the car. A uniform inflow boundary condition is set with $u = u_b = 1$ and a convective outflow condition is used at the opposite boundary in x-direction. Periodic conditions are used in y-direction and free-slip walls are imposed in z-direction. The car is positioned with the front-bumper $3H_{car}$ away from the inlet and with the wheels touching the lower wall at $z = 0$. The particle is discretized with $N_L = 80115$ forcing points and $N_{tri} = 160226$ triangles. Three forcing loops are run through within the improved forcing scheme of the IBM [136]. Adaptive time-stepping is employed with $CFL \approx 0.8$. A weakly unsteady flow is observed for a Reynolds number of $Re_H = u_b H_{car}/\nu = 250$, where the flow is rather poorly resolved with the present moderate resolution. An impression of the flow field is given in Figure 3.4b) by a contour of the instantaneous streamwise velocity and pathlines indicating the recirculation behind the windscreen and the car itself.

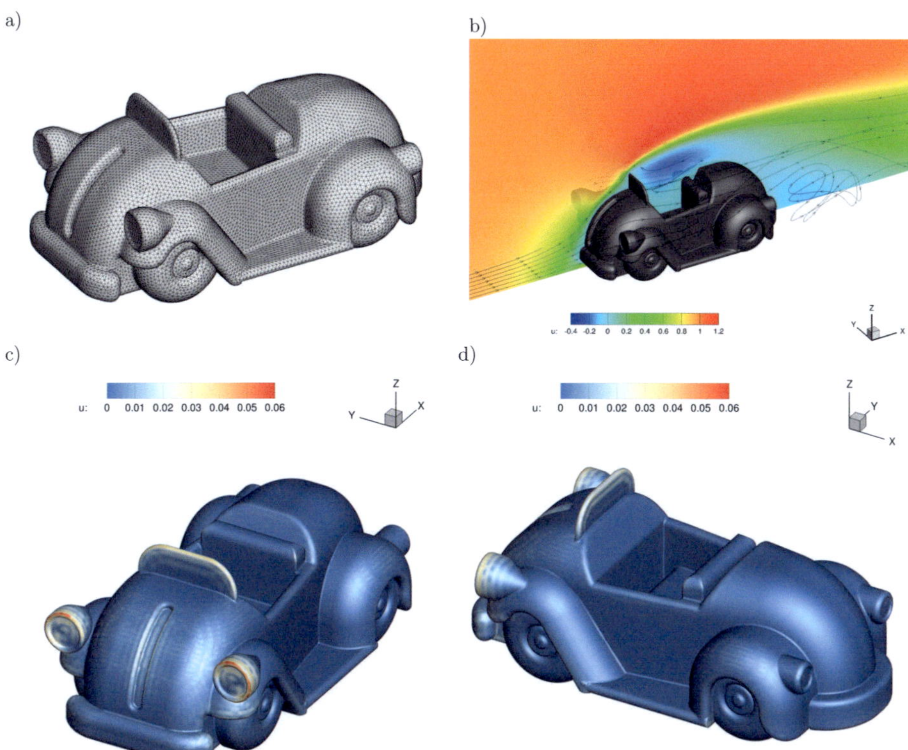

Figure 3.4 Flow around an immersed car model. a) Surface triangulation ($N_{tri} = 41112$). b) Instantaneous streamwise velocity contour and pathlines. c) and d) Absolute value of surface velocity corresponding to error in surface boundary condition enforced by the IBM.

The error in the surface boundary condition is evaluated at the individual forcing points with $\epsilon^{(l)} = \|\mathbf{u}_{fp}^{(l)} - \mathbf{u}_S^{(l)}\|/u_b$ where $\mathbf{u}_S^{(l)} = 0$. The error thus corresponds to the surface velocity

itself in the present setup. A distribution of the error over the surface of the car is shown in Figure 3.4c) and d) where $\mathbf{u}_{fp}^{(l)}$ are attained by an additional interpolation step at the end of the time-step. A slight footprint of the Cartesian grid is visible on the car surface especially in regions of high velocity gradients. The maximum errors occur in stagnation pressure regions in combination with large curvature of the surface, as for the top of the windscreen and the headlights. A global maximum error of $\max(\epsilon) = 5.3\%$ was computed. The quadratic average of the error, $\epsilon_{RMS} = 0.19\%$, is substantially lower. The results are in good agreement with the errors of the IBM reported in [136] for the flow around a cylinder with moderately increased maximum errors.
The simulations were run in parallel on 16 processes. Given the large number of forcing points and the non-optimal load balance with respect to the particle, the performance is still very decent. About 12% of the total computation time are spent in particle routines.
In conclusion, the boundary condition along the very complex surface is enforced within acceptable tolerances. The extension of the IBM, hence, provides a sound foundation for generic particle shapes.

3.1.5 The flow over periodic hills

Introduction and problem definition

The scope of this test case is to validate the current IBM implementation for a generic, curved, outer geometry. The flow over periodic hills in a plane channel is well documented and provides interesting features as flow separation from a curved surface with subsequent reattachment both varying in space and time, a strong shear layer, recirculation, acceleration and wall proximity [73, 24]. The configuration is depicted in Figure 3.5. The domain extents are $\mathbf{L} = (9.0,\ 3.036,\ 4.5)\ h$, with h being the height of the hill [266]. The exact geometry of the hill may be found in the ERCOFTAC database (case 9.2 or case 81 depending on the host server) and was originally studied experimentally in [4]. Periodic boundary conditions are used in streamwise and spanwise direction, whereas no-slip walls bound the domain in y-direction.
The flow over periodic hills has been studied extensively in [24] for a large range of Reynolds numbers, $100 \leq Re_h \leq 10595$, where $Re_h = u_b h/\nu$ with u_b being the bulk velocity at the crest of the hill. In this reference, DNS and highly resolved LES were conducted with two independent codes (MGLET and LESSOC) and experimental data was provided from PIV measurements. The largest Reynolds number, $Re_h = 10595$, was studied in [73] by LES with the focus on subgrid scale modeling and turbulence mechanisms associated with the specific features of this configuration.
In the present study, a DNS was performed for $Re_h = 2800$ with an equidistant grid of $\mathbf{N} = (512,\ 173,\ 256)$, $N_L = 63503$ Lagrangian forcing points and adaptive time step size yielding $CFL = 0.8$. The hill was placed centered at $L_x/2$ and the x-coordinate, designating the position of the profiles in Figure 3.7, origins at the crest of the hill. The geometry is sketched in Figure 3.5, where the forcing points representing the hill and their triangulation are coarsened by a factor of four. Instantaneous contours of the velocity component u/u_b are shown in this figure as well, illustrating the irregular recirculation zone and the turbulent character of the flow in streamwise and spanwise direction.

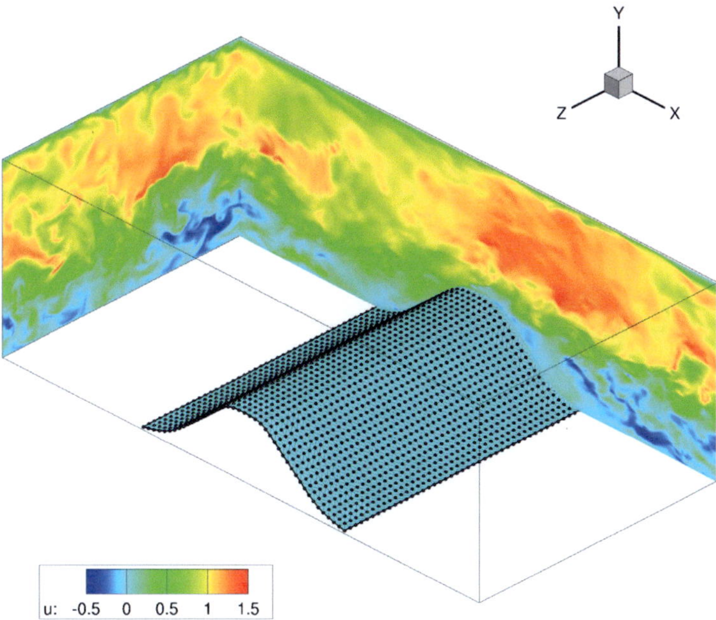

Figure 3.5 Periodic hill, contours show instantaneous streamwise velocity u/u_b for $Re_h = 2800$. Forcing points and triangulation on the hill are coarsened by a factor of 4.

Results for the flow over periodic hills

An instantaneous contour of the vorticity component ω_z is displayed in Figure 3.6a illustrating the size of the turbulent scales. It also gives an impression of the separating boundary layer and the shear layer feeding the production of turbulent kinetic energy. In order to check the grid resolution apart from the wall, the WALE subgrid scale model was implemented to determine the turbulent eddy viscosity ν_t (see appendix D). The simulation was continued for a short period of about two flow through times and ν_t was calculated, but not considered on the rhs of the Navier-Stokes equation, i.e. there was no feedback of the subgrid scale model. The turbulent eddy viscosity ν_t is plotted relative to the kinematic viscosity in Figure 3.6b. There are two distinct peaks of ν_t in each profile, one in the shear layer and one near the top wall which are also apparent in [72]. The fraction of ν_t to the kinematic viscosity does not exceed 7% and is quite low in the entire domain. Therefore the grid resolution should be sufficient to resolve the very major part of the scales in the inner flow. The wall resolution is estimated in an integral way as no local wall shear stress has been determined on the hill. The global momentum balance yields $\langle\tau_w\rangle \approx \langle f_x\rangle V_f/\left(2A_{surf}\right)$ with V_f being the total fluid volume (without the hill), A_{surf} the total surface area consisting of the area of the hill, the lower channel wall apart from the hill and the upper wall. The spatially constant volume force f_x is adjusted in time by a PI-controller to yield the specified u_b. The average non-dimensional wall distance of the first grid point is then computed to be $\langle y^+\rangle \approx \sqrt{\langle\tau_w\rangle/\rho_f}\, y_{st,1}/\nu = 2.2$, which is only slightly above the recommendation for

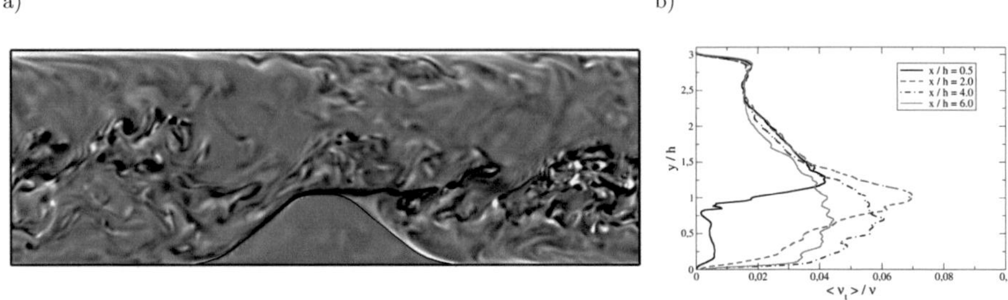

Figure 3.6 a) Instantaneous contour of plane-normal vorticity ω_z for $Re_h = 2800$. b) Short time average of turbulent eddy viscosity ν_t as a fraction of kinematic viscosity obtained by WALE-model from post-processing, i.e. no feedback to simulation.

wall-resolved simulations of $y^+ \lesssim 1$ [72].

Averaged profiles of the velocity component $\langle u \rangle / u_b$ are plotted in Figure 3.7 at different streamwise locations and compared to the reference data from [24]. Very good agreement is found at all locations. The first profile at $x/h = 0.5$ is located shortly after the flow separates from the hill, note that there is an artificial flow within the hill driven by the IBM. The second and third profile at $x/h = 2.0$ and $x/h = 4.0$ show the recirculation zone behind the hill, whereas the flow has reattached at $x/h = 6.0$. The mean reattachment location is at $x_R/h = 5.32$, which is in excellent agreement with the value of 5.41 given in [24]. Statistics were gathered for about 24 flow-through times. The mean velocity profiles were sufficiently converged after this time. Turbulent statistics would need further averaging and they are not reported here as this is beyond the scope of the present study.

The computation time associated with the IBM was less than 0.5% of the total computation time for the present test case.

Concluding remarks for the hill test case

With the present results for the mean flow being in good agreement with the reference data [24], it has been shown that the current IBM implementation is applicable for curved, outer geometries. However, there are open questions that need to be addressed when pursuing to higher Reynolds numbers. The equidistant grid is an issue in terms of near wall resolution. Therefore the IBM needs to be adapted to a stretched grid or a local refinement with hanging nodes has to be realized (Appendix I). The aspect of wall functions and subgrid scale modeling needs to be elucidated in combination with the IBM.

The local shear stress at the embedded body could be determined from the wall normal velocity gradient using an extra interpolation step within the IBM by

$$\left.\frac{\partial u_t}{\partial n}\right|_w \approx \frac{\Delta u_t}{\Delta n} = \frac{(u_t^i(\mathbf{x}_{fp} + \Delta n\, \mathbf{n}_{fp}) - u_t^i(\mathbf{x}_{fp}))}{\Delta n}, \tag{3.8}$$

with $\mathbf{n}_{fp}$ being the unit normal vector at the forcing point, and u_t^i the interpolated, streamwise tangential velocity.

A sound basis for particle laden flows in curved, realistic geometries is available.

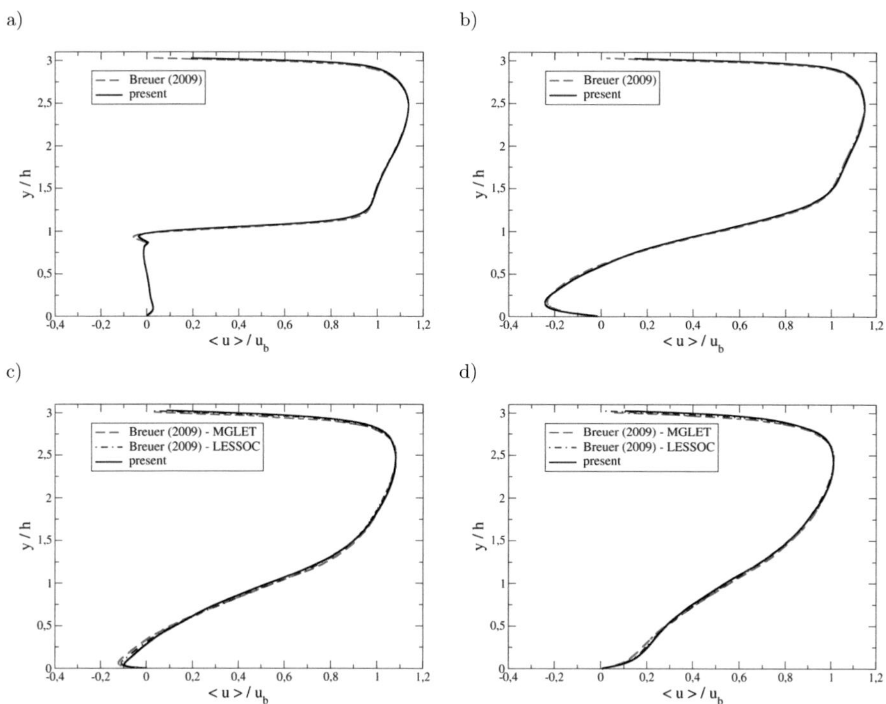

Figure 3.7 Mean velocity $\langle u \rangle / u_b$ at different streamwise locations for $Re_h = 2800$: a) $x/h = 0.5$, b) $x/h = 2.0$, c) $x/h = 4.0$, d) $x/h = 6.0$.

3.2 Non-spherical particles

3.2.1 Introduction to Euler's theorem of rigid body rotation

The motion of a rigid body in three-dimensional space can be decomposed into a translation, e.g., of the body's center of mass and a rotation around this origin. Accordingly, different coordinate systems are introduced to describe the particle motion [212, 91, 148]. First, there is the global laboratory system as the native environment for the simulations. The computational domain, the boundary conditions, the fluid flow, and the particle trajectories are studied in this system. Second, there is the laboratory-parallel system. It is particle-dependent and its origin is located in the center of mass of a specific particle with its axes being parallel to the laboratory system. The particle undergoes a sole rotation in this coordinate system. The description of the translation of non-spherical particles does not differ from the previous notation and is thus not further discussed here. Finally, a body-fixed coordinate system is introduced whose origin is identical to the one of the laboratory-parallel system, but its axes are aligned with the principal axes of the particle at each instant in time. The quantities expressed in such a body-fixed coordinate system are denoted with a prime, e.g. x'.
Euler's theorem of rigid body rotation states that any sequence of rotations can be equivalently described by a single rotation of this body by an angle about one axis [91, 148, 259]. The rotation of a rigid object is a linear similarity transformation and for the respective transformation matrix holds $det\left(\mathbf{A}\right) \equiv 1$, i.e. there is no compression or stretching of the body. The application of this transformation on a vector can be interpreted either as a rotation of the vector itself creating a new vector in the same coordinate system, or on the other hand as a rotation of the coordinate system in the opposite rotational sense [91]. The inverse transform is indicated by $\mathbf{A}^{-1}$ and due to the orthogonality of the transformation matrix the inverse matrix equals the transpose, $\mathbf{A}^{-1} = \mathbf{A}^{T}$.
A three-dimensional rotation by an angle ϕ around a unit axis $\mathbf{a}$ through the origin is described by

$$\mathbf{A}_{\mathbf{a},\phi} = \begin{pmatrix} a_{\mathrm{x}}^2\left(1-\cos\phi\right)+\cos\phi & a_{\mathrm{x}}a_{\mathrm{y}}\left(1-\cos\phi\right)-a_{\mathrm{z}}\sin\phi & a_{\mathrm{x}}a_{\mathrm{z}}\left(1-\cos\phi\right)+a_{\mathrm{y}}\sin\phi \\ a_{\mathrm{y}}a_{\mathrm{x}}\left(1-\cos\phi\right)+a_{\mathrm{z}}\sin\phi & a_{\mathrm{y}}^2\left(1-\cos\phi\right)+\cos\phi & a_{\mathrm{y}}a_{\mathrm{z}}\left(1-\cos\phi\right)-a_{\mathrm{x}}\sin\phi \\ a_{\mathrm{z}}a_{\mathrm{x}}\left(1-\cos\phi\right)-a_{\mathrm{y}}\sin\phi & a_{\mathrm{z}}a_{\mathrm{y}}\left(1-\cos\phi\right)+a_{\mathrm{x}}\sin\phi & a_{\mathrm{z}}^2\left(1-\cos\phi\right)+\cos\phi \end{pmatrix}, \tag{3.9}$$

which simplifies for a rotation axis aligned with one of the axes of the laboratory-parallel system to one of the three-dimensional rotation matrices

$$\mathbf{A}_{\mathrm{x}} = \begin{pmatrix} 1 & 0 & 0 \\ 0 & \cos\phi_{\mathrm{x}} & -\sin\phi_{\mathrm{x}} \\ 0 & \sin\phi_{\mathrm{x}} & \cos\phi_{\mathrm{x}} \end{pmatrix}, \quad \mathbf{A}_{\mathrm{y}} = \begin{pmatrix} \cos\phi_{\mathrm{y}} & 0 & \sin\phi_{\mathrm{y}} \\ 0 & 1 & 0 \\ -\sin\phi_{\mathrm{y}} & 0 & \cos\phi_{\mathrm{y}} \end{pmatrix}, \quad \mathbf{A}_{\mathrm{z}} = \begin{pmatrix} \cos\phi_{\mathrm{z}} & -\sin\phi_{\mathrm{z}} & 0 \\ \sin\phi_{\mathrm{z}} & \cos\phi_{\mathrm{z}} & 0 \\ 0 & 0 & 1 \end{pmatrix}. \tag{3.10}$$

An active transformation rotates a vector; it changes its components while the vector norm remains unchanged. A passive rotation on the other hand describes a rotation of the coordinate system and is achieved by rotating with a negative rotation angle in the above rotation matrices which is equivalent to switching the sign of the sine entries. The same vector is represented in different coordinate systems. The formalism of active and passive rotation is the same [91].
The rotational orientation of a body in three-dimensional space can be parametrized in manifold ways where all of those are based on Euler theorem of rigid body rotation stated above

[259]. A helpful, comprehensive survey is given in [51]. Classically used are three Euler angles which describe three successive rotations around pre-defined axes. In the zyx-convention of Euler angles, the laboratory-parallel coordinate system is first rotated by an angle ϕ_3 around the z-axis, then by an angle ϕ_2 around the intermediate y-axis, and subsequently by an angle ϕ_1 around the final x'-axis to obtain the body-fixed coordinate system. The order of the subsequent finite rotations must not be interchanged as the associated transformations are not commutative. The definition also implies that the laboratory-parallel system and the body-fixed system coincide if all Euler angles are zero. There are various definitions of Euler angles, the x-convention (or 3-1-3 or zxz) seems to be the most common one in physics [148]. Here, the zyx-convention is chosen as it is more intuitive to use and rather widespread in engineering applications. Defined this way, the Euler angles can be interpreted as yaw, pitch and roll of an aircraft for example, as explained in the 'Deutsche Luftfahrtnorm' [75]. The matrix of the full transformation in the zyx-convention is given by

$$\mathbf{A} = \begin{pmatrix} \cos\phi_2\cos\phi_3 & \cos\phi_2\sin\phi_3 & -\sin\phi_2 \\ \sin\phi_1\sin\phi_2\cos\phi_3 - \cos\phi_1\sin\phi_3 & \cos\phi_1\cos\phi_3 + \sin\phi_1\sin\phi_2\sin\phi_3 & \sin\phi_1\cos\phi_2 \\ \cos\phi_1\sin\phi_2\cos\phi_3 + \sin\phi_1\sin\phi_3 & \cos\phi_1\sin\phi_2\sin\phi_3 - \sin\phi_1\cos\phi_3 & \cos\phi_1\cos\phi_2 \end{pmatrix} . \tag{3.11}$$

The transformation of the vector $\mathbf{x}$ in the laboratory-parallel system to its representation $\mathbf{x}'$ in the body-fixed coordinate system is obtained via

$$\mathbf{x}' = \mathbf{A}\cdot\mathbf{x} . \tag{3.12}$$

The nomenclature for the Euler angles varies throughout the literature where the present notation was adopted from [96] and $\phi_3 = \phi$, $\phi_2 = \theta$, $\phi_1 = \psi$ provides the useful conversion to the nomenclature of Goldstein et al. [91][1], a textbook on classical mechanics often referenced in the present context. Accordingly, $\boldsymbol{\phi}_p = (\phi_1, \phi_2, \phi_3)^T$ lists the angles describing the particle orientation, where the particle index p is omitted when the respective components are addressed. Note that $\boldsymbol{\phi}_p$ is not a vector in a physical sense [91], but a list of the Euler angles.

The angular velocity vector is denoted by $\boldsymbol{\omega}_p = (\omega_x, \omega_y, \omega_z)^T$ in the laboratory-parallel system and its body-fixed counterpart is $\boldsymbol{\omega}_p'$. The relation between the components of the angular velocity and the temporal derivatives of the Euler angles, $\dot{\boldsymbol{\phi}}_p$, is given by [91, 96]

$$\boldsymbol{\omega}_p = \begin{pmatrix} \cos\phi_2\cos\phi_3 & \sin\phi_3 & 0 \\ \cos\phi_2\sin\phi_3 & \cos\phi_3 & 0 \\ -\sin\phi_2 & 0 & 1 \end{pmatrix}\cdot\begin{pmatrix} \dot\phi_1 \\ \dot\phi_2 \\ \dot\phi_3 \end{pmatrix} , \quad \boldsymbol{\omega}_p' = \begin{pmatrix} 1 & 0 & -\sin\phi_2 \\ 0 & \cos\phi_1 & \sin\phi_1\cos\phi_2 \\ 0 & -\sin\phi_1 & \cos\phi_1\cos\phi_2 \end{pmatrix}\cdot\begin{pmatrix} \dot\phi_1 \\ \dot\phi_2 \\ \dot\phi_3 \end{pmatrix} . \tag{3.13}$$

The above formulae can be substantially simplified for small angles, infinitesimal rotations or the rotation around a single principal axis [91]. The numerical treatment of the most general case is, however, problematic. A singular mapping is found for $\cos\phi_2 = 0$ [259], often referred to as Gimbal lock, and numerical errors in the evaluation of (3.13) become large if this state is approached. In the present applications, there is mostly a single dominant rotation axis, as e.g. for the zig-zag of a bubble or the interaction of an ellipsoid with an inclined wall. A solution to the numerical issues is employing Euler parameters as introduced in Section 3.2.5 below.

[1] Note that there are two flaws in the transformation matrix in zyx-convention as given in the book, which were reported to the publisher by the present author.

3.2.2 Angular momentum equation of the particle

In contrast to a sphere, the moment of inertia is represented by a tensor for particles of more complex shape. A torque, $\mathbf{M}_p$, acting on the particle causes a temporal change of its angular momentum. The angular equation of motion of a non-spherical particle in the laboratory-parallel coordinate system reads

$$\frac{d\left(\mathbf{I}_p\boldsymbol{\omega}_p\right)}{dt} = \rho_f \oint_S \mathbf{r} \times (\boldsymbol{\tau} \cdot \mathbf{n}_S)\, dS = \mathbf{M}_p \;, \tag{3.14}$$

with the symmetric moment of inertia tensor

$$\mathbf{I}_p = \begin{pmatrix} I_{\mathtt{xx}} & I_{\mathtt{xy}} & I_{\mathtt{xz}} \\ I_{\mathtt{yx}} & I_{\mathtt{yy}} & I_{\mathtt{yz}} \\ I_{\mathtt{zx}} & I_{\mathtt{zy}} & I_{\mathtt{zz}} \end{pmatrix} = \begin{pmatrix} \int \rho_p\,(\mathtt{y}^2+\mathtt{z}^2)\,dV_p & -\int \rho_p\,\mathtt{x}\,\mathtt{y}\,dV_p & -\int \rho_p\,\mathtt{x}\,\mathtt{z}\,dV_p \\ I_{\mathtt{yx}} = I_{\mathtt{xy}} & \int \rho_p\,(\mathtt{x}^2+\mathtt{z}^2)\,dV_p & -\int \rho_p\,\mathtt{y}\,\mathtt{z}\,dV_p \\ I_{\mathtt{zx}} = I_{\mathtt{xz}} & I_{\mathtt{zy}} = I_{\mathtt{yz}} & \int \rho_p\,(\mathtt{x}^2+\mathtt{y}^2)\,dV_p \end{pmatrix} , \tag{3.15}$$

where $\mathtt{x}$, $\mathtt{y}$, $\mathtt{z}$ denote coordinates with respect to the center of mass, i.e. coordinates in the laboratory-parallel system. Here, contributions to $\mathbf{M}_p$ originating from collisions are not considered.

In the body-fixed reference frame, the moment of inertia is constant. Since the coordinate axes are aligned with the body's principal axes at each instant in time, the moment of inertia tensor, $\mathbf{I}'_p$, holds only the principal moments of inertia and off-diagonal entries are zero,

$$\mathbf{I}'_p = \begin{pmatrix} I_{x'x'} & 0 & 0 \\ 0 & I_{y'y'} & 0 \\ 0 & 0 & I_{z'z'} \end{pmatrix} , \tag{3.16}$$

with the respective transformation [91, 148] being given by

$$\mathbf{I}_p = \mathbf{A} \cdot \mathbf{I}'_p \cdot \mathbf{A}^T \;. \tag{3.17}$$

For an tri-axial ellipsoid, the principal moments of inertia are given by

$$I_{x'x'} = \frac{1}{5} m_p \left(b^2 + c^2\right) , \quad I_{y'y'} = \frac{1}{5} m_p \left(a^2 + c^2\right) , \quad I_{z'z'} = \frac{1}{5} m_p \left(a^2 + b^2\right) , \tag{3.18}$$

with m_p being the particle mass and a, b, c the semi-axes.

The angular equation of motion expressed in the body-fixed reference frame is addressed in Section 3.2.5 (equation (3.30)). Next, the computation of the torque, which acts on the particle and determines the right hand side of (3.14), is discussed in the context of the IBM.

3.2.3 Evaluation of the force and torque acting on the particle

The net force and torque acting from the fluid on the particle are determined by integrating the hydrodynamic stresses including pressure over the particle surface. The evaluation of this surface integral is, however, numerically challenging and can be avoided within the IBM by transferring the surface integral to a volume integral of the underlying 'artificial' fluid. An in-detail discussion of the approach can be found in [136, 134]. Here, the approach is recapitulated with some remarks towards non-spherical particles and their orientation. The balance of the fluid's angular momentum in a volume V with a surface S reads

$$\frac{d}{dt} \int_V \rho_f \mathbf{r} \times \mathbf{u}\, dV = \rho_f \oint_S \mathbf{r} \times (\boldsymbol{\tau} \cdot \mathbf{n}_S)\, dS + \int_V \rho_f \mathbf{r} \times \mathbf{f}\, dV \;. \tag{3.19}$$

If the balance is applied to a volume bounded by the particle surface, i.e. the particle volume V_p, the right hand side of the particle angular momentum equation (3.14) can be re-written as

$$\frac{d\left(\mathbf{I}_p\boldsymbol{\omega}_p\right)}{dt} = \frac{d}{dt}\int_{V_p} \rho_f \mathbf{r} \times \mathbf{u}\, dV - \int_{V_p} \rho_f \mathbf{r} \times \mathbf{f}\, dV\,. \tag{3.20}$$

The integral over the Eulerian forces can be replaced by a summation over the Lagrangian forcing points as apparent from (3.2)

$$\int_{V_p} \rho_f \mathbf{r} \times \mathbf{f}\, dV = \sum_{l=1}^{N_L} \rho_f \left(\mathbf{x}_{fp}^{(l)} - \mathbf{x}_p\right) \times \mathbf{f}_{fp}^{(l)}\, \Delta V_L^{(l)}\,. \tag{3.21}$$

This term is thus directly available from the IBM forcing procedure ensuring the boundary condition at the particle surface. For the computation of the remaining term in (3.20), the temporal derivative is processed within the particle time-integration scheme and the volume integral is numerically evaluated via

$$\int_{V_p} \rho_f \mathbf{r} \times \mathbf{u}\, dV = \sum_{i,j,k} \rho_f \left(\mathbf{x}_{i,j,k} - \mathbf{x}_p\right) \times \mathbf{u}_{i,j,k}\, \alpha_{i,j,k} V_{i,j,k}^{cell}\,, \tag{3.22}$$

where $\alpha_{i,j,k}$ denotes the particle volume fraction, i.e. the portion of the finite Eulerian cell occupied by the particle,

$$\alpha_{i,j,k} = \frac{V_{i,j,k}^{p}}{V_{i,j,k}^{cell}}. \tag{3.23}$$

The particle volume fraction is efficiently computed from

$$\alpha_{i,j,k} = \frac{\sum_{m=1}^{8} \left|\min\left(\phi_m,\, 0\right)\right|}{\sum_{m=1}^{8} \left|\phi_m\right|}\,, \tag{3.24}$$

employing the signed-distance level-set function of the particle, ϕ, computed at the $m = 1, \ldots, 8$ corner points of the cell [136]. The signed-distance level-set function is negative inside the particle and positive outside, while the distance is measured normal to the surface. The particle surface itself is indicated by the zero level-set. Figure 3.8a) provides a sketch illustrating the concept. The approach is very efficient and showed second order convergence [136]. The signed-distance level-set function for a rotated ellipsoid is given by

$$\phi_{i,j,k}^{ellipsoid} = \sqrt{\left(\frac{x'_{i,j,k}}{a}\right)^2 + \left(\frac{y'_{i,j,k}}{b}\right)^2 + \left(\frac{z'_{i,j,k}}{c}\right)^2} - 1, \tag{3.25}$$

with the coordinates of the Cartesian cell transformed to the body-fixed frame

$$\mathbf{x}'_{i,j,k} = \mathbf{A} \cdot \left(\mathbf{x}_{i,j,k} - \mathbf{x}_p\right). \tag{3.26}$$

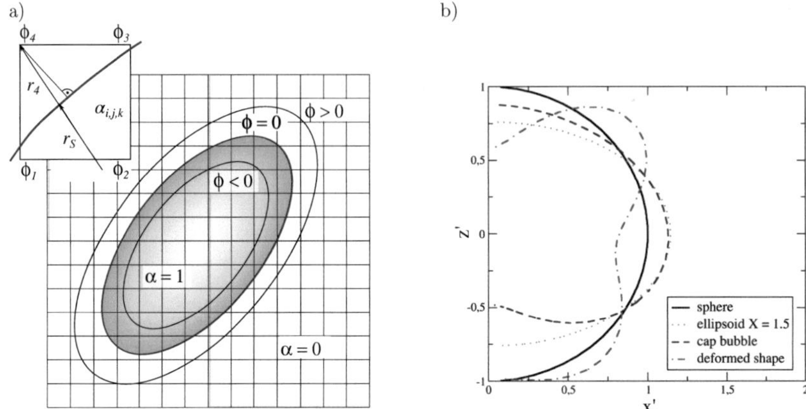

Figure 3.8 Evaluation of particle volume fraction, $\alpha_{i,j,k}$. a) Schematic sketch of signed-distance approach. b) Axisymmetric shapes for evaluation summarized in Table 3.2.

The signed-distance level-set function can be obtained for particles of basically arbitrary shape [253]. In Chapter 4, complex particle shapes are represented by spherical harmonic functions. The evaluation of the signed, surface-normal distance is somewhat cumbersome for these complex, analytical shapes. A comparison of the computation of the particle volume fraction swapping the signed surface-normal distance by the signed radial distance $r_m - r_S$ in equation 3.24, showed negligible deviations for the spheroidal shapes tested. The difference of these distance measures is illustrated in Figure 3.8a). The computational effort decreases when using the signed radial distance. Table 3.2 lists the numerical, relative error in the computation of the particle volume, V_p, determined from $\sum_{i,j,k} \alpha_{i,j,k} V_{i,j,k}^{cell}$ for the axisymmetric shapes shown in Figure 3.8b). The original level-set approach was used for the spherical and ellipsoidal particle, while the radial distance was employed for the cap-shaped particle and the deformed shape. The numerical error is unaltered and also second order convergence is achieved. As a further validation (not shown), the moment of inertia tensor was computed using a numerical evaluation of the volume integrals in 3.15 employing the particle volume fraction and comparing it to the reference solution from (3.17) for an ellipsoid tilted around all three axes.

$\frac{d_{eq}}{\Delta x}$	sphere	p	ellipsoid, $X = 1.5$	p	cap	p	deformed shape	p
10	-1.46E-02		-1.33E-02		-1.30E-02		-1.22E-02	
20	-3.50E-03	2.1	-3.17E-03	2.1	-3.42E-03	1.9	-2.71E-03	2.2
40	-8.12E-04	2.1	-7.42E-04	2.1	-8.26E-04	2.1	-6.72E-04	2.0
80	-1.87E-04	2.1	-2.02E-04	1.9	-2.00E-04	2.0	-1.60E-04	2.1

Table 3.2 Numerical, relative error in volume, V_p, determined from $\sum \alpha_{i,j,k} V_{i,j,k}^{cell}$ and order of convergence p with increasing the resolution of the Cartesian grid.

In summary, the torque acting on the non-spherical particle is obtained in a very efficient way and the equation of motion for the rotation can be solved as discussed in the next paragraphs. The determination of the net force acting on the non-spherical particle for the translational momentum equation works in an analogous way accounting for the staggered grid [136, 134].

3.2.4 Solution in laboratory-parallel system

As in the original method for a sphere [134, 136], the angular momentum equation of the particle is solved in the laboratory-parallel system. This seems intuitive and is straightforward based on the original implementation of the IBM. The rate of change of the angular momentum in the k-th Runge-Kutta sub-step is obtained by the finite difference

$$(\mathbf{I}_p \cdot \boldsymbol{\omega}_p)^{k+1} = (\mathbf{I}_p \cdot \boldsymbol{\omega}_p)^k + 2\alpha^k \Delta t \ \mathbf{M}_p^k , \tag{3.27}$$

with α^k being coefficients of the present low-storage Runge-Kutta scheme. In order to solve the problem, one has to deal with the change of the moment of inertia tensor, $\mathbf{I}_p$, in time. This is done by the following iterative procedure:

0) Compute once the right hand side of 3.27 including the current torque and moment of inertia tensor in the laboratory-parallel system.
1) Formulate an initial guess for $\mathbf{I}_p^{k+1}$. Since for a sphere $\mathbf{I}_p(t) = const.$ using $\mathbf{I}_p^k$ seems appropriate for spheroidal particles.

`do` $j = 1, j_{max}$

2) Solve the linear system of equation posed by (3.27) for $\boldsymbol{\omega}_p^{k+1}$. A direct solution by means of Gaussian elimination is conducted.
3) Compute the rate of the Euler angles, $\dot{\boldsymbol{\phi}}_p^{k+1}$, or Euler parameters from (3.13) or (3.31), respectively.
4) Determine the new orientation of the particle at $k+1$, e.g. using the trapezoidal rule for the new Euler angles,

$$\boldsymbol{\phi}_p^{k+1} = \frac{1}{2}\left(\dot{\boldsymbol{\phi}}_p^k + \dot{\boldsymbol{\phi}}_p^{k+1}\right) 2\alpha^k \Delta t \, .$$

5) Update the rotation matrix $\mathbf{A}$ and moment of inertia tensor $\mathbf{I}_p^{k+1}$ by (3.17).
6) Exit loop if specified residual criterion with respect to j-iteration is fulfilled

`end do`

Usually only a very few iterations are necessary for convergence. For spheroidal particles with moderate aspect ratios and given the CFL-limit of the explicit time integration, the temporal change of the inertial tensor is quite small. However, $\|\boldsymbol{\omega}_p \cdot d\mathbf{I}_p/dt\|$ can become important in the more general case, as reported in Table 3.11e) for an initially tumbling elongated ellipsoid. A more elegant way of solving the angular momentum equation is using the body-fixed coordinate system as discussed in the next Section employing Euler parameters as a parametrization of the rotation matrix.

3.2.5 Solution employing Euler parameters in body-fixed system

The body-fixed system is referred to as a coordinate system centered in $\mathbf{x}_p$ with its axes aligned with the principal axes of the particle. The rotation is described employing Euler parameters. With this approach, a rotation is described by four parameters which can be directly derived from Euler's angle-axis theorem. The four parameters are grouped to a quaternion with $q = (q_0, q_1, q_2, q_3) = (q_0, \mathbf{q})$ [302, 243, 259]. The first parameter is a scalar and the other three build a vector in three dimensional space and thus a quaternion can be interpreted as angle and axis. Euler parameters form a unit quaternion and hence the subsidiary condition $q_0^2 + q_1^2 + q_2^2 + q_3^2 = 1 = \|q\|^2$ must hold. The multiplication of two quaternions, $q \circ p$, is defined as [302, 243, 259]

$$q \circ p = (q_0 p_0 - \mathbf{q} \cdot \mathbf{p},\, q_0 \mathbf{p} + p_0 \mathbf{q} + \mathbf{q} \times \mathbf{p}) \,. \tag{3.28}$$

The product is non-commutative. Sequences of rotations can be expressed by quaternion multiplications in a compact way. Here, we stick to the rotation matrix formalism. The transformation of a vector from the laboratory-parallel coordinate system to the body fixed coordinate system is still given by $\mathbf{x}' = \mathbf{A} \cdot \mathbf{x}$, where the orthogonal rotation matrix now reads in terms of Euler parameters [51]

$$\mathbf{A} = \begin{pmatrix} q_0^2 + q_1^2 - q_2^2 - q_3^2 & 2\,(q_1 q_2 + q_0 q_3) & 2\,(q_1 q_3 - q_0 q_2) \\ 2\,(q_2 q_1 - q_0 q_3) & q_0^2 - q_1^2 + q_2^2 - q_3^2 & 2\,(q_2 q_3 + q_0 q_1) \\ 2\,(q_3 q_1 + q_0 q_2) & 2\,(q_3 q_2 - q_0 q_1) & q_0^2 - q_1^2 - q_2^2 + q_3^2 \end{pmatrix} . \tag{3.29}$$

Applied to the torque vector acting on the particle, the transformation yields the torque in the body-fixed coordinate system, $\mathbf{M}_p' = \mathbf{A} \cdot\ \mathbf{M}_p$. This can be used to solve Euler's angular momentum equation in the body-fixed reference frame [91, 148, 243],

$$\begin{aligned} I_{x'x'} \dot{\omega}_{x'} - (I_{y'y'} - I_{z'z'})\, \omega_{y'} \omega_{z'} &= M_{x'} \\ I_{y'y'} \dot{\omega}_{y'} - (I_{z'z'} - I_{x'x'})\, \omega_{z'} \omega_{x'} &= M_{y'} \\ I_{z'z'} \dot{\omega}_{z'} - (I_{x'x'} - I_{y'y'})\, \omega_{x'} \omega_{y'} &= M_{z'} \end{aligned} \tag{3.30}$$

to obtain $\dot{\boldsymbol{\omega}}_p'$. Obviously, the moment of inertia tensor in the body-fixed frame is constant in time and no iterative procedure is necessary. Integration in time with the present Runge-Kutta scheme yields $\boldsymbol{\omega}_p'$ which is transformed to the laboratory-parallel system giving $\boldsymbol{\omega}_p$ to be used in the IBM. The time-derivatives of the Euler parameters are then given by

$$\dot{q} = \frac{1}{2} q \circ \omega_p' = \frac{1}{2} \omega_p \circ q. \tag{3.31}$$

with $\omega_p' = (0, \omega_{x'}, \omega_{y'}, \omega_{z'})$. Finally, the Euler parameters and the respective rotation matrix are updated integrating $\dot{q}$ in time employing the trapezoidal rule. The numerical integration introduces errors leading to a violation of the unit constraint for the Euler parameters. These are thus renormalized by coordinate projection, i.e. dividing by $\|q\|$ [243].

The approach is very efficient and eludes the usage of sine or cosine functions. It represents an always non-singular mapping avoiding the possibility of a Gimbal lock. The method is used in classical physics, robotics, aerospace dynamics, computer graphics, but also for simulations of particulate flows [315, 313]. Euler parameters should represent the preferential description for future work, especially for systems with many non-spherical particles.

3.2.6 Application to the rotation of an ellipsoid

Prescribed rotation

The method describing the particle rotation and orientation was applied to a variety of generic test cases and problems of physical interest for validation purposes. First, simple elementary rotations around the principal axes of the non-spherical particle were prescribed and superpositioned with a given translational path, e.g. to mimic a zig-zag trajectory of a particle and study the flow field (not shown). Here, a prescribed off-axis rotation is presented where $\boldsymbol{\omega}_p = (\pi,\, 0,\, -2\pi)\,T^{-1}$ was assigned in the laboratory-parallel system for the period T. It is rather difficult to find visualizations of time-sequences for intricate rotations. The test case is therefore inspired by a jump from the slope-style discipline in free-skiing called 'cork5'. The body undergoes one full rotation around the $\mathbf{z}$-axis (360°) and a half rotation (180°) around the $\mathbf{x}$-axis and it thus faces backwards after the cycle. A prolate ellipsoid with semi-axes $a > b = c$ and aspect ratio $X = 2$ is used. Figure 3.9 shows a three-dimensional visualization of the rotation. For illustration purposes, the equi-spaced time-instants were placed along a virtual parabola in a xz-plane. At time $t = 0$, the long axis $2a$ aligned with the $\mathbf{x}$-axis and is indicated by the forcing point $\mathbf{x}_{fp,a}$. This specific point is highlighted during the laps of time in Figure 3.9a) and further a body-fixed direction (of view) is indicated by an arrow at the start and the end. The temporal history of $\mathbf{x}_{fp,a} - \mathbf{x}_p$ and its path are provided in Figure 3.9b) and c).

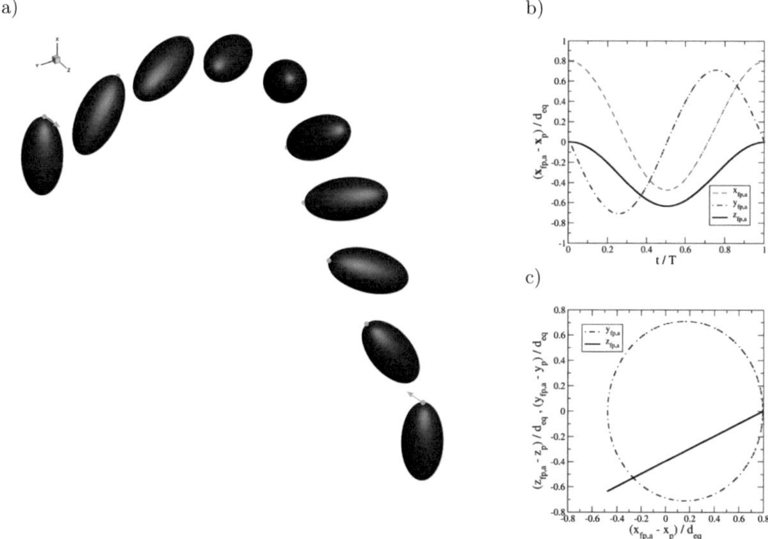

Figure 3.9 Prescribed rotation of a prolate ellipsoid. a) Three-dimensional visualization along virtual parabola in a xz-plane. b) Temporal history of $\mathbf{x}_{fp,a} - \mathbf{x}_p$ describing the relative position of a long semi-axis. c) Path of point $\mathbf{x}_{fp,a} - \mathbf{x}_p$.

Ellipsoid in shear flow

The fluid-particle coupling is examined in this section studying the rotational behavior of an ellipsoid in shear flow. A linear shear flow, given by the velocity profile $u(y) = s(y - y_p)$ with the shear parameter s, is realized by two counter-moving walls bounding the computational domain in the y-direction as sketched in Figure 3.10a). The non-dimensional numbers characterizing the problem are the shear Reynolds number, $Re_s = 4\,s\,a^2/\nu$, the confinement ratio $L_y/a = 8$, the aspect ratio $X = a/b = 2$, and the density ratio π_ρ. Two of the parameters are fixed in this study as indicated and a prolate ellipsoid is studied. A computational domain of $\mathbf{L} = (6.4,\, 6.4,\, 6.4)\, d_{eq}$ was used and discretized equidistantly with $\mathbf{N} = (128,\, 128,\, 128)$. The remaining boundary conditions are chosen as periodic. The time-step is adjusted to yield $CFL = 0.5$, keeping in mind that the viscous terms are treated implicitly. These numerical parameters were used in all simulations presented below.

Jeffery orbit

The rotation of a neutrally buoyant ellipsoid, $\pi_\rho = 1$, is considered under Stokes flow conditions, $Re_s \ll 1$. For these prerequisites, an analytical solution can be derived known as Jeffery orbit [119]. A small principal axis of the ellipsoid and the vorticity axis of the flow are aligned, the ellipsoid rotates around this axis and thus $\phi_3 = \phi_z$. From the analytical solution [119], the angular velocity of the particle can be expressed as a function of the particle inclination,

$$\omega_z(\phi_z) = -\frac{s}{2}\left(1 - \frac{X^2 - 1}{X^2 + 1}\cos\left(2\phi_z\right)\right), \tag{3.32}$$

i.e. the particle rotates fastest when oriented vertical and its long axis is parallel to the y-axis, and it rotates slowest when oriented horizontal and its long axis points in the x-direction. Figure 3.10c) provides a comparison of the analytical solution with present simulation results obtained with $Re_s = 0.5$. Very good agreement is found with a slight overprediction of the angular velocity at small ϕ_z. Figure 3.10b) shows the temporal evolution of the absolute value of the particle's angular velocity with solution in body-fixed and laboratory-parallel reference frame. Both solutions coincide with a small performance plus for the quaternion approach in the body-fixed system. The period of the orientation angle is given by $T(\phi_z) = 2\pi/s\ (X + X^{-1})$ [315]. It was underestimated by 3.6% with the present simulations.

The effect of particle inertia

The influence of particle inertia, as well as the impact of the vicinity of a solid wall on the motion of ellipsoidal particles was studied in [85, 86, 26]. A boundary-element method is employed for the solution in the Stokes flow regime. The authors use an inertial parameter based on the particle relaxation time, τ, and the shear parameter which reads $\tau\, s = \frac{4}{3}\pi \frac{\rho_p a^2}{\mu_f}\, s$. Figure 3.10d) shows the angular velocity versus the particle orientation for various inertial parameters τs and a comparison to data from Gavze & Shapiro [86]. For an inertial ellipsoid, there are two values of ω_z for each argument $\cos(2\phi_z)$ corresponding to a left-right asymmetry with respect to the vertical orientation. The slope of the graph agrees with the value of the theoretical limit given by the Jeffery orbit for small inertial parameters ($\tau\, s = 10$). With increasing $\tau\, s$, the average slope decreases ($\tau\, s = 100$) and approaches zero for very high inertia ($\tau\, s = 1000$), while $-\omega_z/s$ tends towards 0.5. The difference between the two branches of ω_z increases, runs through a global maximum and again decreases as the particle inertia is increased. The comparison to the reference data is conducted for an inertial

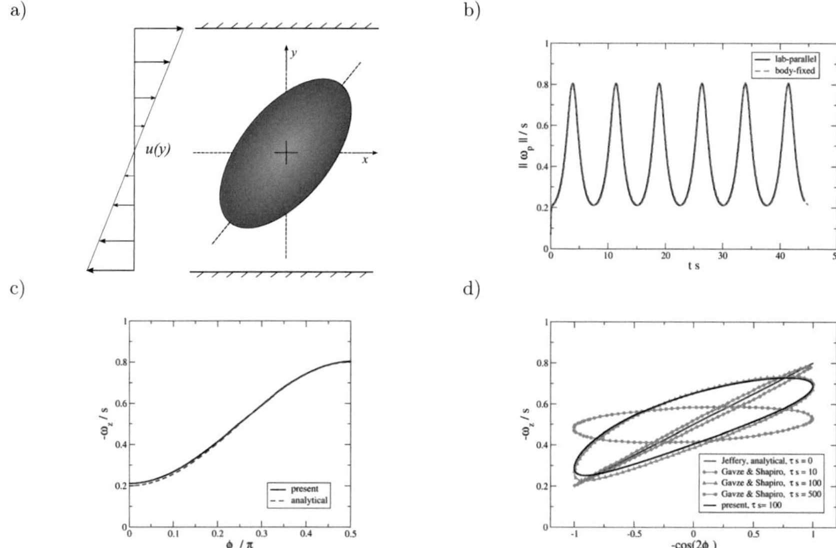

Figure 3.10 Rotation of an ellipsoid in shear flow. a) Configuration. b) and c) Jeffery orbit of neutrally buoyant prolate ellipsoid, $X = 2$: b) Temporal evolution of the absolute value of the particle's angular velocity with solution in body-fixed and laboratory-parallel reference frame. c) Angular velocity versus particle orientation and comparison to analytical solution. d) Inertial ellipsoid, angular velocity versus particle orientation for various inertial parameters τs and comparison to data from Gavze & Shapiro [86].

parameter of $\tau\, s = 100$, while the shear Reynolds number is kept at $Re_s = 0.5$ in the present simulation. The simulation is run in the absence of gravity as in the reference [86]. Very good agreement is found for the rotational behavior of an inertial ellipsoid as apparent from Figure 3.10d). A minor overprediction of the angular velocity is present at small $-\cos(2\phi_z)$ and the graph is slightly shifted towards higher inertial parameters, corresponding e.g. to a somewhat larger particle. This might by related to the IBM and the employed delta function to force the boundary condition at the interface. A similar effect is discussed in Section 3.3.4. In conclusion, the impact of the particle inertia is well captured for the case considered.

Rotational behavior for finite Reynolds numbers
The rotational behavior of ellipsoidal, neutrally buoyant particles in Couette flow was studied for finite Reynolds numbers in [216] using a lattice Boltzmann method. Three regimes were found for prolate ellipsoids, i.e. the major axis is the axis of revolution. For low to moderate Reynolds numbers, the ellipsoid rotates around its minor axis, i.e. parallel to the vorticity vector of the flow, as it was shown above for the Jeffery orbit. In the intermediate range of Reynolds numbers, an additional stable precession was found where the mean angle between the major axis and the vorticity vector increases with Reynolds number. Finally for high Reynolds numbers, the prolate ellipsoid rotates around its major axis with a constant rate. Further regimes and the effect of initial orientation were recently documented in [116].

Figure 3.11a)-d) shows instantaneous orientations of a prolate ellipsoid in shear flow in

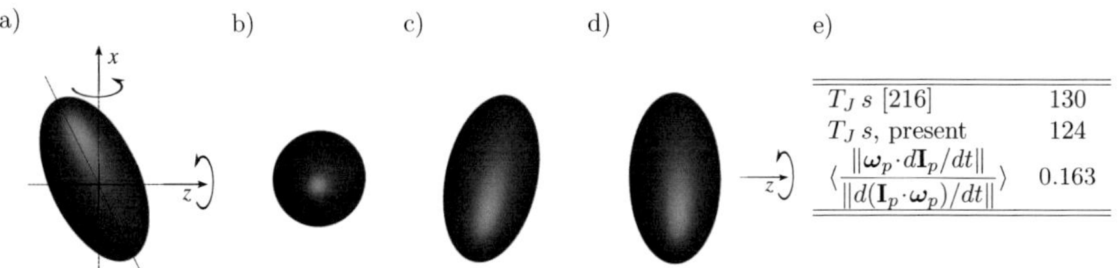

$T_J\, s$ [216]	130
$T_J\, s$, present	124
$\langle \frac{\Vert\omega_p \cdot d\mathbf{I}_p/dt\Vert}{\Vert d(\mathbf{I}_p \cdot \omega_p)/dt\Vert} \rangle$	0.163

Figure 3.11 Rotational states of ellipsoids in shear flow. a)-d) Prolate ellipsoid at low Re_s: a) Early inclination, b) and c) intermediate states, d) final, stable rotation with small semi-axis parallel to the vorticity vector.

the low Reynolds number regime from a present simulation. The simulation was conducted for $Re_s = 32$ and the ellipsoid was initially oriented in terms of Euler angles with $\phi_{p,0} = (0,\, 0.25,\, 0.5)\,\pi$. The long semi-axis is initially inclined with respect to the y-axis by 45° in the yz-plane. Only a qualitative comparison to the observations reported in [216] is performed here, because the full set of initial conditions could not be obtained from the reference and thus a slightly different setup was chosen.

The present results on the rotational behavior are in agreement with the reference. The ellipsoid performs a periodic rotation around the z-axis as well as an additional precession. The inclination angle of the rotation axis with respect to the vorticity vector of the flow and thus the precession are gradually reduced until the small semi-axis of the ellipsoid becomes parallel to the z-axis and the stable orbit is reached. A time scale T_J is defined measuring the time until the angle between the small semi-axis and the z-axis drops under 5° and a Jeffery-like orbit is reached. Table 3.11e) lists this time scale normalized with the shear parameter, s, and a comparison to the data approximated from [216]. The time scales are in good agreement. Further, the table reports time-average of the ratio of the vector norm of $\|\boldsymbol{\omega}_p \cdot d\mathbf{I}_p/dt\|$ and $\|d(\mathbf{I}_p \cdot \boldsymbol{\omega}_p)/dt\|$ obtained during the initial transient towards the stable orbit in $(0,\, 0.3T_J)$. The denominator $\|d(\mathbf{I}_p \cdot \boldsymbol{\omega}_p)/dt\|$ equals $\|\mathbf{M}_p\|$ which is always non-zero in the present case. Experimental observations in laminar pipe flow support the observations concerning the stable orbit in the low Re_s regime. It was found in [131] for suspended rods that the long axis is oriented in planes passing through the axis of the tube.

3.2.7 Turbulent open channel flow with ellipsoidal roughness elements

Introduction and problem definition

One of the central questions concerning turbulent flows over rough walls is to quantify the influence of the roughness on the turbulent structure [33, 241, 220]. Different roughness types or shapes of the roughness elements alter wall-bounded turbulence in distinct ways [123].
The goal of this section is to provide first steps and a feasibility study towards the application of more general particle shapes in the representation of rough walls, sediment beds, or porous media.
The majority of the work on phase-resolved simulations of multiphase flow concentrates on spherical particles throughout. Heitkam et al. study the stability and therefore probability of occurrence of spherical packing structures [101]. The turbulent flow over a rough bed in an open channel is investigated in [295, 294, 297, 33] with respect to sediment erosion processes. In [33] the focus lies on forces acting on individual elements of the fixed sediment bed additionally to the influence of the roughness on the turbulence. Further, Vowinckel et al. [295, 294, 297] study many mobile particles moving and colliding on top of the regular sediment bed and which then form sediment structures like ripples and ridges.
In the present work, turbulent open channel flow over a rough bed is studied. The flow is driven by a volume force f_x which is adjusted in time by a PI-controller to give a constant bulk Reynolds number of $Re_b = 2900 = u_b H/\nu$, where $H = 1$ m is the free height over the sediment bed and $u_b = 1$ m/s is the bulk velocity. Periodic boundary conditions are applied in x- and z-direction and a free slip condition is set at the upper boundary as in [134, 33, 297, 312]. A no slip wall is situated at the lower boundary underneath the roughness elements. The domain extends $\mathbf{L} = (6,\, 1.1,\, 3)\, H$ which is about the lower limit to get decorrelated turbulent structures in streamwise and spanwise direction [134]. The y-coordinate counts from the top of the regularly packed ellipsoids forming the sediment bed, i.e. the lower boundary is situated at $y = -2a$.
The number of roughness elements forming the sediment bed is $N_p = 3600$ of fixed, oblate ellipsoids with an aspect ratio of $X = 2$, and a long semi-axis of $a = 0.05H$.
If the roughness elements do not extend over the viscous sublayer of the turbulent boundary layer, the roughness is classified as smooth. When reaching into the buffer layer, where both viscous and turbulent stresses are affected, the flow is denoted as transitionally rough. In the fully rough regime, the mean flow and the turbulent statistics are altered significantly by the roughness, which then extends well into the logarithmic layer. Thus according to [194, 123], the flow is expected to be within the transitionally rough regime with relative roughness heights of about twenty times the viscous length scale.
Different orientations of the ellipsoids are studied as shown in Figure 3.12:

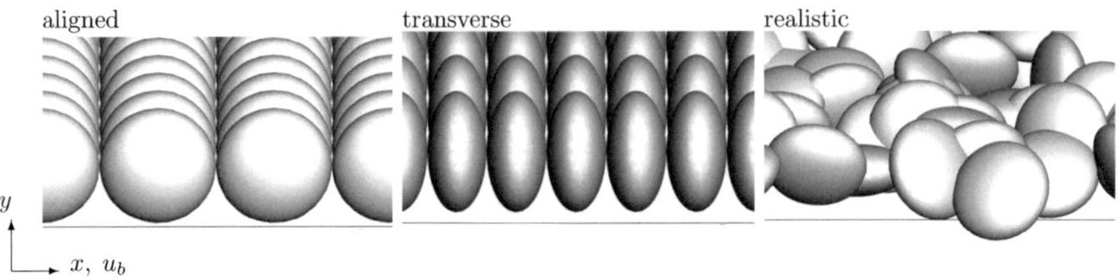

Figure 3.12 Orientation of ellipsoidal roughness elements in sediment bed with flow from left to right.

- *aligned* orientation, small semi-axis pointing in spanwise direction
- *transverse* orientation, small semi-axis pointing in streamwise direction
- *realistic* orientation from sedimentation of ellipsoids

The two regular packings (aligned, transverse) consist of a single layer of ellipsoids in rectangular arrangement. The more realistic orientation is obtained from a precursor simulation of freely, sedimenting ellipsoids. In this precursor simulation, three horizontal layers of ellipsoids are initially placed in aligned orientation around $\langle y_p \rangle \approx 0.7L_y$ and then settle with density ratio $\pi_\rho = 2.0$ in a laminar channel with no-slip walls at the upper and lower boundary. The reason for this changed setup is that higher shear rates yield larger rotation rates and therefore larger diversity in orientation at low sedimenting heights. Only a subset of the original size of the computational domain with $L_x = L_z = 1.5$ is considered with a corresponding fraction of ellipsoidal particles and the resulting distribution of ellipsoids is then repeated periodically to obtain the sediment bed in the large computational domain. In this pre-cursor simulation inter-particle and particle-wall collisions were accounted for employing collision model CM-1 described in Section 7.4. To demonstrate the validity of the ellipsoid collisions, the interaction of a single ellipsoid with an inclined wall was studied and the results are presented in the subsequent paragraph. A collision model with only a weak repelling potential is employed in the pre-cursor simulation of the sedimenting particles which yields an overlap of ellipsoids in the final distribution (see Figure 3.12) mimicking spheroidal roughness elements in comparison to fully accurate ellipsoids. This overlapping allows for an average vertical particle position of $\langle 2y_p \rangle / H \approx 0$ which is then used in the computation with fixed ellipsoids in the large computational domain. The pre-cursor simulation was stopped as this average vertical particle position was reached, i.e. before the particles came fully to rest. The average roughness height is therefore comparable to the simulations with aligned and transverse orientation in regular packings displayed in Figure 3.12. Nevertheless, the sediment bed is characterized by a higher heterogeneity of porosity as there are several holes in the sediment and in contrast some spots with two ellipsoids on top of each other. The average porosity λ_{por} of the packings is listed in the table of Figure 3.14 and is highest for the realistic orientation. It is evaluated in the interval $y/H \in (-0.1,\ 0)$ according to the definition $\lambda_{por} = \langle 1 - V^p_{i,j,k} / V^{cell}_{i,j,k} \rangle$ and characterizes the fraction of fluid volume contained in a cell at position (i, j, k) of the Cartesian grid. For the realistic packing, the orientation of the ellipsoids is almost random with a small preference towards the transverse orientation. Note that in contrast to the rather stochastic distribution of roughness elements used here, sediment structures like ridges, ripples or dunes form on larger scales of real sediment beds with moving particles [297].

Rigid ellipsoid interacting with an inclined wall in viscous fluid

The intention of this section is to provide a qualitative impression of the interaction of a rigid ellipsoid with an inclined wall in viscous fluid. With respect to the numerics, the test case serves as an examination of the routines involved to calculate the particle-wall distance and the contact point, as well as the collision forces and respective moments outlined in paragraphs 7.2.2 and 7.4. The detailed study of inter-particle and particle-wall collisions with non-spherical particles is beyond the scope of this work. However, a check for reasonable and sensible results is undertaken here to ensure adequate results in the simulation runs with multiple particles in bounded domains presented in this text.
A sketch of the configuration is given in Figure 3.13. It comprises a rigid ellipsoid first settling in viscous fluid and then interacting repeatedly with a tilted flat plate. The inclination of the wall is realized employing a gravity vector pointing downwards and tilted by 45° with respect to the y-axis. The domain with extents $\mathbf{L} = (25.6,\, 6.4,\, 6.4)\, d_{eq}$ is bounded by no-slip walls in y-direction and periodic boundary conditions are applied in x- and z-direction. An equidistant grid with $\mathbf{N} = (512,\, 128,\, 128)$ is employed. The spatial resolution therefore corresponds to $d_{eq}/\Delta x = 20$ and a constant time step of $\Delta t = 1 \cdot 10^{-3}$ is applied.
The oblate ellipsoid of aspect ratio $X = 2$ is heavier than the surrounding fluid with $\pi_\rho = 8$. A Galilei number of $G = 250$ characterizes the ratio of buoyancy to viscous forces. Only wall-normal collision modeling is used where the forces are obtained from a quadratic repelling potential with $k_{col.n} = 100$ (CM-1 in Section 7.4). The particle-wall distance is computed based on the ellipsoid contact function described in Section 7.2.2.
An initial particle position $\mathbf{x}_{p,0} = (0.8,\, 5.0,\, 3.2)$ is set and the ellipsoid is oriented with its small semi-axis b parallel to gravity. Both, the ellipsoid and the fluid, are initially at rest. During the simulation, the translation in z and the rotation around the x- and y-coordinate axes are switched off to reduce the complexity of this generic test case. After settling on a straight trajectory, the ellipsoid collides with the inclined wall. The collision moment leads to a rotation of the ellipsoid following the slope of the wall and to an immediate second wall contact (Figure 3.13a)) as expected from similar simulations in [310]. The Stokes numbers characterizing the first, wall-normal collisions is obtained as $St = \pi_\rho d_{eq} |\mathbf{u}_{p,n}^{in}|/(9\nu) = 126$. Interpreted by physical means, the influence of viscous dissipation is rather small and a corresponding spherical particle would retain about 70% of its kinetic energy calculated from the normal approach velocity $u_{p,n}^{in}$ [135, 126, 125, 92]. A very similar value is observed here for the wall-normal rebound of the ellipsoidal particle with $|\mathbf{u}_{p,n}^{out}| = 0.63|\mathbf{u}_{p,n}^{in}|$. Wynn [310] reports a value of $|\mathbf{u}_{p,n}^{out}| \approx 0.72|\mathbf{u}_{p,n}^{in}|$ for an ellipsoid of the same aspect ratio, $X = 2$, and angle of incidence on the wall, 45°, but with the simulations being conducted without viscous effects and the results being averaged over all impact orientations of the ellipsoid. For ellipsoidal particles, the normal rebound velocity may be even higher than the normal impact velocity due to transfer of energy from the tangential direction [310] and a change in orientation of the ellipsoid. For illustration, an ellipsoid in cross-flow with its small semi-axis b parallel to the mean flow experiences a larger drag than the same ellipsoid having its long semi-axis a pointing towards the incident flow as shown in Section 3.4.3 and [316].
After an intermediate transient, the further course of the numerical experiment is characterized by the ellipsoid sliding along the wall and performing successive backward flips as indicated in Figure 3.13b). The insets Figure 3.13c) and d) give an impression of the flow

field by contours of the absolute value of velocity where c) shows an instant shortly after impact and d) one during the lift-off for a backflip. As the leading edge of the ellipsoid is raised with respect to the wall, more fluid squeezes in between the wall and the ellipsoid hull. As a consequence the particle experiences a lift force which moves it away from the wall and a backward rotation follows. A famous and very similar example from the world of motor sport is the backflip-accident of a Mercedes GTR at the 24 Hours of Le Mans in 1999.

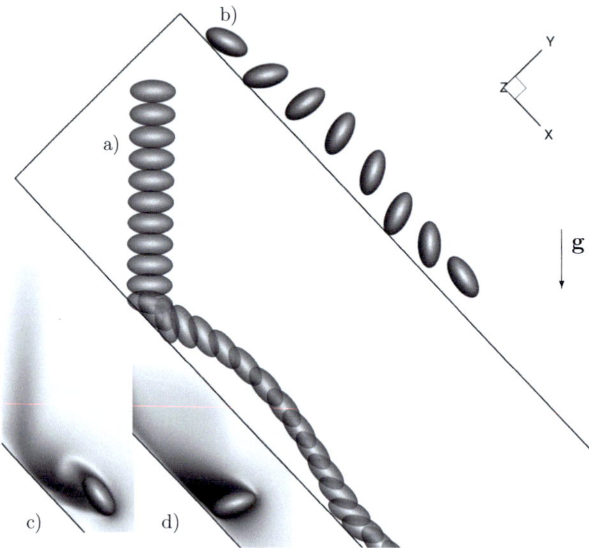

Figure 3.13 Rigid ellipsoid interacting with an inclined wall. a) Ellipsoid trajectory around impact. b) Trajectory during second passage of domain with ellipsoid performing backward flips. c) - d) Impression of flow field by contour of absolute value of velocity with color scale in interval $(0, u_g)$: c) shortly after first impact, d) during lift-off for backflip within second passage.

In conclusion, the goals of the test were fulfilled. A correct prediction of the contact point was obtained for different orientations of the ellipsoid and also in case of wall penetration. The particle-wall collisions appear physically sound and so does the particle trajectory. A numerically stable run was realized. A reduced spring stiffness of the collision model by a factor of ten to $k_{col.n} = 10$ (not shown) yielded higher values for the divergence of the velocity field due to larger particle-wall penetration. The overall physics were not significantly affected and a very similar first rebound and successive backflips in the second passage of the domain were observed. With respect to collisions of non-spherical particles, substantially more work needs to be invested into the improvement and validation of the collision modeling, as well as the interpretation of the involved physics. Special attention should be given to the transfer of translational kinetic energy to rotational kinetic energy during the collision. Quantitative comparison to experiments or other simulations should be considered in the future even though there is only scarce data for rigid ellipsoids.

Results on rough wall, turbulent open channel flow

An overview of the different runs for the simulation of turbulent channel flow over different roughness elements is given in Table 3.14. The computational domain was discretized with an equidistant grid of $\mathbf{N} = (1024, 188, 512)$ yielding $y^+ = \Delta y / l_\tau = 1.5$ for the largest Re_τ obtained with the realistic sediment bed. The time-step size was variable and adjusted according to $CFL = 1.0$. Table 3.14 also lists the relative roughness height H^+_{rough} corresponding to the average height of the roughness elements (twice the long semi-axis for the ellipsoids and the diameter for the spheres) normalized with the viscous length scale l_τ of the turbulent channel flow. The equivalent sand-grain roughness usually used for rough walls is not defined consistently for the studied cases, since a general description of three-dimensional roughness is rather difficult [23, 241, 298, 70]. The results obtained with ellipsoidal roughness elements are compared to the smooth channel solution and to a simulation presented in [134] with spherical roughness elements in hexagonal packing.

run	Re_τ	y^+	H^+_{rough}	λ_{por}
transverse	205	1.2	20.5	0.476
aligned	202	1.2	20.2	0.476
realistic	259	1.5	25.9	0.584
spheres [134]	202	1.2	23.3	0.395
smooth	184	1.0	0	1

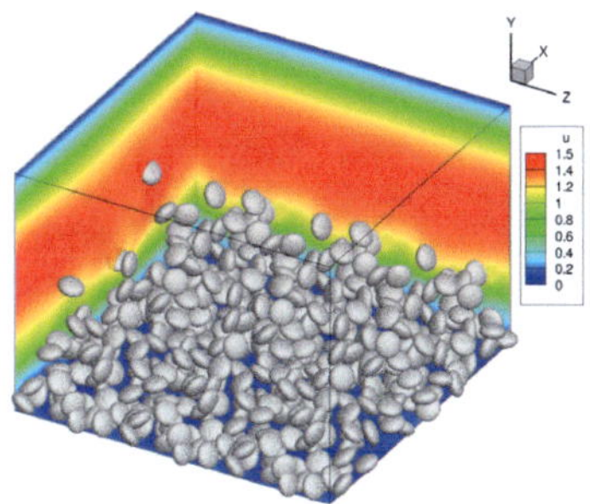

Figure 3.14 Overview of runs listing friction Reynolds number, Re_τ, average y-plus, H^+_{rough}, relative roughness height, H^+_{rough}, and average porosity of the bed, λ_{por}. Instantaneous view on sedimenting ellipsoids in laminar channel to obtain a realistic bed with contours of streamwise velocity u.

All rough wall configurations studied here lead to an increase in drag compared to the smooth wall reference. The regular packings with aligned and transverse orientation of the ellipsoidal roughness elements do not differ significantly and give an increase of $Re_\tau = u_\tau H/\nu$ of about 11% in relation to smooth wall turbulent channel flow at $Re_\tau = 184$. This increase obtained with ellipsoids of aspect ratio $X = 2$ also agrees well with the outcome for spherical roughness elements obtained in [134]. The effect of particle shape is therefore negligible for the parameters studied here in regular packing. The realistic sediment bed with almost randomly oriented and irregularly distributed ellipsoids yields an increase of 41% in Re_τ compared to the smooth configuration and and increase of 27% compared to the sediment bed consisting of regularly packed ellipsoids (see table in Figure 3.14). The average wall shear stress which defines the friction velocity $u_\tau = \sqrt{\langle \tau_w \rangle / \rho_f}$ was determined from the global momentum balance $\langle \tau_w \rangle \approx \langle f_x \rangle H$ therefore including the pressure loss due to interstitial flow and due to the no-slip wall below the ellipsoids. A more detailed discussion on the friction velocity and the positioning of a virtual wall for the flow over roughness elements is given in [33].
Figure 3.15 shows instantaneous velocity fields u for the three cases and gives an impression

of the respective sediment beds. The horizontal contour in these plots is placed at $y/H = 0.05$ corresponding approximately to the maximum in the streamwise velocity fluctuations shown in the graphs of Figure 3.17. The two regular packings of ellipsoids show the typical high speed and low speed streaks extending almost along the entire streamwise size of the computational domain. Qualitatively, the streamwise size of the structures decreases for the realistic bed as Re_τ increases. Figure 3.16 shows complementary data and an iso-contour of $u/u_b = 0.2$. Here, a significant difference is visible between the aligned and transverse orientation of ellipsoids. In the aligned case, the iso-surface has a wavy character strongly correlated to the low speed streaks whereas in the transverse case, streamwise bands situated in the spanwise gaps between the ellipsoids stretch straight along the full length of the domain. The realistic bed yields a more heterogeneous and uneven iso-surface with also some smooth regions.

The mean velocity profiles and the Reynolds stresses are given in Figure 3.17. The average streamwise velocity $\langle u \rangle / u_b$ becomes close to zero at the top of the single sediment layer for all regular packings. However, the realistic bed has a non-zero contribution down to $y \approx -a$, i.e. due to the increased porosity more interstitial flow is observable. The average velocity profiles and all Reynolds stress profiles collapse for the regular packings of spherical and ellipsoidal roughness elements with slightly larger fluctuations compared to the smooth wall case. For the realistic case, a significant increase in the velocity fluctuations of all components is observed. Especially the wall-normal contribution to the turbulent kinetic energy approximately doubles compared to the smooth wall case. With increasing Re_τ the maxima in the Reynolds stresses are located closer to the wall.

Conclusions and outlook for rough wall, turbulent open channel flow

Phase-resolving simulations of turbulent, open channel flow over rough walls were performed in the transitionally rough regime. Ellipsoidal roughness elements were employed in contrast to sediment beds consisting of spheres often used in the literature. An increase of turbulent fluctuations and Re_τ is found for all roughness configurations compared to the smooth wall case. A negligible impact of the shape and orientation of the roughness elements on the turbulence statistics is observed for regularly packed ellipsoids of aspect ratio $X = 2$ and spheres with a similar roughness height. In contrast, an irregular distribution of ellipsoidal roughness elements, modeling a more realistic sediment bed, yields a significant increase of Re_τ at a comparable absolute roughness height. The shape of the roughness elements is expected to have a marked influence when turning towards less spheroidal shapes within the transitionally rough regime. Sharp shapes in aligned orientation with a transverse spacing, as for instance found on the skin of fast sharks, are called riblets and can even yield a drag reduction [79, 78, 11, 10].

A simple local, geometric argument in the alignment of spheres promotes a bias in the structure of large scale clustering [101]. It is also shown that the type of collision modeling and the treatment of particle contact have a significant impact on large scale particle structures forming on top of the sediment bed [294, 297, 134]. Hence, similar small scale effects concerning the particle shape can be expected to have large effects on sediment motion, incipient sediment erosion etc. The given implementation enables the study of basically arbitrary particle shapes within the proposed immersed boundary method.

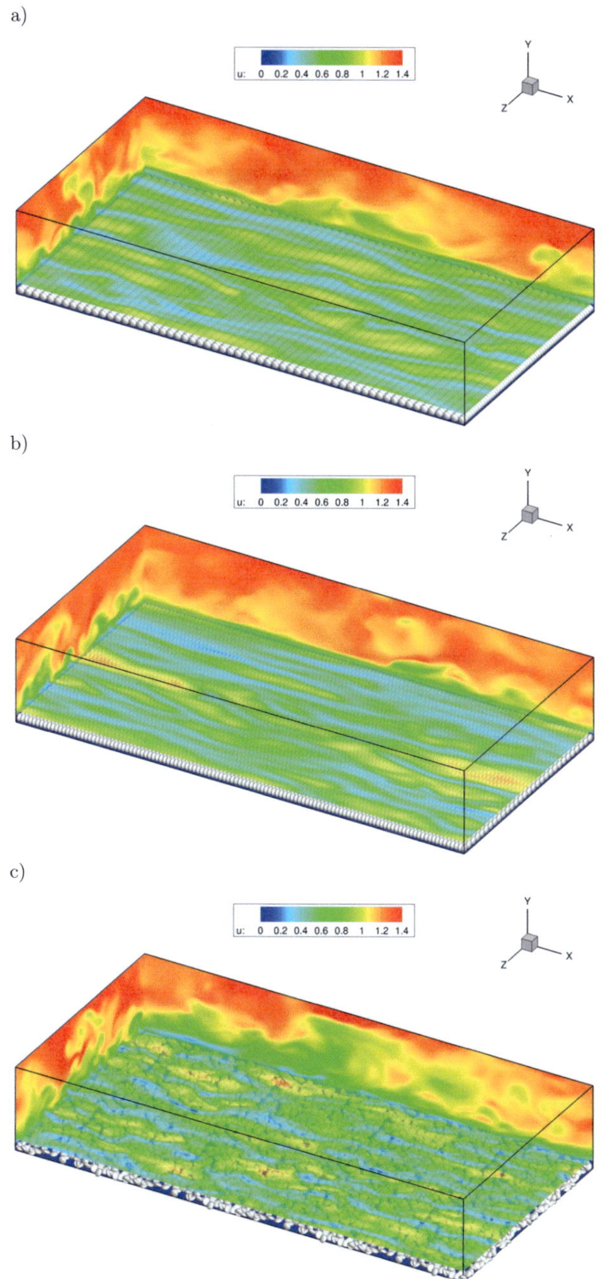

Figure 3.15 Instantaneous velocity field u for a) aligned, b) transverse and c) realistic orientation. Horizontal contour situated at $y/H = 0.05$.

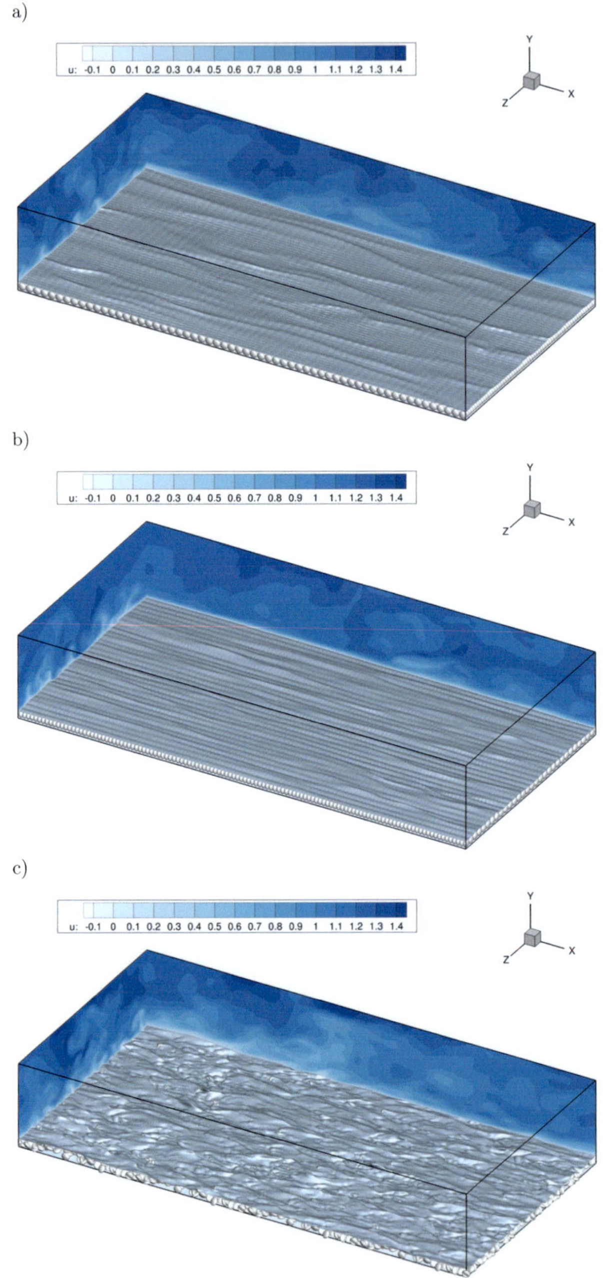

Figure 3.16 Instantaneous velocity field u and iso-contour $u/u_b = 0.2$ for a) aligned, b) transverse and c) realistic orientation.

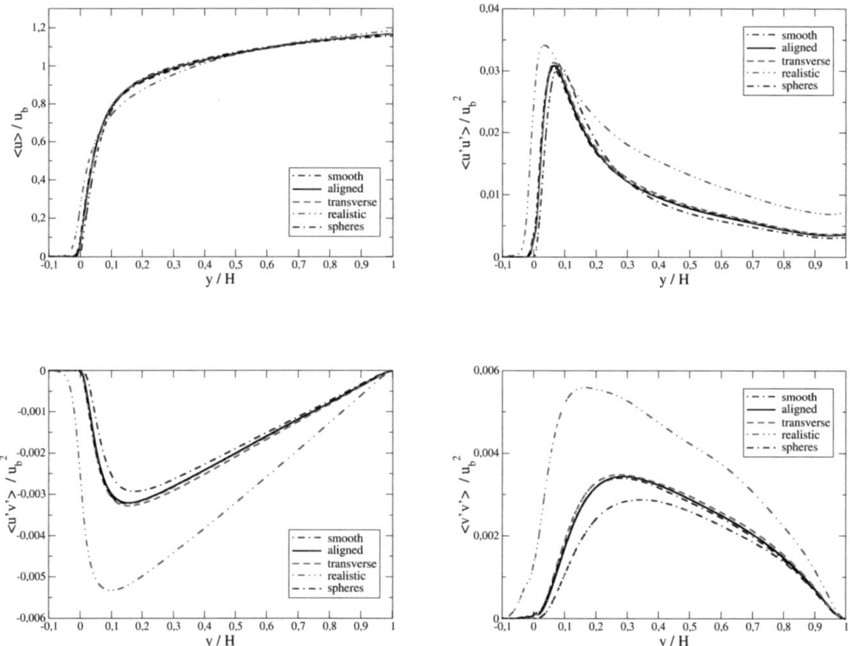

Figure 3.17 Mean velocity profile and Reynolds stresses for the three cases aligned, transverse and realistic in comparison to a smooth channel and data for spherical roughness elements from [134].

3.3 Virtual mass concept for very light particles

3.3.1 Introduction

One of the main objectives of the present work is the simulation of bubble dynamics employing a variant of the IBM developed in [134, 136]. For particles with a density markedly larger than that of the surrounding liquid, the particle dynamics are mainly governed by the inertia of the immersed body itself. For bubbles on the other hand, the immersed object has a vanishing density and hence the crucial inertia is determined by the attached fluid mass which is accelerated with the bubble. This effect is generally addressed as added mass. The dynamics of a heavy, settling sphere [181] served as a test case for the validation of the present IBM in [134, 136]. It is, however, pointed out in [130, 129] that rising, light, solid spheres do not obey the laws for free settling and their average drag deviates from the standard drag curve obtained for a fixed sphere [37]. Light spheres behave more like gas bubbles in contaminated systems [128]. The regimes of freely rising and falling spheres and respective wake visualizations are discussed in [286, 287, 288]. A very extensive experimental study and a review on the dynamics and wakes for the motion of a sphere in quiescent fluid is provided in [113]. The study reveals that heavy falling spheres always settle on a dominantly rectilinear trajectory, possibly oblique with low amplitude oscillations. Below a density ratio of $\pi_\rho \lesssim 0.4$, spheres follow a zig-zag trajectory for $Re_t \in (260,\, 1550)$, whereas the critical density ratio is somewhat higher, $\pi_\rho \lesssim 0.6$, at larger Reynolds numbers, $Re_t > 1550$.
Numerical simulations of light spherical particles are rather rare. A fictitious domain method applicable to a wide range of density ratios in particulate flows was presented in [7]. Jenny et al. [122] performed a systematic study of a freely moving sphere over a wide range of density ratios. At the low end of the density ratio, the authors report that their numerical results for rigid spheres are in close agreement to literature data for spherical bubbles [309]. Although the experiments [113] and the numerical predictions [122] contain certain discrepancies [63], there is an agreement that above a critical Reynolds number only light spheres undergo substantial path oscillations, including a regime of periodic zig-zag motion. Further, there is consensus that the dynamics of the particle motion are significantly different for heavy and light spheres.
Low density ratios ($\pi_\rho \lesssim 0.4$) are inaccessible by the original IBM [134, 136]. In the following Section 3.3.2, a method is developed to overcome this limitation and to access the bubble-like regime. The corresponding extension of the IBM, suggested in Section 3.3.3 obeys the following properties:

- Applicability to arbitrary density ratios, especially $\pi_\rho \ll 1$.
- Multiple particles in bounded domains can be handled.
- Accuracy, efficiency and robustness of the present IBM [136] are retained keeping the explicit time integration and coupling scheme.

The improvement is based on a modified time integration scheme for the particle momentum equation. It is termed 'virtual mass approach'. The corresponding virtual mass is of purely numerical nature and shall not be confused with the physical added mass. The modified time integration scheme is first applied to an analytical model problem. The extended IBM is then validated against DNS data for the initial acceleration of a sphere towards its terminal velocity in quiescent liquid for various density ratios. The current section is part of the manuscript [249].

3.3.2 Development of the numerical method

Discussion of the problem with low particle densities

In a Lagrangian, phase-resolving simulation, the translational and rotational motion of a single spherical particle can be described by its linear and angular momentum equation formulated in the laboratory system

$$m_p \frac{d\mathbf{u}_p}{dt} = \rho_f \int_S \boldsymbol{\tau} \cdot \mathbf{n} \, dS + (\rho_p - \rho_f) \, V_p \, \mathbf{g} \,, \tag{3.33a}$$

$$I_p \frac{d\boldsymbol{\omega}_p}{dt} = \rho_f \int_S \mathbf{r} \times (\boldsymbol{\tau} \cdot \mathbf{n}) \, dS \,, \tag{3.33b}$$

as recalled from the introductory Section 1.2.2.
The Basset-Boussinesq-Oseen (BBO) equation describes the particle motion,

$$m_p \frac{d\mathbf{u}_p}{dt} = \sum_i \mathbf{F}_i = \mathbf{F}_D + \; \mathbf{F}_M + \mathbf{F}_L + \mathbf{F}_B + \mathbf{F}_{AM} + \mathbf{F}_P + \mathbf{F}_{BH} + \dots \,, \tag{3.34}$$

where $\mathbf{F}_D$ denotes the drag force, $\mathbf{F}_M$ the Magnus force, $\mathbf{F}_L$ the lift force, $\mathbf{F}_B$ buoyancy, $\mathbf{F}_{AM}$ the added mass force, $\mathbf{F}_P$ the pressure gradient force and $\mathbf{F}_{BH}$ the Basset history force. For point particles, i.e. no resolution of the fluid-particle-interface is attained, the BBO-equation provides the numerical framework. All forces are evaluated using appropriate correlations also employing information from the underlying fluid. For a discussion of those correlations see, e.g., Clift et al. [37].
In the present context, the added mass force is of specific interest which is the force necessary to accelerate the fluid surrounding the particle,

$$\mathbf{F}_{AM} = m_{AM} \left(\frac{D\mathbf{u}}{Dt} - \frac{d\mathbf{u}_p}{dt} \right) = C_{AM} \, \rho_f \, V_p \frac{d\mathbf{u}_{rel}}{dt} \,, \tag{3.35}$$

with C_{AM} being the added mass coefficient and $\mathbf{u}_{rel}$ being the relative velocity between the particle and the fluid. The added mass coefficient depends primarily on the geometry of the individual particle and on the vicinity of other particles or bounding walls [257]. It does not depend on the particle density, the acceleration, the Reynolds number or the boundary condition at the particle surface (no-slip or free-slip) [186]. For a single sphere in an unbounded domain and irrotational flow, one can derive $C_{AM} = 0.5$ [37, 114]. This value for C_{AM} seems also valid under relaxed conditions and is frequently used [299, 170, 155]. As a result of the inertia of the surrounding fluid, the acceleration of a very light particle remains finite.
In a phase-resolving DNS, the surface integrals in equation (3.33) are evaluated directly from the flow field yielding all forces and torques acting on the particle in an integral manner. For example, the added mass force is inherently included in the surface integrals in (3.33). It is then basically possible to access, e.g., the individual forces stemming from different physical mechanisms which enter equation (3.34).

The solution of (3.33) becomes problematic when $\pi_\rho \to 0$ and thus $m_p \to 0$ as the left hand side (lhs) of equation (3.33) becomes singular. It has been pointed out that the IBM of Uhlmann [282] with explicit coupling becomes unstable for spherical particles with $\pi_\rho \lesssim 1.2$, i.e. already for spheres heavier than the surrounding fluid. The stability range was extended by Kempe et al. to enable the simulation of moderately light particles down to $\pi_\rho \approx 0.4$ [134, 136]. Interestingly, this density ratio corresponds closely to the threshold for substantial path oscillations obtained in the experiments [113] discussed in the introduction. With the original formulation of the IBM, a lower density ratio presents with immediate strong oscillations in $\mathbf{u}_p$ and $\boldsymbol{\omega}_p$ which lead to the divergence of the solution within the first few time steps of the run. The scenario is basically insensitive to refinement of the spatial and temporal resolution. For the ascent of a sphere in quiescent liquid, it was derived in [120, 122] that explicit time integration must be expected to become unstable for $\pi_\rho < 0.5$. For light spheres, these authors use implicit time integration of the particle momentum equation and implicit coupling to the fluid phase [121, 120, 122]. More details are provided in Section 3.3.4 below.

The problems occurring at low density ratio ,π_ρ, can further be avoided by dealing with the true added mass or the added mass tensor in the more general case of non-spherical particles. A large body of work on the motion of particles freely rising or falling in fluids is based on the solution of the generalized Kirchhoff equations [63, 186]. In this framework, the momentum conservation of the combined fluid-particle system is considered with applicability to general shapes. The dynamics and paths of an oblate ellipsoid were studied in [187] and the motion of finite height cylinders was studied in [62, 67]. As a prerequisite, the added mass tensor of the particle needs to be obtained from theoretical considerations or once from the short-term response of the particle-fluid system to a given body acceleration [168, 186]. There are certain limitations to this approach. The studies are conducted in an unbounded domain corresponding to, e.g., the condition that the fluid is at rest at infinity. Only single, non-deformable particles are considered since for a temporally varying bubble shape or for an approaching second particle, the added mass changes in time.

In the simulation of point bubbles in complex, turbulent flows based on equation (3.34), the added mass force, $\mathbf{F}_{AM}$, can simply be shifted to the lhs enabling stable time integration. The added mass coefficient is assumed constant or obtained from a correlation, and the relative velocity, $\mathbf{u}_{rel}$, is computed using the fluid velocity, $\mathbf{u}$, of the cell which comprises the particle. For phase-resolving simulations in turbulent, bounded flows, a plausible idea could hence be the evaluation of the actual added mass force, splitting it from the integral force acting on the particle and also shifting it to the lhs of equation (3.33a). However, $\mathbf{u}_{rel}$ is not easily obtained in a general flow field since it is conceptually hard to define the fluid velocity at the position of the particle. Furthermore, the coefficient C_{AM} is difficult to assess in the general case of multiple particles of complex, possibly time-dependent shape in bounded domains. Such an approach can hardly be formulated in a consistent way compatible with the requirements of a DNS. Therefore, a simple, but effective alternative is developed in the next section.

Basic idea

In the following, the concept is discussed for the translational momentum equation of a spherical particle in quiescent fluid. It will be apparent subsequently that the idea is also applicable to the equation governing the rotational motion, to more general particle shapes and to complex flows in bounded geometries.

First, a virtual force $\mathbf{F}_v$ is added to both sides of (3.33a),

$$m_p \frac{d\mathbf{u}_p}{dt} + \mathbf{F}_v = \rho_f \int_S \boldsymbol{\tau} \cdot \mathbf{n}\, dS + (\rho_p - \rho_f)\, V_p\, \mathbf{g} + \mathbf{F}_v\,, \tag{3.36}$$

where the virtual force is expressed similarly to the added mass force as

$$\mathbf{F}_v = C_v\, \rho_f\, V_p \frac{d\mathbf{u}_p}{dt}\,. \tag{3.37}$$

Note that the particle velocity, $\mathbf{u}_p$, is used in this definition and not the relative velocity. The ambiguous evaluation of the latter is not necessary. This yields

$$(\rho_p + C_v\, \rho_f)\, V_p \frac{d\mathbf{u}_p}{dt} = \rho_f \int_S \boldsymbol{\tau} \cdot \mathbf{n}\, dS + (\rho_p - \rho_f)\, V_p\, \mathbf{g} + \mathbf{F}_v\,. \tag{3.38}$$

This formally removes the singularity on the lhs when $\rho_p \to 0$ for $C_v > 0$. The second step is a suitable discretization in time of $\mathbf{F}_v = C_v\, \rho_f\, V_p\, d\mathbf{u}_p/dt$ on the right hand side (rhs). It has then to be shown that consistency and stability are achieved for arbitrary π_ρ with the resulting full temporal discretization.
Note that $\mathbf{F}_v$ does not need to have physical meaning, but is a purely mathematical term designed to stabilize the temporal integration. In the literature, the added mass effect is sometimes also addressed as "virtual mass effect". Here, the term "virtual mass force" is employed to denote the numerical stabilization mechanism described above, since the definition (3.37) shows some similarity to the actual added mass force. It should however not be confused with the true physical added mass force, $\mathbf{F}_{AM}$, in (3.35). The latter is resolved directly by the IBM, but cannot be separated easily from the other forces in a complex flow.
In the following, the numerical solution of the original problem (3.33a) and the modified problem (3.38) are addressed. A general time integration scheme evolves the solution for $\mathbf{u}_p$ from time level t^n to time level $t^{n+1} = t^n + \Delta t$ by $\mathbf{u}_p^{n+1} = \mathbf{u}_p^n + \Delta t\, \boldsymbol{\psi}_p^*$ employing the increment function $\boldsymbol{\psi}_p^*$, which depends on the chosen scheme. One main goal is to retain the order of the time integration scheme when the virtual mass concept is applied. The order of convergence q remains unchanged when substituting the increment function $\boldsymbol{\psi}_p^*$ by another increment function,

$$\boldsymbol{\psi}_p = \boldsymbol{\psi}_p^* + \mathcal{O}\left(\Delta t^q\right)\,, \tag{3.39}$$

in the discrete evolution of $\mathbf{u}_p$ [50] yielding

$$\mathbf{u}_p^{n+1} = \mathbf{u}_p^n + \Delta t\, \boldsymbol{\psi}_p\,. \tag{3.40}$$

Further goals to be met are stability, accuracy and ease of implementation. Several time schemes have been developed on this basis and will be presented in the next sections.

Description of the generic test case

To assess the virtual mass concept, a simple test case is introduced for which an analytical solution exists. It features a sphere sedimenting or rising in quiescent, unbounded fluid. Stokes flow conditions are assumed to prevail ($Re \ll 1$). Dropping the index p for the particle velocity, the translational particle momentum equation reads

$$m_p \dot{u}(t) = F_{D,Stokes} + F_B\,, \tag{3.41}$$

with the Stokesian drag force and the buoyancy force,

$$F_{D,Stokes} = -6\pi r_p \mu_f \, u \;, \tag{3.42}$$

$$F_B = \frac{4}{3}\pi r_p^3 g \left(\rho_p - \rho_f\right) . \tag{3.43}$$

In (3.41) and below, the temporal derivative is denoted by a dot. This setting, together with the initial condition $u(t=0)=0$, will be addressed as 'Stokes sphere problem' in the following. Equation (3.41) sets a suitable scene to develop and assess different time schemes. Adding $F_v = C_v m_v \dot{u}$ on both sides of (3.41) yields the modified problem including the virtual mass,

$$(m_p + m_v)\,\dot{u}(t) = F_{D,Stokes} + F_B + F_v \; . \tag{3.44}$$

This can be rewritten as

$$\dot{u}(t) = f(u(t), \dot{u}(t)) = c_u u + c_g g + c_a \dot{u}(t)\,, \tag{3.45}$$

where

$$c_u = -\frac{9\mu_f}{2r_p^2\left(\rho_p + C_v\rho_f\right)}\,, \quad c_g = -\frac{\rho_p - \rho_f}{\left(\rho_p + C_v\rho_f\right)}\,, \quad c_a = -\frac{C_v\rho_f}{\left(\rho_p + C_v\rho_f\right)}\,. \tag{3.46}$$

In the following, the abbreviation

$$f_v(t) = c_a \, \dot{u}(t) \tag{3.47}$$

is used. The original initial value problem with the initial condition for the particle velocity,

$$\dot{u}(t) = f(u(t)) \qquad \text{with} \qquad u_0 = u(t=0)\,, \tag{3.48}$$

is now replaced by the following modified initial value problem using the virtual mass approach,

$$\dot{u}(t) = f(u(t), \dot{u}(t)) \quad \text{with} \quad u_0 = u(t=0), \quad \dot{u}_0 = \dot{u}(t=0), \tag{3.49}$$

requiring an initial condition for the velocity and the acceleration.
The analytical solution of the problem (3.41) for $u_0 = 0$ is given by

$$u_{ref}(t) = u_\infty \left(1 - \exp(-\lambda\, t)\right) , \tag{3.50a}$$

$$\dot{u}_{ref}(t) = \lambda \, u_\infty \, \exp(-\lambda\, t)\,, \tag{3.50b}$$

where

$$u_\infty = \frac{2\left(\rho_p - \rho_f\right) r_p^2 g}{9\mu_f}\,, \quad \lambda = \frac{9\mu_f}{2r_p^2\rho_p}\,. \tag{3.51}$$

The analytical solution of this problem will be used as a reference below. It is plotted in Figure 3.18 for density ratios $\pi_\rho = \rho_p/\rho_f = 0.1$, 0.5 and 2.0 using the physical properties $\mu_f = 0.001$, $\rho_f = 1000$, $g = 9.81$ and $r_p = 0.0007$. These properties are used throughout for the Stokes sphere test case. As this section primarily deals with purely numerical issues, the units are omitted.

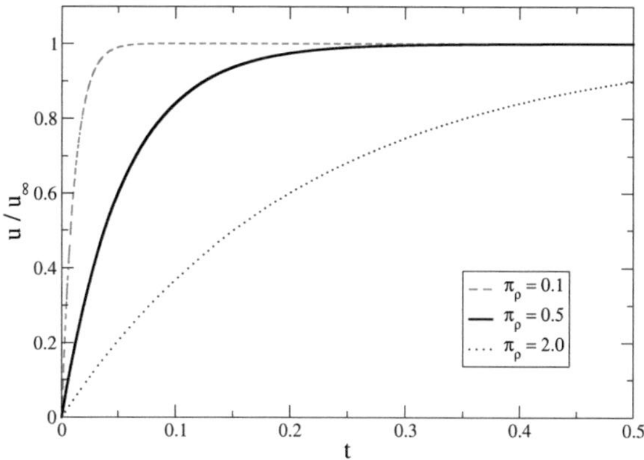

Figure 3.18 Exact solution of Stokes sphere problem (3.41). Vertical particle velocity over time for different density ratios π_ρ.

Parameters of the numerical experiments

To illustrate the following discussion and to motivate certain algorithmic developments, the original problem (3.41) and the modified problem (3.45) are solved with different time integration schemes. The physical properties were chosen as stated above with a density ratio of $\pi_\rho = 0.5$. The temporal discretization was conducted with $\Delta t = t_{end}/N_{\Delta t}$ and $t_{end} = 0.5$ for the number of time steps, $N_{\Delta t}$, provided in Table 3.3.
A virtual mass coefficient of $C_v = 0.5$ was used in all schemes denoted with the label *vm*. The order of convergence of the different schemes to follow is addressed by the maximum error compared to the reference solution given by (3.50),

$$\epsilon_{max} = \max\{|u(t) - u_{ref}(t)|\}, \tag{3.52}$$

The order q of the schemes is then numerically calculated from

$$q = \frac{1}{\ln 2} \ln \frac{\epsilon_{max}(2\Delta t)}{\epsilon_{max}(\Delta t)}, \tag{3.53}$$

using the two runs with the finest time step, $N_{\Delta t} = 1600$ and $N_{\Delta t} = 3200$.
The numerical results for the Stokes sphere problem obtained with the different time schemes defined below are assembles in Table 3.3.

Table 3.3 Stokes sphere problem. Maximum error ϵ_{max} in $u(t)$ for different time schemes with labels as defined in the text. Parameters $\pi_\rho = 0.5$, $C_v = 0.5$, $t_{end} = 0.5$.

Scheme	$N_{\Delta t} = 400$	800	1600	3200	**Order** q
AB2	3.99E-05	1.04E-05	2.65E-06	6.68E-07	1.99
AB2-vm	1.14E-04	2.94E-05	7.44E-06	1.87E-06	1.99
AM3-vm	1.51E-07	1.90E-08	2.38E-09	2.99E-10	3.00
CN-vm	1.61E-05	4.10E-06	1.01E-06	2.43E-07	2.06
LF-vm	6.50E-05	1.68E-05	4.25E-06	1.07E-06	1.99
RK3	1.01E-07	1.25E-08	1.56E-09	1.94E-10	3.00
RK3-a-vm	1.43E-07	1.84E-08	2.34E-09	2.94E-10	3.00
RK3-1-vm	2.13E-03	1.10E-03	5.56E-04	2.80E-04	0.99
RK3-1-AM3-vm	5.79E-04	3.00E-04	1.53E-04	7.72E-05	0.99
LF-RK3-vm	3.95E-05	1.02E-05	2.58E-06	6.50E-07	1.99
AB2-RK3-vm	4.59E-05	1.18E-05	3.00E-06	7.55E-07	1.99
Lag-RK3c-vm	4.10E-06	1.11E-06	2.87E-07	7.31E-08	1.97
Lag-RK3tp-vm	9.45E-05	2.42E-05	6.13E-06	1.54E-06	1.99

Multistep methods

Multistep methods construct a higher order approximation by using additional data points which have already been computed [68, 50]. For comparison, the original problem is solved here using a second-order accurate Adams-Bashforth scheme. The abbreviation *AB2* denotes the explicit two-step Adams-Bashforth scheme for constant time step size

$$f^n = c_u u^n + c_g g \,, \tag{3.54}$$

$$u^{n+1} = u^n + \frac{\Delta t}{2}\left(3f^n - f^{n-1}\right) . \tag{3.55}$$

For the Stokes sphere test case (3.41), second order accuracy is obtained as illustrated by the result in Table 3.3.

For the modified problem, employing a second order backward finite difference formulation [68] for the acceleration $\dot{u}^n$ from (3.45), yields the following explicit Adams-Bashforth scheme, *AB2-vm*:

$$f_v^n = c_a \frac{3u^n - 4u^{n-1} + u^{n-2}}{2\Delta t} \,, \tag{3.56}$$

$$f^n = c_u u^n + c_g g + f_v^n \,, \tag{3.57}$$

$$u^{n+1} = u^n + \frac{\Delta t}{2}\left(3f^n - f^{n-1}\right) . \tag{3.58}$$

With this two-step scheme with constant time step size, again second order accuracy is obtained for the Stokes sphere problem. Only a slightly larger maximum error is apparent from Table 3.3 compared to the original scheme AB2.

If the data point t^{n+1} is incorporated into the computation, an implicit formulation is obtained. The implicit three-step Adams-Moulton method, abbreviated as *AM3-vm*, reads

$$f_v^n = c_a \frac{11u^n - 18u^{n-1} + 9u^{n-2} - 2u^{n-3}}{6\Delta t} \,, \tag{3.59}$$

$$f^n = c_u u^n + c_g g + f_v^n \,, \tag{3.60}$$

$$u^{n+1} = u^n + \frac{\Delta t}{12}\left(5f^{n+1} + 8f^n - f^{n-1}\right), \tag{3.61}$$

with constant time step size. For this scheme, an iterative procedure is required as the rhs contains values at t^{n+1}. A third order backward finite difference formulation [68] for the acceleration $\dot{u}^n$ is used and overall 3rd order accuracy is attained for the analytical test case. With the third order scheme, the absolute maximum error is substantially smaller compared to the two-step methods (Table 3.3). Additional multistep methods including the virtual mass contribution, such as the Crank-Nicolson scheme (*CN-vm*) and the Leap-Frog scheme (*LF-vm*) are listed in the Appendix F .

With the virtual mass approach, the original order of the method is retained using backward finite differences of the same order for $\dot{u}^n$ on the rhs in the test case considered. Multistep methods are easy to implement and are quite efficient in CPU time per time step. The downside is the difficult startup, as values for $t < 0$ are not available within the first time steps. For the chosen test case, the analytical solution was used for the first four time steps, throughout. Further disadvantages are a rather large demand on memory resources for saving the values at the old time levels and an increased complexity when using variable time step size.

Runge-Kutta methods

Runge-Kutta methods use specific sub-step values in the time interval $[t^n,\ t^{n+1}]$ to construct a higher order approximation [50]. In contrast to the multistep methods, no additional values from older time steps are necessary. Therefore, the startup at $t = 0$ is straightforward. Furthermore, a variable time step size is easily accounted for. An s-stage, explicit Runge-Kutta method reads as

$$u^{n+1} = u^n + \Delta t \sum_{i=1}^{s} b_i k_i\,, \tag{3.62}$$

where for $i = 1, \ldots, s$,

$$k_i = f\left(t^n + c_s\Delta t,\ u^n + \Delta t \sum_{j=1}^{i-1} a_{ij} k_j\right). \tag{3.63}$$

Higher order accuracy is achieved by proper weighting of the increments k_i with the weights b_i in (3.62). The increments are obtained at sub-step time levels $t^n + c_i\Delta t$ employing nested information from the previous sub-steps weighted with a_{ij} where $a_{ij} = 0$ for $j \geq i$. Figure 3.19 lists the coefficients of the present three-stage Runge-Kutta method in form of the Butcher scheme [50]. In Appendix G, the conditions on the set of coefficients are re-derived for which the scheme is consistent and the higher order error terms vanish so that the three-stage scheme is third order accurate.

$$\begin{array}{c|ccc} c_1 & & & \\ c_2 & a_{21} & & \\ c_3 & a_{31} & a_{32} & \\ \hline & b_1 & b_2 & b_3 \end{array} \qquad\qquad \begin{array}{c|ccc} 0 & & & \\ \frac{8}{15} & \frac{8}{15} & & \\ \frac{2}{3} & \frac{1}{4} & \frac{5}{12} & \\ \hline & \frac{1}{4} & 0 & \frac{3}{4} \end{array}$$

Figure 3.19 Butcher scheme of the coefficients for a three-stage explicit Runge-Kutta method. The right graph contains the values used in the present work [308, 134].

In the original method implemented in our code PRIME, a three-stage Runge-Kutta method is employed with implicit treatment of the viscous terms by a Crank-Nicolson method [136, 134]. Overall second order accuracy in time is achieved [219, 282, 134]. More specifically, a low storage Runge-Kutta scheme is used [136] leading to a different nomenclature for the coefficients [308, 137, 305]. In simple words, the final weighted summation of the increments in (3.62) is re-written to allow a successive evaluation without storing intermediate values. For simplicity, the Butcher notation is kept here representing a general Runge-Kutta scheme. It is hence desirable to develop a virtual mass variant of this scheme which is at least of second-order accuracy.
With respect to the implementation of the virtual mass concept into the PRIME code, it is hence desirable to develop a virtual mass variant of this scheme which is at least of second-order accuracy.
For comparison, a standard three-step Runge-Kutta scheme (*RK3*) is employed. It reads for the original unmodified problem (3.41)

$$k_1 = c_u u^n + c_g g \,, \tag{3.64}$$
$$k_2 = c_u \left(u^n + a_{21}\Delta t\, k_1\right) + c_g g \,, \tag{3.65}$$
$$k_3 = c_u \left(u^n + a_{31}\Delta t\, k_1 + a_{32}\Delta t\, k_2\right) + c_g g \,. \tag{3.66}$$

With the coefficients from Figure 3.19, third order convergence is achieved for the Stokes sphere test case (Table 3.3).
Now the virtual mass contribution shall be included to enable lower particle density ratios. Proceeding as for the multistep methods by calculating $\dot{u}^n$ with a third-order backward finite difference formula at t^n and using this for f_v^n in all Runge-Kutta sub-steps yields the scheme

$$f_v^n = c_a \frac{11u^n - 18u^{n-1} + 9u^{n-2} - 2u^{n-3}}{6\Delta t} \,, \tag{3.67}$$

$$k_1 = c_u u^n + c_g g + f_v^n \,, \tag{3.68}$$
$$k_2 = c_u \left(u^n + a_{21}\Delta t k_1\right) + c_g g + f_v^n \,, \tag{3.69}$$
$$k_3 = c_u \left(u^n + a_{31}\Delta t k_1 + a_{32}\Delta t k_2\right) + c_g g + f_v^n \,, \tag{3.70}$$

$$u^{n+1} = u^n + \Delta t \left(b_1 k_1 + b_2 k_2 + b_3 k_3\right) \,, \tag{3.71}$$

named *RK3-1-vm* here. It turns out that only first order convergence is obtained for the Stokes sphere test case (Table 3.3).
Just using the temporal derivative $\dot{u}^n$ hence is insufficient for a higher order scheme. A more accurate prediction of $\dot{u}$ is therefore needed at the sub-step levels $c_i\Delta t$. The derivation of the original Runge-Kutta coefficients is only valid for the unmodified problem $\dot{u} = f(u(t))$ [50]. If the additional argument $\dot{u}$ is introduced into the rhs, the coefficients k_i have to be changed to achieve optimal order. A Runge-Kutta scheme can basically be seen as a predictor corrector scheme on the time interval $(t^n,\ t^{n+1})$ [68]. For instance, a two-stage Runge-Kutta scheme uses an explicit Euler predictor to calculate $u_{pred}^{n+1/2} = u^n + \frac{1}{2}\Delta t\, f\left(t^n,\ u^n\right)$. This prediction is then corrected using the midpoint rule for $u^{n+1} = u^n + \frac{1}{2}\Delta t\, f\left(t^{n+1/2},\ u_{pred}^{n+1/2}\right)$. The predictor is only first-order accurate and therefore cannot be used to accurately calculate $\dot{u}^{n+1/2}$ at the sub-step time level.
In the following, we illustrate that in principle the order of the Runge-Kutta scheme can be

retained with the virtual mass approach. This is done by using the analytical solution of the problem (3.50) to determine $f_{v,ref}\left(t^n + c_i \Delta t\right)$ in the second and third sub-step yielding

$$f_v^n = c_a \frac{11u^n - 18u^{n-1} + 9u^{n-2} - 2u^{n-3}}{6\Delta t} , \tag{3.72}$$

$$k_1 = c_u u^n + c_g g + f_v^n , \tag{3.73}$$

$$k_2 = c_u \left(u^n + a_{21}\Delta t k_1\right) + c_g g + f_{v,ref}\left(t^n + c_2 \Delta t\right) , \tag{3.74}$$

$$k_3 = c_u \left(u^n + a_{31}\Delta t k_1 + a_{32}\Delta t k_2\right) + c_g g + f_{v,ref}\left(t^n + c_3 \Delta t\right) , \tag{3.75}$$

termed *RK3-a-vm* here. Indeed, third order convergence and maximum errors comparable to the original scheme *RK3* are achieved for the test problem (Table 3.3), which highlights the need for an accurate prediction of $\dot{u}(t)$ at the sub-step time levels. This, however, is difficult since the sub-steps themselves are only of lower order, i.e. an accurate prediction of u is not directly available for a higher order interpolation of $\dot{u}(t)$.
As a solution to this problem, an additional higher order predictor step over the whole time step is suggested here to enable an accurate approximation of $\dot{u}(t)$ on the entire interval $(t^n,\ t^{n+1})$ as detailed in the next paragraph.

Predictor-corrector schemes

It is again emphasized, the goal is to achieve a scheme for the virtual mass approach with second order convergence as previously with the unmodified method. It is desired that the three-stage Runge-Kutta scheme currently implemented in the code PRIME can be kept. Supplemental steps in the time integration scheme are a common procedure to add specific features like lower absolute errors and automatic error control or increased stability [68].
One possibility is to combine an Adams-Bashforth-predictor and the trapezoidal rule for the virtual force which is then applied in the Runge-Kutta sub-steps:

$$f_v^n = c_a \frac{3u^n - 4u^{n-1} + u^{n-2}}{2\Delta t} , \tag{3.76}$$

$$f^n = c_u u^n + c_g g + f_v^n , \tag{3.77}$$

$$u_{pred}^{n+1} = u^n + \frac{\Delta t}{2}\left(3f^n - f^{n-1}\right) , \tag{3.78}$$

$$f_{v,pred}^{n+1} = c_a \frac{3u_{pred}^{n+1} - 4u^n + u^{n-1}}{2\Delta t} , \tag{3.79}$$

$$\tilde{f}_v^{n+\frac{1}{2}} = \frac{1}{2}\left(f_v^n + f_{v,pred}^{n+1}\right) , \tag{3.80}$$

$$k_1 = c_u u^n + c_g g + \tilde{f}_v^{n+\frac{1}{2}} , \tag{3.81}$$

$$k_2 = c_u \left(u^n + a_{21}\Delta t k_1\right) + c_g g + \tilde{f}_v^{n+\frac{1}{2}} , \tag{3.82}$$

$$k_3 = c_u \left(u^n + a_{31}\Delta t k_1 + a_{32}\Delta t k_2\right) + c_g g + \tilde{f}_v^{n+\frac{1}{2}} , \tag{3.83}$$

$$u^{n+1} = u^n + \Delta t \left(b_1 k_1 + b_2 k_2 + b_3 k_3\right) . \tag{3.84}$$

The scheme is named *AB2-RK3-vm* here and yields second order convergence as illustrated in Table 3.3. The predictor is evaluated at time level t^n and is used subsequently in all Runge-Kutta sub-steps. In a similar fashion, other multistep methods may be employed as a predictor, illustrated for example by the scheme *LF-RK3-vm* in Appendix F. On the contrary, the scheme *RK3-1-AM3-vm*, also specified in Appendix F, demonstrates that using a higher order corrector after the Runge-Kutta scheme still only yields first-order, because the predicted u^{n+1}_{pred} is only first order accurate and so is $f_{v,pred}$. All these methods are fully explicit, and no inner iterations are necessary. A reduction of the absolute error was achieved compared to the sole application of the predictor schemes for the Stokes sphere test case considered (Table 3.3).
The realization remains complicated for variable time step size. This can be circumvented by modifying the predictor step. The usage of a Lagrange polynomial, P_N, of order N is proposed to predict f_v. With u given at arbitrary, but different instants in time, t_j, $j = 0, \ldots, N$, the interpolation reads

$$u(t) \approx P_N(t) = \sum_{j=0}^{N} L_j(t)\, u(t_j) \quad \text{where} \quad L_j(t) = \prod_{i=0,\, i\neq j}^{N} \frac{t - t_i}{t_j - t_i}, \tag{3.85}$$

and the first derivative is obtained by

$$\dot{u}(t) \approx P_N'(t) = \sum_{j=0}^{N} L_j'(t)\, u(t_j) \quad \text{where} \quad L_j'(t) = \frac{\sum_{i=0,\, i\neq j}^{N} \prod_{k=0,\, k\neq i}^{N} (t - t_k)}{\prod_{i=0,\, i\neq j}^{N} (t_j - t_i)}. \tag{3.86}$$

A second order polynomial, interpolating three adjacent points (t^n, u^n), (t^{n-1}, u^{n-1}) and (t^{n-2}, u^{n-2}), is then given by

$$\begin{aligned} u(t) \approx P_2(t;\, u^n,\, u^{n-1},\, u^{n-2}) =& \frac{(t - t^{n-1})(t - t^{n-2})}{(t^n - t^{n-1})(t^n - t^{n-2})} u^n \\ &+ \frac{(t - t^{n})(t - t^{n-2})}{(t^{n-1} - t^{n})(t^{n-1} - t^{n-2})} u^{n-1} \\ &+ \frac{(t - t^{n})(t - t^{n-1})}{(t^{n-2} - t^{n})(t^{n-2} - t^{n-1})} u^{n-2}, \end{aligned} \tag{3.87}$$

and its first derivative reads

$$\begin{aligned} \dot{u}(t) \approx P_2'(t;\, u^n,\, u^{n-1},\, u^{n-2}) =& \frac{2t - t^{n-1} - t^{n-2}}{(t^n - t^{n-1})(t^n - t^{n-2})} u^n \\ &+ \frac{2t - t^{n} - t^{n-2}}{(t^{n-1} - t^{n})(t^{n-1} - t^{n-2})} u^{n-1} \\ &+ \frac{2t - t^{n} - t^{n-1}}{(t^{n-2} - t^{n})(t^{n-2} - t^{n-1})} u^{n-2}, \end{aligned} \tag{3.88}$$

where the additional arguments after the semicolon are introduced to indicate which values are being interpolated. Employing a second order Lagrange extrapolation for $f_{v,pred}(t^{n+1})$ and making use of the trapezoidal rule for the virtual force $\widetilde{f_v}^{n+\frac{1}{2}}$ in the Runge-Kutta sub-steps yields the following scheme, labeled *Lag-RK3tp-vm*,

$$f_{v,pred}\left(t^{n+1}\right) = c_a P_2'\left(t^n;\, u^n,\, u^{n-1},\, u^{n-2}\right), \tag{3.89}$$

$$\tilde{f}_v^{n+\frac{1}{2}} = \frac{1}{2}\left(f_v^n + f_{v,pred}^{n+1}\right) , \tag{3.90}$$

$$k_1 = c_u u^n + c_g g + \tilde{f}_v^{n+\frac{1}{2}} , \tag{3.91}$$

$$k_2 = c_u \left(u^n + a_{21}\Delta t\, k_1\right) + c_g g + \tilde{f}_v^{n+\frac{1}{2}} , \tag{3.92}$$

$$k_3 = c_u \left(u^n + a_{31}\Delta t\, k_1 + a_{32}\Delta t\, k_2\right) + c_g g + \tilde{f}_v^{n+\frac{1}{2}} , \tag{3.93}$$

$$u^{n+1} = u^n + \Delta t \left(b_1 k_1 + b_2 k_2 + b_3 k_3\right) . \tag{3.94}$$

The implementation is fairly easy and $\Delta t = t^{n+1} - t^n$ is allowed to change in time. The desired second order convergence is obtained with slightly larger maximum errors compared to the scheme *AB2-RK3* (Table 3.3). A reduction of the error can be achieved by using the second order Lagrange approximation at the sub-step time levels $f_v\left(t^n + c_i\Delta t\right) = c_a P_2'\left(t^n + c_i\Delta t\right)$ leading to scheme *Lag-RK3c-vm* described in Appendix F.

Conclusions for the generic test case

Table 3.3 lists the maximum error obtained with the different time integration schemes and the convergence order deduced from these data. The results illustrate that the virtual mass, introduced into the Newtonian equation of motion of a particle to remove the singularity for $m_p \to 0$, is operational. The virtual mass concept is applicable in a straightforward manner to all multistep methods studied here (*AB2-vm, AM3-vm, CN-vm, LF-vm*) yielding the same order of accuracy as without virtual force. In the present tests, a moderately increased absolute error was observed, as apparent, e.g., by comparing the results of *AB2* and *AB2-vm*. For the Runge-Kutta schemes, an accurate prediction of $\dot{u}$ is necessary at substep time levels to conserve the order of the scheme which is evident, e.g., from the results of the schemes *RK3-a-vm* and *RK3-1-vm*. This prediction of $\dot{u}$ for the Runge-Kutta method can be achieved using a multistep scheme as a predictor and the trapezoidal rule for f_v, which is then used in the Runge-Kutta scheme as shown for schemes *LF-RK3-vm* and *AB2-RK3-vm*. Stability issues might arise from the predictor scheme. The estimate of $\dot{u}$ for the Runge-Kutta method can also be obtained by extrapolation with a Lagrange polynomial which gives the desired second order convergence of schemes *Lag-RK3tp-vm* and *Lag-RK3c-vm*. Furthermore, the additional data gathered in Appendix H show that second-order convergence is also obtained when using the L_1-error instead of the maximum error. The order of convergence remains unchanged if a virtual mass coefficient of $C_v = 1.0$ is used yielding slightly larger absolute errors compared to the reference solution for the Stokes sphere test case. The same observations hold for a density ratio of $\pi_\rho = 0.1$ with somewhat higher errors obtained for the same temporal discretization due to the larger acceleration in this case as apparent from Figure 3.18. The choice of an appropriate initial condition will be addressed below in Section 3.3.3. The more general application of the concept to the phase-resolving simulation of particle motion for an arbitrary density ratio is discussed in the next section.

3.3.3 Phase-resolving simulation of particle motion for an arbitrary density ratio

Implementation in PRIME

In order to allow for very low particle-to-fluid density ratios, the scheme *Lag-RK3tp-vm* was implemented in the multiphase code PRIME [134, 136]. Note that the used Runge-

Kutta method is a low storage scheme. The corresponding coefficients α^k are taken from [308, 137, 282]. With respect to the original implementation, only minor modifications are necessary. The virtual mass approach alters the linear momentum equation of the particle (3.95) and the angular momentum equation (3.98), enabling the simulation of very light particles.
The translational motion of the particle center is obtained for Runge-Kutta step k with $k = 1, 2, 3$, in the time interval $[t^n, t^{n+1}]$ from

$$\frac{\mathbf{u}_p^k - \mathbf{u}_p^{k-1}}{\Delta t} = \frac{\rho_f}{V_p\,(\rho_p + C_v \rho_f)} \left[\int_S \boldsymbol{\tau} \cdot \mathbf{n}\, ds \right]^k + 2\alpha^k \frac{\rho_p - \rho_f}{\rho_p + C_v \rho_f} \mathbf{g} + \frac{2\alpha^k}{V_p\,(\rho_p + C_v \rho_f)} \tilde{\boldsymbol{F}}_v^{n+\frac{1}{2}}, \tag{3.95}$$

where

$$\tilde{\boldsymbol{F}}_v^{n+\frac{1}{2}} = \frac{1}{2} \left(\boldsymbol{F}_v^n + \boldsymbol{F}_{v,pred}^{n+1} \right) . \tag{3.96}$$

The force vector $\boldsymbol{F}_{v,pred}^{n+1}$ has the components

$$\boldsymbol{F}_{v,pred}^{n+1} \cdot \mathbf{e}_\beta = C_v\, \rho_f\, V_p\, \dot{\boldsymbol{u}}_p^{n+1} \cdot \mathbf{e}_\beta = C_v\, \rho_f\, V_p\, P_2'(t^{n+1};\, u_{p,\beta}^n,\, u_{p,\beta}^{n-1},\, u_{p,\beta}^{n-2}) , \tag{3.97}$$

with the index $\beta = 1, 2, 3$ designating the Cartesian components of a vector and $\mathbf{e}_\beta$ the unit vector in the corresponding Cartesian direction, respectively. Each velocity component is hence extrapolated individually and its temporal derivative is computed at t^{n+1}.
In an analogous way as for the linear momentum, the virtual mass concept is also applied for the angular momentum. The rotational motion of a spherical particle is thus computed by solving

$$\frac{\boldsymbol{\omega}_p^k - \boldsymbol{\omega}_p^{k-1}}{\Delta t} = \frac{\rho_f}{I_p + I_v} \left[\int_S \mathbf{r} \times (\boldsymbol{\tau} \cdot \mathbf{n})\; ds \right]^k + \frac{2\alpha^k}{I_p + I_v} \tilde{\boldsymbol{M}}_v^{n+\frac{1}{2}} . \tag{3.98}$$

The virtual moment of inertia in this case is taken to be $I_v = 2/5\, C_{v,\omega} \rho_f V_p r_p^2$, and the rhs of (3.98) determined according to

$$\tilde{\boldsymbol{M}}_v^{n+\frac{1}{2}} = \frac{1}{2} \left(\boldsymbol{M}_v^n + \boldsymbol{M}_{v,pred}^{n+1} \right) , \tag{3.99}$$

with

$$\boldsymbol{M}_{v,pred}^{n+1} \cdot \mathbf{e}_\beta = I_v\, \dot{\boldsymbol{\omega}}_p^{n+1} \cdot \mathbf{e}_\beta = I_v\; P_2'(t^{n+1},\, \omega_{p,\beta}^n;\, \omega_{p,\beta}^{n-1},\, \omega_{p,\beta}^{n-2}) . \tag{3.100}$$

The coefficient $C_{v,\omega}$ is chosen to be $C_{v,\omega} \approx C_v$ here, i.e. the same as for the linear momentum. The surface integrals in squared brackets on the rhs of (3.95) and (3.98) yield the force and torque the fluid exerts on the particle. Their evaluation is discussed in [136] and Section 3.2.3.
Concerning the validation of the virtual mass approach for the angular momentum, simulations were conducted for a sphere in a plane shear flow with and without virtual mass (not shown here). Excellent agreement was found and the outcome is analogous to the one in the next sections for the linear momentum, i.e. the order of convergence is unchanged by the virtual mass approach while enabling low density ratios. Further validation was performed for spheroidal particles accounting for a virtual mass tensor.
With respect to the parallelization, only the master-master communication needs to be altered. Once a particle leaves the local domain of a process, its short history of $\mathbf{u}_p$ and $\boldsymbol{\omega}_p$, i.e. $(u_{p,\beta}^n, u_{p,\beta}^{n-1}, u_{p,\beta}^{n-2})$ and $(\omega_{p,\beta}^n, \omega_{p,\beta}^{n-1}, \omega_{p,\beta}^{n-2})$, is transferred to the new master-process by MPI-communication. The parallel performance and the high efficiency of the method remain basically unchanged no matter whether heavy or light particles are considered.

Specification of the test problem

The test problem consists of a sphere moving under the action of gravity at finite Reynolds number. A single sphere ascending or sedimenting in quiescent fluid is characterized by the Galilei number, G, and the density ratio, π_ρ, with

$$G = \frac{\sqrt{|\pi_\rho - 1|\, g\, d_p^3}}{\nu}, \qquad \pi_\rho = \frac{\rho_p}{\rho_f}. \tag{3.101}$$

The Galilei number is chosen to $G = 170$ where $d_p = 1$ and $\mathbf{g} = (-10,\, 0,\, 0)$ are used here. The fluid density is set to $\rho_f = 1$ here. First, a density ratio of $\pi_\rho = 0.5$ is chosen to enable a comparison of the original method with the modified virtual mass scheme. Lower density ratios lead to instabilities with the original scheme of [136]. Second, the modified scheme with virtual mass is demonstrated to be applicable for a density ratio close to zero, $\pi_\rho = 0.001$. The convergence behavior is studied as well for this case.
The purpose here is validation rather than analysis of the physical phenomena, so that only the initial phase of the trajectory is simulated. This can be accomplished using a cubic domain with $\mathbf{L} = (6.4,\, 6.4,\, 6.4)\, d_p$. It is discretized equidistantly with $\mathbf{N} = (N_x,\, N_y,\, N_z)$ cells with the number of points being varied later on to assess convergence. The spacing of the equidistant grid is $\Delta x = \Delta y = \Delta z = h$. Periodic boundary conditions are applied in all three directions. With the fluid and the sphere initially at rest, the simulation is run until $t_e = 0.2$, i.e. only the beginning of the acceleration phase is considered. For the virtual mass approach (*vm*), $C_v = 0.5$ is used, whereas the original method, i.e. no virtual mass (*nov*), refers to the original scheme of [136].

Initialization

Similar to the considered multistep method, the scheme *Lag-RK3tp-vm* based on the Lagrange polynomial requires special concern at startup. The virtual mass approach needs initial values for $(u^n_{p,\beta},\, u^{n-1}_{p,\beta},\, u^{n-2}_{p,\beta})$ and $(\omega^n_{p,\beta},\, \omega^{n-1}_{p,\beta},\, \omega^{n-2}_{p,\beta})$ at $t = 0$, i.e. for $n = 0$.
While $u^n_{p,\beta} = 0$ and $\omega^n_{p,\beta} = 0$ are given by the formulation of the test case where the particle is initially at rest in quiescent fluid, the additional values defining the translational and angular acceleration are not that clear. For the present problem, the sphere initially experiences no torque from the fluid and all components of the angular acceleration can be set to zero. In the present setup, rotational motion in general is negligible [2]. Wake transition and the connected path instability can lead to rotation in the later stages, though, which are beyond the scope of this study.
Concerning the translational acceleration, two different choices for the initial values in the determination of the virtual force are discussed. First, $\tilde{\mathbf{F}}_v(t = 0) = 0$ is used on the rhs of (3.95), i.e. $u^{n-1}_{p,\beta} = u^{n-2}_{p,\beta} = 0$. This corresponds physically to an initially increased inertia of the particle. Second, the analytical solution for the physical added mass force is used to determine the initial acceleration, $\dot{u}_{p,0}$, under the action of gravity neglecting other forces. The added mass is defined as $m_{AM} = C_{AM}\, \rho_f V_p$ with $C_{AM} = 0.5$ for a sphere, so that

$$(m_p + m_{AM})\, \dot{\mathbf{u}}_{p,0} = \mathbf{F}_B, \tag{3.102}$$

$$\dot{\mathbf{u}}_{p,0} = \frac{(\rho_f - \rho_p)}{(\rho_p + C_{AM}\rho_f)}\mathbf{g}. \tag{3.103}$$

The influence of the initialization is studied by comparison to the reference solution without virtual mass modeling (nov), where no additional initial condition for the acceleration is necessary. The relative error is thus defined as $\epsilon_{ref}(t) = \|\mathbf{u}_p - \mathbf{u}_{p,nov}\| / \|\mathbf{u}_{p,nov}\|$. Table 3.4 lists the maximum error $\max\{\epsilon_{rel}\}$, the time average $\langle\epsilon_{rel}\rangle$, and the error at the end of the time interval $\epsilon_{rel}(t_e)$ with $t \in [\Delta t, t_e]$ and $t_e = 0.2$.

Table 3.4 Influence of the initialization on the relative error in $u_p(t)$ for $\pi_\rho = 0.5$, where the reference is the solution without virtual mass (nov).

Initialization	$d_p/\Delta x$	Δt	$\max\{\epsilon_{rel}\}$	$\langle\epsilon_{rel}\rangle$	$\epsilon_{rel}(t_e)$
$\tilde{\mathbf{F}}_v(t=0) = 0$ on rhs of (3.95)	40	1.0E-03	3.96E-01	3.05E-02	5.64E-03
Analyt. added mass (3.103)	40	1.0E-03	9.30E-02	2.56E-03	4.59E-04

The maximum error occurs for the initial time step and the error introduced by the chosen initialization vanishes with time. The initialization with the analytical solution for the added mass force yields lower errors for the present setup and is thus retained throughout this study. The issue of the initial condition is studied here for completeness to set the scene for the subsequent tests. In the simulation of an experiment where light particles or bubbles are introduced into a flow, strong physical modeling is required for the startup in any case totally overwhelming the issue of choosing $\tilde{\boldsymbol{F}}_v(t=0) = 0$ [247].

Discretization error

To determine the overall order of accuracy of the modified method, the temporal and spatial discretization are studied together, i.e. the time step and the grid spacing are refined maintaining $\Delta t/\Delta x = const.$ to avoid dominance of the spatial error in case of small time steps. The number of forcing points on the surface of the sphere is increased as well with $N_L \sim 1/\Delta x^2$. For the refinement study, a time step of $\Delta t = 5\cdot 10^{-4}$ was chosen together with $N_x^3 = 128^3$. The convergence is studied for the original method (nov) and the modification with virtual mass (*vm*) individually. For both, a reference solution was determined with $N_x^3 = 512^3$. This corresponds to a spatial resolution of $d_p/\Delta x = 80$, i.e. eighty gridpoints over the diameter of the sphere.
The relative error compared to the respective reference is defined by $\epsilon = \|\mathbf{u}_p - \mathbf{u}_{p,ref}\| / \|\mathbf{u}_{p,ref}\|$ and is calculated at $t_e = 0.2$. The results of the refinement study are listed in Table 3.5 and illustrated in Figure 3.20. The convergence order q is determined by a formula analogous to (3.53).

The results show that the solution with the new time scheme is virtually the same as the one obtained with the unmodified method. Practicly the same relative errors and almost identical values for the particle velocity are computed for all numerical resolutions, e.g. $u_p(t_e) = 6.1683 \cdot 10^{-1}$ (vm) and $u_p(t_e) = 6.1625 \cdot 10^{-1}$ (nov) for $d_p/\Delta x = 20$.
The order of convergence is the same in both cases. Using the two finest grids, a convergence order of 1.9 is obtained.

The major benefit of the virtual mass approach is that it allows simulations with very low particle densities. As in the previous section, a rising sphere with $G = 170$ is considered (where $d_p = 1$, $\|\mathbf{g}\| = 10$), but now with a substantially reduced particle-to-fluid density ratio of $\pi_\rho = 0.001$. This case is no more accessible with the original method. All other

Table 3.5 Convergence order q of the method with virtual mass (vm) compared to the original method (nov) for $G = 170$ and $\pi_\rho = 0.5$. Relative error ϵ in $\mathbf{u}_p(t)$ at $t_e = 0.2$.

				nov		*vm*	
N_x	$d_p/\Delta x$	$N_{\Delta t}$	N_L	$\epsilon(t_e)$	q	$\epsilon(t_e)$	q
64	10	200	315	2.9205E-01		2.9140E-01	
96	15	300	708	1.8060E-01	1.2	1.7928E-01	1.2
128	20	400	1258	1.2100E-01	1.4	1.2017E-01	1.4
192	30	600	2828	6.3604E-02	1.6	6.3307E-02	1.6
256	40	800	5028	3.6751E-02	1.9	3.6751E-02	1.9
512	80	1600	20107	Reference		Reference	

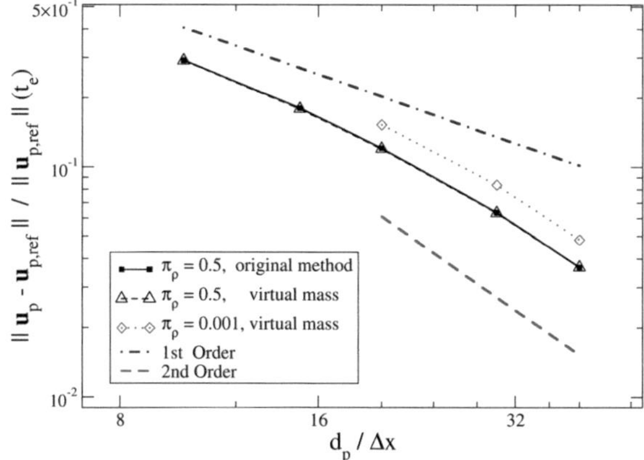

Figure 3.20 Convergence of the original method and the new method with virtual mass for the physical parameters $G = 170$ and $\pi_\rho = 0.5$. Furthermore, convergence of the virtual mass approach for the case of a very light particle, $G = 170$ and $\pi_\rho = 0.001$, which is inaccessible by the original method.

numerical parameters remain as described above. The change in the particle dynamics with π_ρ during the acceleration towards the terminal velocity is, e.g., apparent in Figure 3.21. Here, again the first stage of this transient until $t_e = 0.2$ is of interest and the reduction of the discretization error is studied with simultaneous refinement in space and time. Figure 3.20 shows the relative error in the particle velocity at time t_e over the grid resolution for $\pi_\rho = 0.001$ and $\pi_\rho = 0.5$, where the reference solutions are obtained with $d_p/\Delta x = 80$. The errors cannot be compared between the cases due to the faster acceleration of the light particle. From the simulations, close to second order convergence is also obtained for very low particle densities. The convergence properties of the original IBM are hence retained for very light spheres.

Comments on stability

The simulations concerning the discretization error were performed for small time step size with the intention to accurately determine the order of the new scheme. From the practical experience of many simulations with various setups, it appears that the stability properties of the original IBM are retained with the virtual mass approach for very light particles. The requirement on the time step size remains unchanged and time steps up to $CFL \approx 1.0$ can be used. A more rigorous stability analysis for the motion of a sphere governed by equations (3.33) or (3.34) would be desirable for the future. The virtual mass concept has been applied for the simulation of single bubble motion at high Reynolds numbers [244, 246] employing deformable bubble shapes. Simulations with multiple particles have been performed as well, with no stability problems stemming from the virtual mass approach. In these configurations, high accelerations, also of alternating sign, are present related, for instance, to particle-wake interaction, or particle-wall and inter-particle collisions. Forces and moments resulting from collisions need to be modeled and enter as additional terms on the rhs of (3.95) and (3.98) [135].

The modified method enabled, e.g., the study of a bubble chain [247] or studies on the clustering of spherical bubbles and the formation of wet metal foam [102, 101]. In contrast to the present study in quiescent liquid, many spherical and ellipsoidal light particles in turbulent channel flow were examined with the virtual mass approach in [236, 238, 237] using an adaptive time step size to yield $CFL \approx 1.0$.

3.3.4 Validation with implicit, highly resolved spectral simulations

A quantitative validation of the predictions with the present method is provided for the motion of a sphere under the action of gravity in viscous fluid with $G = 170$ and density ratios $\pi_\rho = 0.001$, 0.5 and 2.0 (where again $d_p = 1$, $g = \|\mathbf{g}\| = 10$, and $\rho_f = 1$ are used here). With respect to the specification of the test problem in Section 3.3.3, the computational domain is enlarged in the direction of particle motion to $\mathbf{L} = (12.8,\, 6.4,\, 6.4)\; d_p$ in order to assess the terminal velocity of the particle. It is discretized with an equidistant grid of $\mathbf{N} = (512,\, 256,\, 256)$ cells, corresponding to $d_p/\Delta x = 40$, and a time step of $\Delta t = 1 \cdot 10^{-3}$ is used. The sphere is initially at rest in quiescent fluid and positioned at $\mathbf{x}_{p,0} = (0.6,\, 3.2,\, 3.2)\; d_p$. The validation data was received from private communication with Mathieu Jenny, the author of [121, 120, 122]. In these studies, the motion of the sphere is simulated in a reference frame moving with the sphere, and a body fitted grid is employed. A spectral discretization in space of high accuracy is used and implicit time integration of the particle momentum equations is applied. With $\pi_\rho \to 0$, the solution becomes independent of the particle density as only the inertia of the surrounding fluid determines the dynamics. The data obtained for $\pi_\rho = 0.001$ with the present approach can hence be compared to the validation data with $\pi_\rho = 0$. For the comparison, the gravitational velocity and time scale are utilized as reference values

$$u_g = \sqrt{|\rho_p - \rho_f|\, g\, d_p}\,, \qquad t_g = \sqrt{\frac{d_p}{|\rho_p - \rho_f|\, g}}\,. \tag{3.104}$$

The corresponding reference length is the particle diameter d_p. Figure 3.21 shows the temporal evolution of the particle velocity for the three density ratios considered and the comparison to the validation data. In these tests, very good agreement is achieved concerning the initial velocity and acceleration for all values of π_ρ considered, including the very light sphere. The results agree well with those of the validation data indicating that the dynamics of the motion of the sphere are predicted correctly by the present method.

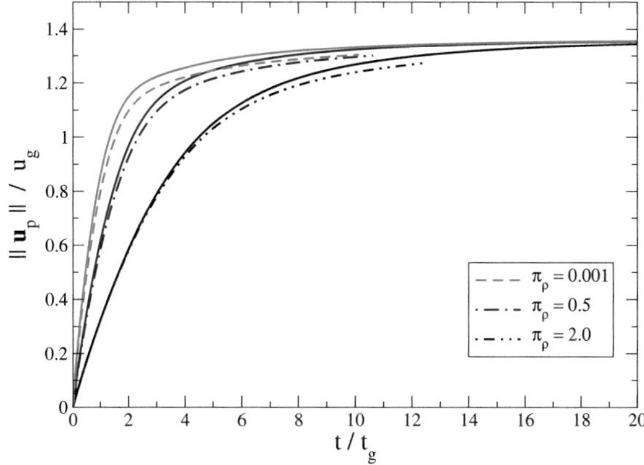

Figure 3.21 Temporal evolution of the particle velocity for $G = 170$ and density ratios $\pi_\rho = 0.001$, 0.5 and 2.0 obtained with the present method (broken lines) with comparison to data from author of [121, 120, 122] (continuous lines).

At later stages, however, a slight underprediction of the terminal velocity is observed, independent of π_ρ. The relative error in time with respect to the validation data from Jenny (index J) is defined by $\epsilon_J(t) = (u_p/u_g - u_J/u_{g,J}) / (u_J/u_{g,J})$. Here, only the dominant velocity component is considered with the transverse components being basically zero in the present case. The time when the particle has covered a distance of $L_x - d_p$ is denoted by t_{max} and it marks the end point of the graphs in Figure 3.21. The relative error in the particle velocity with respect to the validation data at t_{max} is approximately -2% for all density ratios considered and is given in Table 3.6. A cross-comparison concerning the terminal velocity of the experimental data of Mordant and Pinton [181], Veldhuis et al. [286, 287, 288] and Horowitz [113] with the present numerical and validation data, as well as from Uhlmann [282] and Kempe [134], and the standard drag curve of Clift [37] is very difficult because of the path instabilities that have a significant impact on the sphere velocity after the initial transient. The errors between the different experiments are thus larger than the present deviation and a 2% error is indeed an excellent result. Nevertheless, it appears that the IBM of [282, 136] systematically overpredicts the drag of the sphere and consequently predicts a somewhat lower terminal velocity.

Table 3.6 Relative error in the particle velocity with respect to the validation data at t_{max} for $\pi_\rho = 0.001$, 0.5 and 2.0.

Density ratio π_ρ	2.0	0.5	0.001
Relative error $\epsilon_J(t_{max})$	-2.2 %	-2.0 %	-2.1 %

Discussion of the error in the terminal velocity

Factors like domain size, temporal resolution, spatial resolution, and further numerical parameters like the number of additional forcing loops in the IBM according to [136] were considered as sources of the deviation in the terminal velocity and investigated in additional

simulations not reported in detail here. None of the issues mentioned altered the terminal velocity in these simulations to a substantial extent.
On the other hand, the regularized delta function used for spreading the IBM forces from the Lagrange markers located at the particle surface to the Eulerian grid has an influence on the result. This coincides with the observations in [110, 25]. The function of Roma [225] employed in the IBM has a width of three grid cells which results in a diffuse particle interface as seen by the fluid. As a consequence, the effective particle diameter is marginally increased, so that the IBM tends to overestimate the drag force. It was therefore suggested in [110, 25] to shift the forcing points by a fraction of the grid spacing towards the particle interior. First, the influence of the regularized delta function on the dynamics of a single ascending sphere is shown using two different regularized delta functions. Afterwards, results with adjusted surface marker position are given.

Influence of the regularized delta function

The effect of regularized delta function is quantified by changing the width of the function. The three-dimensional function used in the present IBM [136] is constructed from a one-dimensional function, δ_h^{1D}, according to

$$\delta_h(\mathbf{x} - \mathbf{x}_S) = \delta_h^{1D}(x - x_S)\, \delta_h^{1D}(y - y_S)\, \delta_h^{1D}(z - z_S), \tag{3.105}$$

where $\mathbf{x}_S$ denotes a point on the particle surface. The one-dimensional regularized delta function reads

$$\delta_h^{1D}(x - x_S) = \frac{1}{h}\,\phi(r)\,, \tag{3.106}$$

with the spacing of the equidistant grid h, and $r = (x - x_S)/h$. The continuous function ϕ is given by either one of the following two options:

a) Roma [225]

$$\phi_{Roma}(r) = \begin{cases} \frac{1}{3}\left(1 + \sqrt{-3r^2 + 1}\right), & |r| \leq 0.5 \\ \frac{1}{6}\left(5 - 3|r| - \sqrt{-3(1 - |r|)^2 + 1}\right), & 0.5 \leq |r| \leq 1.5 \\ 0, & \text{otherwise}, \end{cases} \tag{3.107}$$

b) Peskin [208, 209, 174]

$$\phi_{Peskin}(r) = \begin{cases} \frac{1}{4}\left(1 + \cos\left(\frac{\pi r}{2}\right)\right), & |r| \leq 2 \\ 0, & \text{otherwise.} \end{cases} \tag{3.108}$$

The function of Roma has a width of three points; the one of Peskin spreads over four points. The properties of the discrete version of these delta functions are discussed in [225] and [208], respectively. Figure 3.22a) shows a plot of (3.107) and (3.108).

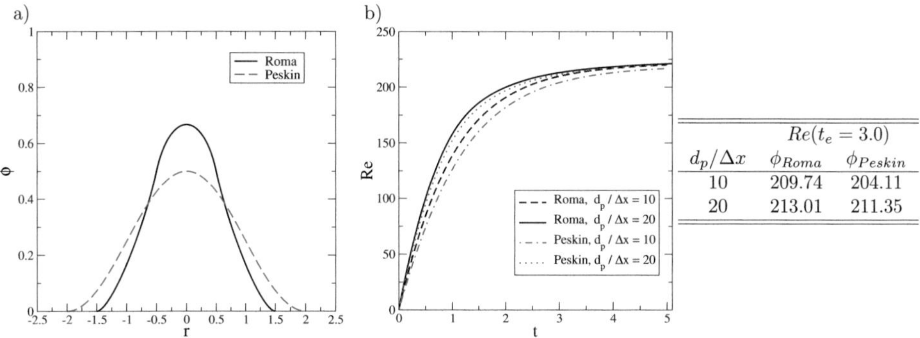

	$Re(t_e = 3.0)$	
$d_p/\Delta x$	ϕ_{Roma}	ϕ_{Peskin}
10	209.74	204.11
20	213.01	211.35

Figure 3.22 a) Regularized delta functions of Roma [225] and Peskin [208]. b) Influence of the width of the regularized delta function on the particle Reynolds number over time for $G = 170$ and $\pi_\rho = 0.5$ using two different spatio-temporal resolutions.

To study the influence of the regularized delta function on the results, additional simulations of the single sphere ascent with $G = 170$ and $\pi_\rho = 0.5$ were conducted successively coarsening the numerical resolution with $\Delta t/\Delta x = const.$ and $\Delta t = 10^{-3}$ for $d_p/\Delta x = 20$. All other parameters are as given in Section 3.3.4 above. Instead of the particle velocity, the instantaneous Reynolds number, $Re = u_p\, d_p/\nu$, is plotted over time in Figure 3.22b) to better illustrate the present regime. The results, given in the respective table, show that a wider regularized delta function yields a slower ascending particle, i.e. an overprediction of the particle drag. The effect is stronger for a coarser grid. The choice of an appropriate regularized delta function for the spreading of the IBM volume forces is a compromise between the smoothing properties and the diffusion of the interface, i.e. a virtually larger particle.

Adjustment of surface marker position

To limit the latter effect, a practical approach is to shift the Lagrangian marker points towards the particle center by a fraction of the grid spacing as proposed in [110, 25]. The impact of the adjustment of the forcing point position on the results for the ascent of a sphere with $G = 170$ is examined for the density ratios $\pi_\rho = 0.5$ and $\pi_\rho = 0.001$. The regularized delta function of Roma was used throughout. A shift of the forcing points towards the interior of the sphere is performed with a radial adjustment distance of $\Delta r_p = -0.3\Delta x$. This results in a smaller effective hydrodynamic diameter generated by the fluid-solid coupling with the IBM. The nominal diameter, d_p, is still used in the momentum equation of the particle.

Simulations were conducted in the same setting as described above. Two grid resolutions were considered for $\pi_\rho = 0.5$.

The influence of the forcing point adjustment on the particle Reynolds number is shown in Figure 3.23 for the two density ratios considered. With forcing point adjustment, the underprediction in the terminal velocity with respect to the reference data of Jenny [121, 122] vanishes for $\pi_\rho = 0.5$ and $\pi_\rho = 0.001$. The present data and the validation data basically collapse even for the coarser grid resolution.

Table 3.7 provides quantitative results concerning the particle Reynolds number and the relative error with respect to the reference data at three instances in time for both density ratios, $\pi_\rho = 0.5$ and $\pi_\rho = 0.001$. The adjustment of the forcing point position does reduce

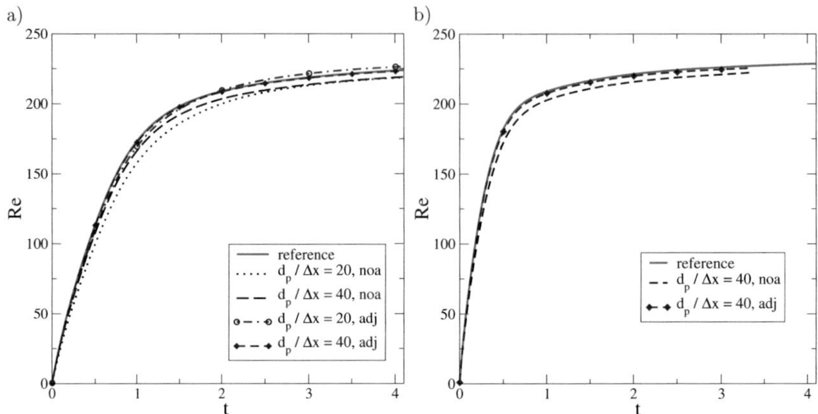

Figure 3.23 Influence of forcing point adjustment on the computed ascent of a single light particle in terms of the particle Reynolds number. a) $\pi_\rho = 0.5$ and b) $\pi_\rho = 0.001$. (noa - no adjustment, adj - adjustment with $\Delta r_p = -0.3\Delta x$) The symbol skip with respect to the time step is 1000. Reference data was provided by the author of [121, 122].

the instantaneous errors for both grid resolutions and density ratios. Very good agreement with the reference data is achieved with relative errors below 1%.

Table 3.7 Influence of forcing point adjustment on the instantaneous Reynolds number and comparison to reference data by Jenny [121] expressed by the relative error ϵ_J. Results for $G = 170$, as well as $\pi_\rho = 0.5$ and $\pi_\rho = 0.001$.

	$\pi_\rho = 0.5$					$\pi_\rho = 0.001$		
d_p/h	20		40		Ref.	40		Ref.
fp adjustment	no	yes	no	yes		no	yes	
$Re(t = 0.2)$	46.68	51.97	51.64	54.03	54.54	99.77	106.96	108.42
ϵ_J $(t = 0.2)$	-0.144	-0.047	-0.053	-0.009		-0.080	-0.013	
$Re(t = 3.0)$	213.01	221.41	213.59	218.23	218.63	220.85	224.46	226.04
ϵ_J $(t = 3.0)$	-0.026	0.013	-0.023	-0.002		-0.022	-0.002	

Even though the adjustment of the forcing point position does improve the agreement with the reference data, it is in general not considered for the simulations with the present IBM. The main reason is the resulting inconsistency of the nominal location of the interface and the position of the forcing points where the boundary condition is enforced. This discrepancy is for instance dubious for the simulation and modeling of particle collisions. A slightly different adjustment would be the correction of the effective diameter in the particle momentum equation in terms of a grid-dependent force correction.

3.3.5 Conclusions and outlook for very light particles

The IBM of [136] faces numerical problems if very light particles are considered. A simple modification of the time scheme of this method is proposed to overcome this limitation, which is named virtual mass approach. For the generic test case of a sphere moving under Stokes flow conditions, it is shown for several time integration schemes that the order of convergence can be retained with the virtual mass approach.
The method was then extended to the three-dimensional situation for phase-resolving simulations and a fully parallel implementation of the method into the code PRIME was conducted. The tests presented demonstrate that the virtual mass approach together with the IBM are capable of predicting the ascent of very light spheres accurately. The approach is straightforward applicable to other particle shapes as well. For heavy and moderately light spheres ($\pi_\rho = 0.5$) the computed solutions is shown to be the same as obtained with the original IBM. Furthermore, a rigorous validation is performed down to a density ratio of $\pi_\rho = 0.001$ by comparison with high-precision data from the literature. A slight deviation in the terminal velocity compared to the validation data is addressed by additional studies on the influence of the delta function for the spreading of the forcing and on the retraction of the forcing point position. The explicit coupling of the original IBM is maintained, so that there is basically no additional computational effort for the extension to very light particles. In this way, the advantages of the IBM, in particular its simplicity and high efficiency are fully retained. The implicit coupling of [121] could be revisited in the future in context of collision modeling or deformable bubbles where stability is an issue. Until the date of this work, taking into account the presentations at the ICMF 2013, the DLES 2013, and the EUROMECH Colloquium 549 on Immersed Boundary Methods, there does not seem to be any other application of an IBM of the present kind to very light particles or bubbles. With respect to the latter, the present approach poses a very efficient alternative to the one-fluid formulation of the Navier-Stokes equation accounting for gas and liquid material properties and jump condition at the interface [215]. Recently, the proposed method was used in several studies of bubbles being modeled as light particles with very satisfactory behavior [102, 236, 237] and is being applied to other problems of this type. Applications to the motion of light particles, as well as individual bubbles and bubble chains will be presented in the next chapters.

3.4 The flow around partially mobile spheroids

3.4.1 Introduction

The scope of this section is to illustrate the progression from a fixed sphere or ellipsoid in uniform cross-flow to a particle moving on a zig-zag trajectory as often encountered in bubbly flows. The flow past a sphere or bluff bodies is a standard problem in fluid mechanics with special emphasis on the forces acting on the embedded body and the wake forming behind it [36]. The motion of non-spherical particles at high Reynolds number was reviewed in [172] and wake-induced paths of free rise or fall were surveyed in [63]. Here, the gap in-between these two cases - fixed or freely moving - is brought in focus by successively releasing specific degrees of freedom with respect to particle shape and motion. At last, the influence of an aligned magnetic field on fixed and oscillating particles in uniform flow is studied.

3.4.2 Fixed sphere

The flow past a fixed sphere is a well known problem and an intensively studied three-dimensional configuration in fluid mechanics. The problem already served as a benchmark within the validation of our code [134] and is briefly revisited here with increased resolution to access the accuracy of the method and to provide a reference with respect to immersed bodies of more complex shape. A single parameter, the Reynolds number $Re = u_\infty d_p/\nu$, describes the problem. Among many other studies, Johnson and Patel [124] provided a detailed description of the flow up to $Re = 300$. Tomboulides and Orszag [273] extended the investigation up to $Re = 1000$.
In contrast to the previous studies, where large computational domains, body fitted grids and highly accurate methods were employed, this study is conducted in rather small computational domains and with the present immersed boundary method. The range of Reynolds numbers is $Re \in [0.5, 1000]$, i.e. spanning from the Stokes regime to a weakly turbulent wake. For the lowest Reynolds number, a cubical domain of edge length $6.4 d_p$ was discretized equidistantly with a resolution of $d_p/\Delta x = 20$. The analytical solution for the Stokes flow around a sphere [38] was imposed as a boundary condition on all boundaries. For all other simulations, the domain extends are $\mathbf{L} = (25.6,\ 12.8,\ 12.8)\, d_p$. A uniform inflow profile with u_∞ and a convective outflow condition are applied in the x-direction. The other boundaries are periodic. This set of boundary conditions is used for all subsequent studied presented in this section. A numerical grid is employed that is equidistant in a patch around the sphere also including the near wake and stretched away from it. Further information on the application of the IBM on such a patch is available from Appendix I. The local resolution is $d_p/\Delta x = 32$ for $Re = 20, 50$ and $d_p/\Delta x = 60$ for $Re \geq 100$, and an adaptive time step size to yield $CFL = 0.8$ is used as for all subsequent simulations presented in this section. The sphere is positioned at $\mathbf{x}_p = (5,\ 6.4,\ 6.4)\, d_p$. The drag coefficients obtained from the simulations are provided in Figure 3.24 and are compared to the correlation of Clift et al. [37] and the Schiller-Naumann correlation, both listed in Table 3.8. In general, good agreement is found. At low Reynolds numbers, the error due to blockage caused by the small lateral extend of the computational domain is dominant, while at high Reynolds numbers the numerical resolution becomes a critical aspect. Further sources of error are discussed in Section 3.3.4.

Re	C_D	C_D [37]	ϵ_{rel}
0.5	50.8	51.5	-1.3%
20	2.93	2.71	8.1%
50	1.67	1.57	6.4%
100	1.13	1.09	3.7%
200	0.809	0.776	4.3%
300	0.685	0.653	4.9%
400	0.632	0.593	6.6%
500	0.573	0.555	3.2%
700	0.526	0.508	3.5%
1000	0.492	0.471	4.5%

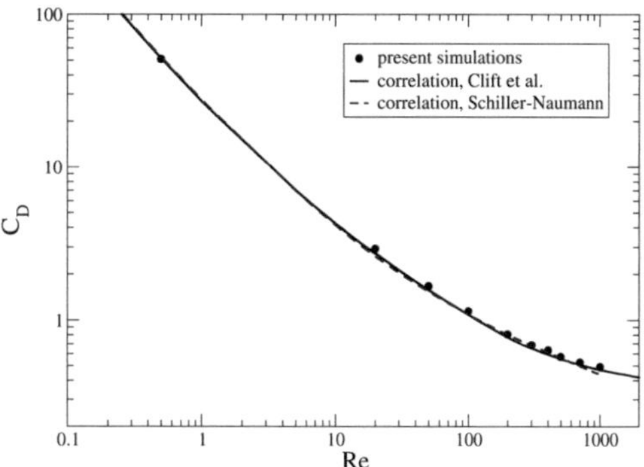

Figure 3.24 Fixed sphere in cross-flow. Drag coefficient C_D versus Reynolds number Re for present simulations and comparison against standard drag curve by Clift et al. [37] and Schiller-Naumann correlation.

Table 3.8 Recommended standard drag curve by Clift et al. [37] where $w_{Re} = \log_{10} Re$ and Schiller-Naumann correlation for C_D^{SN}.

0.01	$< Re \leq$	20	$\log_{10}\left[C_D\, Re/24 - 1\right]$	$= -0.081 + 0.82\, w_{Re} - 0.05 w_{Re}^2$
20	$< Re \leq$	260	$\log_{10}\left[C_D\, Re/24 - 1\right]$	$= -0.7133 + 0.6305 w_{Re}$
260	$< Re \leq$	1500	$\log_{10} C_D$	$= 1.6435 - 1.1242 w_{Re} + 0.1558 w_{Re}^2$
0.01	$< Re \leq$	1000	C_D^{SN}	$= 24/Re\left(1 + 0.15 Re^{0.687}\right)$

The sphere wake is steady, closed and axisymmetric for $Re < 212$. It is characterized by two steady standing vortices making the wake non-axisymmetric in the interval $212 < Re < 274$ [19], before the wake becomes unsteady due to vortex shedding while preserving the reflectional symmetry. The vortices are shed with a single characteristic frequency at $Re = 300$, and a transition to irregular states without reflectional symmetry occurs up to $Re \approx 500$ [273] via quasi-periodic pre-chaotic states [36].

For the present simulations, the Strouhal number, Sr, is in excellent agreement with the reference data from the literature. The non-dimensional frequency associated with the vortex shedding is extracted as $Sr = 0.137$ at $Re = 300$ from the oscillation in the drag coefficient (0.137 [124], 0.136 [273]). It increases monotonously to $Sr = 0.198$ at $Re = 1000$ (0.195 [273]), where the power spectrum of the vertical velocity in a point $5\,d_p$ behind the sphere and $0.4 d_p$ off the axis was used to access the frequency. A further dominant frequency is present in this spectrum at $Sr_2 = 0.342$ associated with the shear layer instability (0.35 [273]). The results also agree very well with the ones reported in the review of experimental investigations given in [231].

For Re = 100, 300, 1000, further simulation were conducted with a reduced size of the computational domain, $\mathbf{L} = (12.8, 6.4, 6.4)\, d_p$, and the results are within 1.5% of the ones reported above.

In summary, good quantitative agreement with available literature data is found, for the drag coefficient and the Strouhal number studied here, over a wide range of Reynolds numbers.

3.4.3 Fixed ellipsoid

In this section, the flow past a fixed, oblate ellipsoid is studied. In comparison to the flow around a sphere, the aspect ratio of the oblate ellipsoid, X, and its orientation in the flow, ϕ_z, enter as additional parameters. Here, we only consider a rotation of the ellipsoid around one of its major axes which is oriented parallel to the z-axis. The Reynolds number, Re_{eq}, is computed based on the sphere volume-equivalent diameter d_{eq}. This length scale is also used in the definition of the drag coefficient, $C_D = F_D / \left(\frac{\rho_f}{2} u_\infty^2 \frac{\pi}{4} d_{eq}^2 \right)$ to avoid scaling issues [111]. Usually the projected area enters the definition of the drag. Hence, the configuration is already substantially more complex and reference data is rare. The data from [316] for a very similar configuration was obtained with a rather coarse resolution, $d_{eq}/\Delta x = 12$. The present simulations are conducted with the setup described above for the sphere. An ellipsoid with $X = 1.25$ is initially placed in its streamlined orientation, $\phi_z = 0°$, for 100 time-units $t\, u_{ref}/d_{eq}$ with $u_{ref} = u_\infty$. Then the ellipsoid is tilted consecutively in steps of 10° remaining in a static orientation for 50 time-units. The configuration is sketched in Figure 3.25a). Three different Reynolds numbers are considered, $Re_{eq} = 100, 300, 1000$.

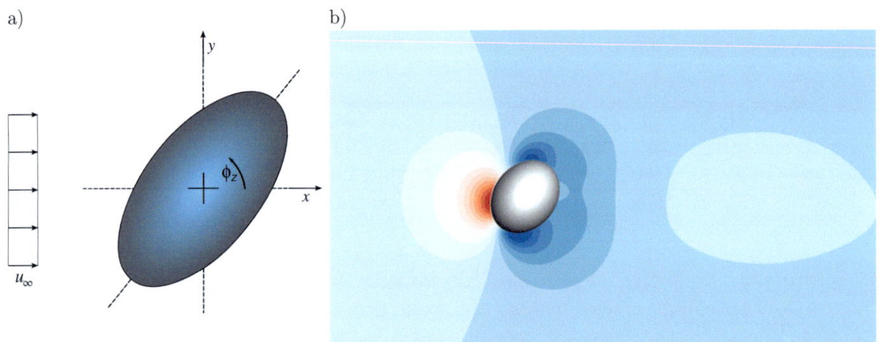

Figure 3.25 Ellipsoid in cross-flow. a) Configuration. b) Qualitative pressure field for $Re_{eq} = 100$, $X = 1.25$ and $\phi_z = 50°$ (close-up).

The temporal evolution of the drag force and the lift force in y-direction are shown in Figure 3.26 for the three Reynolds numbers. The results are normalized with the drag force for the sphere at the same Re_{eq} and obtained with present method. For $Re_{eq} = 100$, temporally constant values for drag and lift are obtained for each inclination, while drag and lift oscillate for the higher Re_{eq}. A zero lift force acts on the particle for the symmetric orientations $\phi_z = 0°$ and $\phi_z = 90°$ for $Re_{eq} = 100$. For the intermediate angles a lift force in the downward direction is obtained where the dependency with respect to ϕ_z is basically symmetric around $\phi_z = 45°$. The maximum absolute lift force in this study was obtained at $\phi_z = 30°$ and $\phi_z = 50°$ with $F_{L,y}/F_{D,sphere} \approx 0.11$. For $Re_{eq} = 300$, a positive mean lift is present for the symmetric orientations associated with the RSP mode of the wake discussed later. For the intermediate angles, the positive wake induced lift is superimposed with the negative lift stemming from the inclination yielding a negative net lift at e.g. $\phi_z = 50°$. The symmetry with respect to $\phi_z = 45°$ is lost. The amplitude of the regular oscillation in the lift force increases with ϕ_z. For $Re_{eq} = 1000$, the oscillations in the lift force are irregular, the symmetry with respect to $\phi_z = 45°$ is roughly regained and again approximately zero mean lift is obtained for the symmetric orientations. The maximum absolute lift force was obtained at $\phi_z = 40°$ with $F_{L,y}/F_{D,sphere} \approx 0.16$.

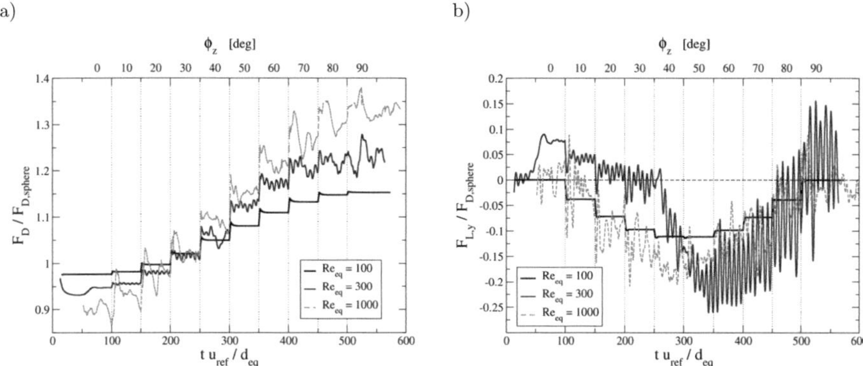

Figure 3.26 Fixed ellipsoid in cross-flow. Influence of static orientation ϕ_z. a) Temporal evolution of normalized drag force. b) Lift component in y-direction.

The time-averaged drag force is shown in Figure 3.27 as a function of the static orientation ϕ_z. The time-average was computed neglecting the first 15 time units of each inclination time interval. In retrospective, longer averaging-intervals would have been beneficial for the two higher Re_{eq}. Compared to the sphere, a lower drag is measured for $\phi_z < 30°$ for all Reynolds numbers. A steeper inclination yields an increase in drag. The dependency nicely follows a $\sin^2 \phi_z$ function based on the two extreme orientations which is indicated by the line plots in the Figure. The spread between the lowest drag at $\phi_z = 0°$ and the highest value at $\phi_z = 90°$ increases with Re_{eq}. For the highest Reynolds number the spread slightly exceeds a value of $0.4F_{D,sphere}$.

Three different semi-empirical correlations are studied for their capability of predicting the drag coefficient for the present configuration. The comparison of simulation data with the predictions from these correlations are gathered in Figure 3.27b).

The ellipsoid used in this study has a sphericity of $\Psi = 0.991$, where the sphericity denotes the ratio between the surface area of the volume-equivalent sphere and the surface area of the present particle. Being geometrically so close to a sphere and considering that at high Re_{eq} the pressure or form drag dominates over the viscous contribution, one might construct a correlation from the well-known correlation for the sphere and a term considering the actual projected area. The first correlation (3.109) thus reads,

$$C_D = C_{D,sphere}^{SN} \Psi_n^{-1} , \tag{3.109}$$

where $C_{D,sphere}^{SN}$ is the drag coefficient of an volume equivalent sphere obtained with the Schiller-Naumann correlation from Table 3.8. The crosswise sphericity, Ψ_n, is the ratio of the cross-sectional area of the volume equivalent sphere and the projected cross-sectional area of the ellipsoid [111]. In the present configuration, the projected shape is an ellipse [290], and one can derive $\Psi_n = a^{-1/3} b^{-2/3} \sqrt{(a^2 - b^2) \sin^2 \phi_z + b^2}$ for an oblate ellipsoid rotated along

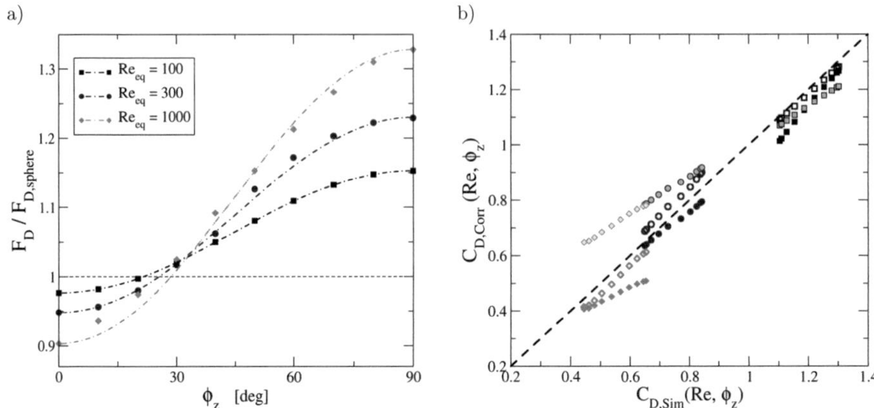

Figure 3.27 Normalized drag force as a function of static orientation ϕ_z for $Re_{eq} = 100, 300, 1000$. a) Time-averaged simulation data (symbols) and $F_{D,0°} + (F_{D,90°} - F_{D,0°}) \sin^2 \phi_z$ (lines). b) Comparison of simulation data $C_{D,Sim}(\phi_z, Re_{eq})$ and predictions from correlations $C_{D,Corr}(\phi_z, Re_{eq})$ (3.109) (filled symbols), (3.110) (empty symbols), (3.111) (light symbols).

one of its major axes. Good agreement is found with the simulation data specifically for $Re_{eq} = 100, 300$. The correlation also predicts the general dependency of the drag force with the inclination correctly, and it reestablishes the drag of a sphere at the specific Reynolds number at $\phi_z \approx 30°$. However, the correlation lacks the Reynolds dependency of the spread in Figure 3.27a) leading to larger deviations at high Re_{eq}.
The second correlation is given by

$$C_D = C_{D,0°}^{corr} + \left(C_{D,90°}^{corr} - C_{D,0°}^{corr}\right) \sin^2 \phi_z \,, \tag{3.110}$$

where $C_{D,0°}^{corr} = 7.57/Re_{eq}^{0.421}$ and $C_{D,90°}^{corr} = 5.61/Re_{eq}^{0.321}$ are the drag coefficients for the extreme orientations and need to be correlated additionally from the data [316]. The $\sin^2 \phi_z$ dependency seems to be a sound fit independent of the Reynolds number. The data in [316] for two different ellipsoids, a fiber and a disc also yield a value for the exponent close to 2. More data points would be necessary to correlate $C_{D,0°}^{corr}$ and $C_{D,90°}^{corr}$ with better accuracy.
The third correlation is taken from [111] and is intended as a 'simple' correlation for a broad spectrum of particle shapes. It was obtained from more than 2000 data points for various shapes and reads

$$C_D = \frac{8}{Re_{eq}} \frac{1}{\sqrt{\Psi_n}} + \frac{16}{Re_{eq}} \frac{1}{\sqrt{\Psi}} + \frac{3}{\sqrt{Re_{eq}}} \frac{1}{\Psi^{3/4}} + 0.42 \cdot 10^{0.4(-log\Psi)^{0.2}} \frac{1}{\Psi_n} \,. \tag{3.111}$$

In the present comparison, this correlation yields the largest deviations, especially at high Reynolds numbers. The average deviations are 8.4% for correlation (3.109), 4.9% for (3.110) and 16.8% for (3.111).
Further correlations for the lift force and as well as the torque acting on the non-spherical particle could be obtained from the present data [316, 172]. This is for now beyond the scope of this study. One should also keep in mind the differences between the static configuration used here and a dynamically changing orientation, which will be considered in the next sections.

3.4.4 Rotary oscillation

In contrast to a fixed, static inclination, the orientation of the ellipsoid in the flow, ϕ_z, is determined by the interaction of the particle with the flow now. The other degrees of freedom concerning rotation and translation are blocked. On top of the Reynolds number, Re_{eq}, and the aspect ratio of the oblate ellipsoid, X, the particle-to-fluid density ratio, π_ρ, enters as an additional parameter to describe the problem.
Two studies were conducted. The first addresses the rotational response of the ellipsoid in cross-flow to a given initial inclination as a function of π_ρ. The second study examines the wake-induced rotary oscillation.

Initial phase

The equilibrium orientation of an oblate ellipsoid in a uniform cross-flow is $\phi_z = 90°$. Figure 3.25b) shows a qualitative pressure field. The surface normal in the center of pressure is not aligned with the center of mass of the particle and the resulting pressure force thus creates a moment on the particle. This moment is always directed towards an orientation with $\phi_z = 90°$, i.e. the small semi-axis is parallel to the mean flow. The also symmetric position $\phi_z = 0°$ does not yield a moment on the particle. However, already small disturbances will create a rotation towards the equilibrium orientation.
An oblate ellipsoid of aspect ratio $X = 1.25$ was initially inclined by $\phi_z = 50°$. The Reynolds number is chosen as $Re_{eq} = 100$, i.e. the wake is steady. Simulations were performed with a domain size of $\mathbf{L} = (12.8,\ 10,\ 10)\, d_{eq}$ and a fixed particle position of $\mathbf{x}_p = (5,\ 5,\ 5)\, d_{eq}$. The local spatial resolution is $d_p/\Delta x = 60$ obtained from a grid with $\mathbf{N} = (512,\ 256,\ 256)$ cells. The simulation is initialized with the developed flow field from one of the previous studies. Figure 3.28 shows the temporal evolution of the orientation ϕ_z towards the equilibrium orientation for various density ratios. The density ratio is varied as $\pi_\rho = 0.001,\ 1,\ 2,\ 4,\ 8$, i.e. covering the range from an air filled particle to an iron particle in water. For the lowest π_ρ, the inertia of the attached fluid dominates, whereas for high π_ρ the particle's inertia itself is significant. With increasing π_ρ, the damping in the amplitude of the oscillation and the frequency decrease. The evolution $\phi_z(t)$ shows some similarity to a damped harmonic

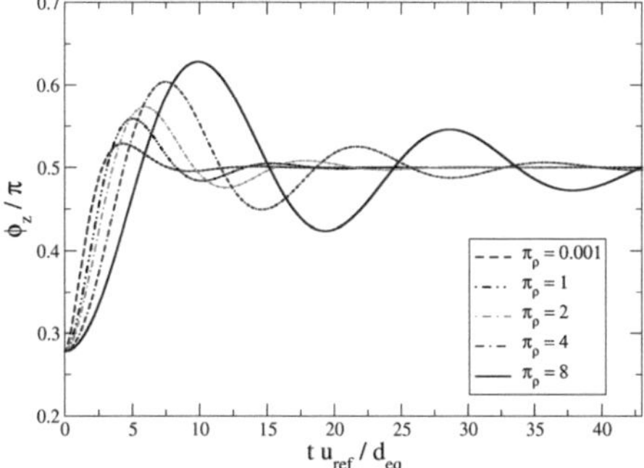

Figure 3.28 Initial phase of temporal evolution of orientation ϕ_z for various density ratios, π_ρ, and $Re_{eq} = 100$, $X = 1.25$, $\phi_{z,0} = 50°$.

oscillation as for a spring-mass-damper system. The analogy was studied in the student thesis of Köhler [141] and revealed rather moderate agreement. Especially the damping cannot be easily described, e.g., proportional to the angular velocity with a constant coefficient. The test case, however, gives some insight into the response time of the system which becomes important when studying wake induced oscillations.

Wake-induced rotary oscillation

The wake behind axisymmetric bodies, as spheroids and finite height cylinders, was studied in much detail in [64, 36] employing accurate numerical methods and body fitted grids, as well as large computational domains. Figure 3.29a) is adapted from [36] and displays the wake regimes of oblate spheroids as a function of Reynolds number and aspect ratio, with $1/X = 1$ corresponding to a sphere and $1/X = 0$ to an infinitely thin disk. The names of the wake modes are adopted from [64]. Visualizations of some modes are provided in Figure 3.29b)-g) from present simulations. At the low end of the Reynolds range, the wake is steady, closed and axisymmetric. The *SS mode* is a steady state and characterized by two counter-rotating standing vortices. The visualization in Figure 3.29b) shows the rotational sense of these vortices. They induce a velocity, v_{ind}, as indicated, which results in a constant lift force and a rotational moment, $M_{z,ind}$.
For higher Reynolds numbers, the wakes becomes unsteady as vortex shedding sets in. The critical Reynolds number decreases with $1/X$ as the surface curvature increases and so does the production of vorticity. The *RSB mode* stands for reflectional symmetry braking mode and is characterized by a non-zero mean lift. It was not encountered in the present study, but can be found specifically for a flat disk as visualized in [64, 36]. The *RSP mode* denotes the reflectional symmetry preserving mode, where also a non-zero mean lift is obtained. In contrast, the *SW mode*, short for standing wave mode, has a zero mean lift. It is very similar to the RSP mode, but on top of the reflectional symmetry in the xz-plane there is an average symmetry in the xy-plane [64]. The *RW mode* stands for rotating wave mode and the wake performs an additional helical motion [36], while a rotating, oscillating lateral force acts on the particle. *Irregular* modes are found at the high end of the Reynolds range consisting of a superposition of the previous modes with additional shedding frequencies, helical wake advection and transition to chaotic wakes.
Especially the rotational regimes seem very sensitive, e.g. there are substantial differences in the regime map when considering finite height cylinders compared to spheroids [36] at the same aspect ratios.

The wake-induced, rotary oscillation of an ellipsoid in cross-flow was studied for density neutral particles, $\pi_\rho = 1$, and various Re_{eq}. Two aspect ratios are considered, $X = 1.25$ and $X = 2.5$, respectively.
The numerical parameters of the simulations are $\mathbf{L} = (12.8,\, 6.4,\, 6.4)\, d_{eq}$, $\mathbf{x}_p = (5,\, 3.2,\, 3.2)\, d_{eq}$, $\mathbf{N} = (512,\, 256,\, 256)$ and $d_p/\Delta x = 60$. According to Biot-Savarts law, the velocity induced by a vortex declines away from the generating vortex element and hence the far wake has little, direct effect on the particle. Nevertheless, potential follow-up studies should consider substantially larger computational domains.

The temporal evolution of the particle orientation ϕ_z is shown in Figure 3.30. For $X = 1.25$ and $Re_{eq} = 300$, the ellipsoid oscillates almost regularly around its equilibrium orientation with amplitudes mildly varying around 4.5°. The associated wake pattern is an SW mode with a slight RW mode contribution resulting in the variation in the oscillation. The simulation for $Re_{eq} = 350$ was started from a result file at the lower Re_{eq} as indicated in the graph. With increasing Re_{eq} the wake structures become irregular and the contribution of

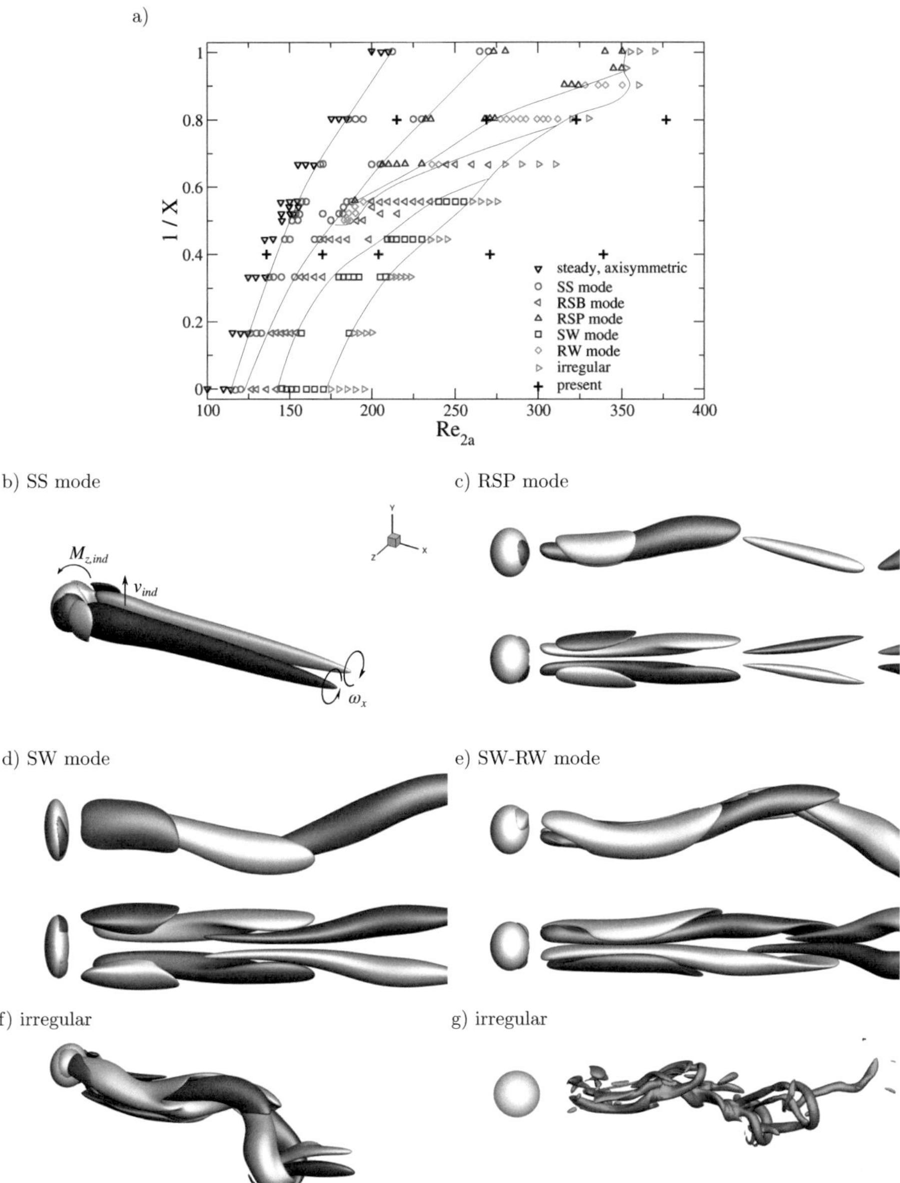

Figure 3.29 Wake forms of an oblate ellipsoid. a) Wake states for fixed spheroids as function of $1/X$ and $Re_{2a} = u_\infty 2a/\nu = X^{1/3} Re_{eq}$, adapted from [36]. b)-f) Wake visualization behind rotationally oscillating ellipsoid by instantaneous iso-contours of streamwise vorticity, $\omega_x d_{eq}/u_\infty = \pm 0.5$ b) SS mode for $X = 1.25$, $Re_{eq} = 200$. c) RSP mode for $X = 1.25$, $Re_{eq} = 250$. d) SW mode for $X = 2.5$, $Re_{eq} = 150$. e) SW-RW mode for $X = 1.25$, $Re_{eq} = 300$. f) irregular for $X = 1.25$, $Re_{eq} = 350$. g) irregular visualized by λ_2-iso-contour for $X = 1.0$, $Re_{eq} = 1000$.

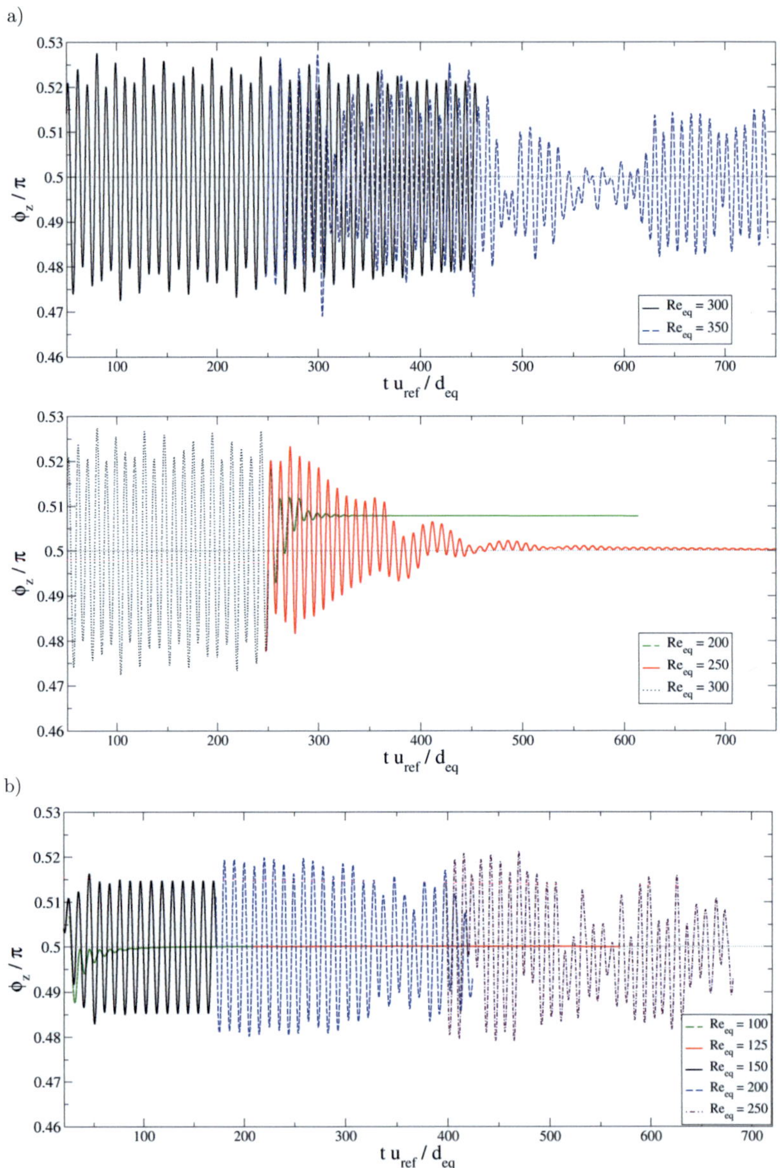

Figure 3.30 Temporal evolution of orientation ϕ_z for various Reynolds numbers, Re_{eq}, and $\pi_\rho = 1$. a) Aspect ratio $X = 1.25$. b) Aspect ratio $X = 2.5$.

the helical, long-wave component increases. Consequently, the oscillation in $\phi_z(t)$ also becomes more non-uniform with strong variations in amplitude. Low amplitudes are associated with vortex pairs contained in a plane parallel to the xy-plane. The run for $Re_{eq} = 250$ was also started from a $Re_{eq} = 300$ result. In this case, the symmetry plane of the wake rotates to a plane perpendicular to the previous state and parallel to the xy-plane. An RSP-mode with non-zero mean lift in the z-direction is obtained and thus no non-zero mean inclination is measured in ϕ_z. For $Re_{eq} = 200$ on the other hand, an SS mode develops and the two steady, counter-rotating vortices, oriented approximately parallel to the xz-plane, lead to a constant inclination of the ellipsoid with $(90 + 1.4)°$.
The results for $X = 2.5$ reveal a perfectly harmonic oscillation for $Re_{eq} = 150$ associated with a clean SW wake mode and again increasing irregularity as the Reynolds number is increased to $Re_{eq} = 200$ and $Re_{eq} = 250$ associated with the wake transition. The amplitudes in the oscillation of $\phi_z(t)$ are lower than for $X = 1.25$. The simulation with $Re_{eq} = 100$ was started from an early result of the $Re_{eq} = 150$ run explaining the initial damped oscillation. Then a steady, axisymmetric wake develops and the ellipsoid is constantly oriented in its equilibrium position. Increasing the Reynolds number to $Re_{eq} = 125$ yields first an SS mode with vortex pairs in the xy-plane which becomes very weakly unsteady towards the end.
The time-averaged drag coefficients and the Strouhal numbers are summarized in Table 3.9. The drag is slightly lower than for the fully fixed case for $X = 1.25$, $Re_{eq} = 300$ and $\phi_z = 90°$. Note that the wake-induced characteristic frequency is very close to the frequency obtained from the initial response studied above for $X = 1.25$, $Re_{eq} = 100$ at the same density ratio, $\pi_\rho = 1$.

Table 3.9 Mean drag $\langle C_D(d_{eq})\rangle$ and Strouhal number $Sr(\phi_z)$ for ellipsoid in cross-flow with free rotation in ϕ_z.

$X = 1.25$, $\pi_\rho = 1$	$Re_{eq} = 200$	250	300	350	
$Sr(\phi_z)$	—	0.103	0.107	0.119	
$\langle C_D\rangle$	0.812	0.815	0.770	0.736	
$X = 2.5$, $\pi_\rho = 1$	$Re_{eq} = 100$	125	150	200	250
$Sr(\phi_z)$	—	—	0.104	0.105	0.112
$\langle C_D\rangle$	1.083	1.021	0.996	0.959	0.924

The configuration was also studied in the diploma thesis of Beetz [12]. The aspect of the numerical resolution was scrutinized and the grid resolution was further refined for the present study. With inclinations less than 5°, the problem would also be well suited for body fitted grids with deforming meshes. A comparison of the loads acting on the rotationally oscillating ellipsoid by means of the generalized Kelvin-Kirchhoff equations [186] and the static correlations of [316] revealed that the latter correlations are rather inaccurate for the predictions of dynamic systems. In summary of the present study, the wake regime and the ellipsoid's oscillation in $\phi_z(t)$ are closely correlated. With respect to the wake regimes of the fixed counterpart [36], a rotational degree of freedom seems to slightly shift the critical Reynolds numbers for regime transitions towards higher values and the wake rotation around an axis parallel to the x-axis seems to be suppressed to some extent. Further studies with larger computational domains and higher numerical accuracy are necessary to provide evidence for this tendency.

3.4.5 Oscillating ellipsoid

Now, additionally a free translation in y_p is allowed, while x_p and z_p remain fixed. As previously, the rotation in ϕ_z is free, while rotations in ϕ_x and ϕ_y are still blocked.
Preliminary experiments of similar configurations were conducted in the diploma thesis of Roßbach [226]. PIV measurements were conducted for the flow around a sphere with $Re \in [3260, 26500]$. The sphere was either fixed, mounted with a steel flat bar allowing an oscillation in one plane, or the sphere was attached to a cord allowing rather free oscillations with $\pi_\rho = 1.3$.
The present study addresses two aspect ratios, $X = 1.25$ and $X = 2.5$, with fixed Reynolds numbers of $Re_{eq} = 300$ and $Re_{eq} = 250$, respectively. Density neutral particles, $\pi_\rho = 1$, as well as very light particles, $\pi_\rho = 0.001$, are considered.
The numerical parameters of the simulations are $\mathbf{L} = (12.8,\, 6.4,\, 6.4)\, d_{eq}$, $\mathbf{x}_p = (5,\, 3.2,\, 3.2)\, d_{eq}$ as for the purely rotary oscillation, but employing an equidistant grid with $\mathbf{N} = (512,\, 256,\, 256)$, and thus $d_p/\Delta x = 40$.
The temporal evolution of the orientation ϕ_z and the particle position y_p is given in Figure 3.31 for $X = 1.25$ and $Re_{eq} = 300$. For both density ratios, the particle undergoes a periodic zig-zag motion characterized by regular oscillations in ϕ_z and y_p. The amplitudes in the oscillation are significantly larger for $\pi_\rho = 0.001$ compared to $\pi_\rho = 1$ and the leading frequency is somewhat higher. For the density neutral particle, a regular modulation with a lower frequency can be observed. Table 3.10 summarizes the major figures related to the particle dynamics. The dynamics of the very light ellipsoid with aspect ratio $X = 1.25$ are indeed comparable to the motion of a single bubble studied in Chapter 5. In comparison to the sole rotational degree of freedom, the amplitude in the oscillation in ϕ_z is approximately tripled at about the same frequency.
For $X = 2.5$ and $Re_{eq} = 250$, the ellipsoid constantly glides sideways with about four percent of the bulk velocity and is inclined about 4° from the equilibrium position in average. Aside from the 'oblique' trajectory in $y_p(t)$ the particle vibrates rather regularly. The frequency of the vibration and the amplitude are again larger for $\pi_\rho = 0.001$ compared to the density neutral particle.

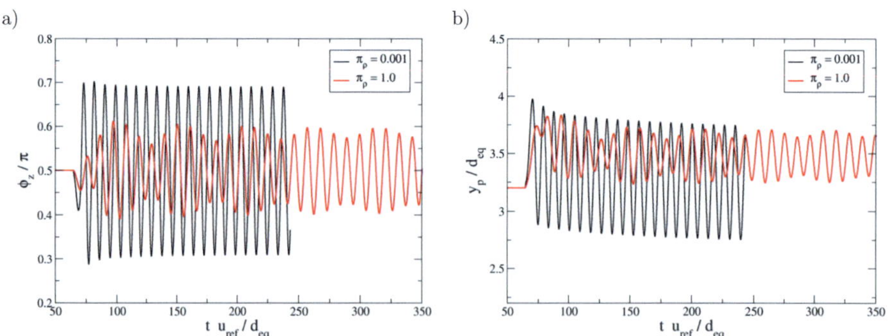

Figure 3.31 Temporal evolution of orientation ϕ_z (a) and of particle position y_p (b) for $\pi_\rho = 1$ and $\pi_\rho = 0.001$, with $Re_{eq} = 300$, $X = 1.25$

The drag or normalized force in x-direction is significantly higher for the transversely oscillating particle than for the mild sole rotation and the completely fixed spheroid. The

Table 3.10 Particle dynamics of ellipsoid in cross-flow with free rotation in ϕ_z and free translation in y_p. The other motional degrees of freedom are blocked.

$X = 1.25$, $Re_{eq} = 300$	$Sr(\phi_z)$	$\langle\phi_z\rangle$	$\lvert\phi_z\rvert_{max} - \langle\phi_z\rangle$	trajectory	Δy_p	$\langle C_D\rangle$
$\pi_\rho = 1$	0.097	90°	17°	zig-zag	$0.43d_{eq}$	0.805
$\pi_\rho = 0.001$	0.119	90°	34°	zig-zag	$1.0d_{eq}$	1.035
$X = 2.5$, $Re_{eq} = 250$	$Sr(\phi_z)$	$\langle\phi_z\rangle$	$\lvert\phi_z\rvert_{max} - \langle\phi_z\rangle$	trajectory	$\langle v_p\rangle/u_\infty$	$\langle C_D\rangle$
$\pi_\rho = 1$	0.114	86°	16°	'oblique', vibrating	−0.041	1.062
$\pi_\rho = 0.001$	0.130	95°	22°	'oblique', vibrating	0.042	1.142

more pronounced the lateral dynamics are, the higher is the increase in drag. This is also apparent from the experimental work in [113, 128] and should be considered for the proper use of correlations.

The mechanism of the drag increase for the oscillating spheroids is presumably similar to the one for an oscillating cylinder [232, 321]. As a thought experiment, a rapidly oscillating cylinder in uniform cross-flow traces out a larger projected area than the fixed one [232], the fluid experiences a virtually larger obstacle and the drag increases. A formulation was derived by von Kármán [292] for the drag of a cylinder expressed as a function of the streamwise and transverse vortex spacing, l_{vs} and h_{vs}, in a staggered vortex street of potential vortices with circulation Γ:

$$C_{D,cyl} = \frac{2\Gamma}{u_\infty^2 d_p}\left[\frac{h_{vs}}{l_{vs}}(u_\infty - 2u_{vs}) + \frac{\Gamma}{2\pi l_{vs}}\right], \tag{3.112}$$

where the propagation speed of the vortex street, u_{vs}, is obtained from

$$u_{vs} = \frac{\Gamma}{2l_{vs}}\tanh\left(\frac{\pi h_{vs}}{l_{vs}}\right).$$

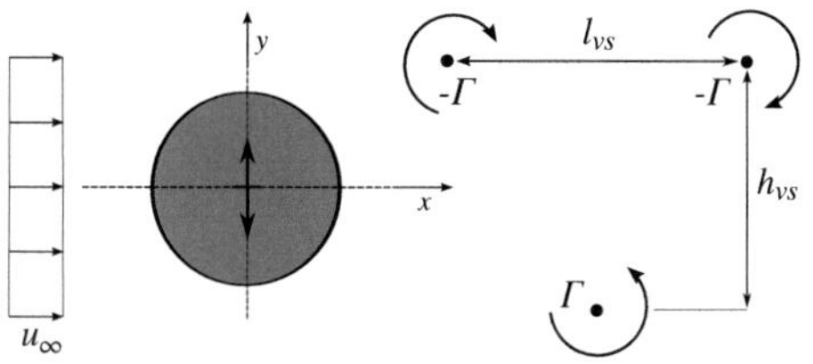

Figure 3.32 compares wake structures for $X = 1.25$, $Re_{eq} = 300$ with only rotatory oscillation and an additional wake-induced oscillation in y_p. With the oscillation in y_p, the streamwise vortex spacing, l_{vs}, decreases and the transverse vortex spacing, h_{vs}, increases. So assuming the circulation associated with a single shed vortex remains constant, equation (3.112) qualitatively yields an increase in drag as observed in the numerical simulation.

3.4.6 Impact of an aligned magnetic field

Fixed spheroids

The influence of an aligned magnetic field on the flow past a fixed spheroid is studied. A few results from [245] are revisited here to illustrate and explain the major effects. In [245], also the flow past a cylinder, for $Re = 100$, was considered and a validation against data from the literature, e.g. [251], was conducted showing very good agreement. Furthermore, a suppression of the von Kármán vortex street by the aligned magnetic field was found

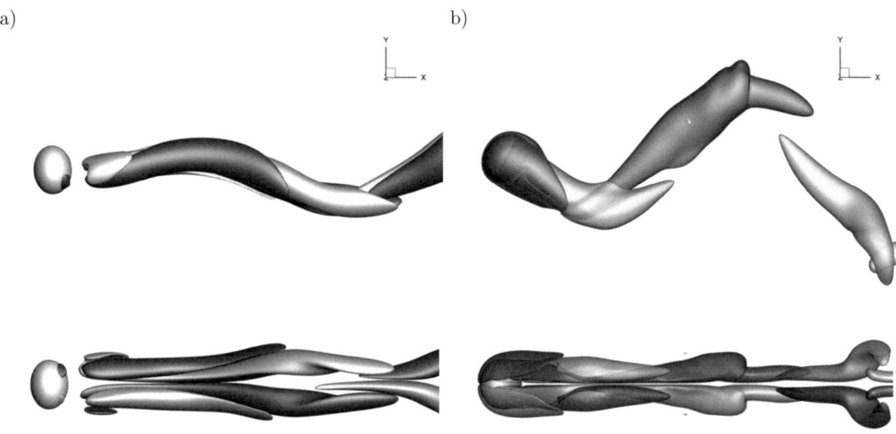

Figure 3.32 Wake structures visualized by iso-contours of streamwise vorticity $\omega_x d_{eq}/u_\infty = \pm 0.5$ for $X = 1.25$, $Re_{eq} = 300$, $\pi_\rho = 0.001$. a) Only rotatory oscillation in ϕ_z. b) Additional oscillation in y_p. Top: xy-view. Bottom: xz-view.

accompanied by an elimination of the fluctuations in lift and a slight decrease in drag for for $N \lesssim 0.5$. A further increase of the magnetic interaction led to a monotonous increase in drag.

The flow past a fixed sphere and a fixed ellipsoid with $X = 2$, $\phi_z = 90°$ are recapitulated here. Simulations were conducted in a domain of extent $\mathbf{L} = (20, 10, 10)\, d_{eq}$ with an equi-spaced grid of $\mathbf{N} = (400, 200, 200)$ cells with the spheroid placed at $\mathbf{x}_p = (5, 5, 5)\, d_{eq}$. A Reynolds number of $Re_{eq} = 200$ was chosen, i.e. an axisymmetric wake without vortex shedding is found for the sphere in the case without magnetic field. Introducing a longitudinal magnetic field, the drag on the sphere increases monotonously and scales with $\sqrt{N}$ for $N > 1$ (Figure 3.33a)), which is in agreement with findings for higher Re and N in [173]. The drag increases basically linearly with N for $N < 1$ (Figure 3.33b)). The above scaling laws are quite insensitive to the numerical resolution and are found on a coarser grid as well with differing absolute values for C_D, though.

Figure 3.33 illustrates the mechanism leading to a drag increase with stronger magnetic interaction. The Lorentz force is proportional to the lateral velocities v and w for a longitudinal magnetic field B_x, so that increasing magnetic interaction leads to a damping of the lateral velocity components. A substantial reduction of the velocity component v is apparent from Figure 3.33 as the magnetic interaction is increased from $N = 0.2$ to $N = 4.0$, also leading to a straightening of the stream lines around the sphere. The recirculation area becomes more conical and the rear pressure is reduced. A larger region of stagnant fluid develops in front of the sphere, which is characterized by an increased front pressure. The modified pressure distribution is the main reason for an increase in drag with increasing magnetic interaction parameter.

For the oblate ellipsoid, an SW mode is observed for the wake and C_D and C_L oscillate with $C_D(d_{eq}) = 1.36$ and $\langle C_L \rangle \approx 0$ in the case without magnetic field. With increasing N, the streamwise vorticity is reduced, as discussed in detail in Chapter 5 for a single rising bubble. A wake transition towards an RSP mode for $N = 0.2$, an SS mode for $N = 0.5$ and a steady

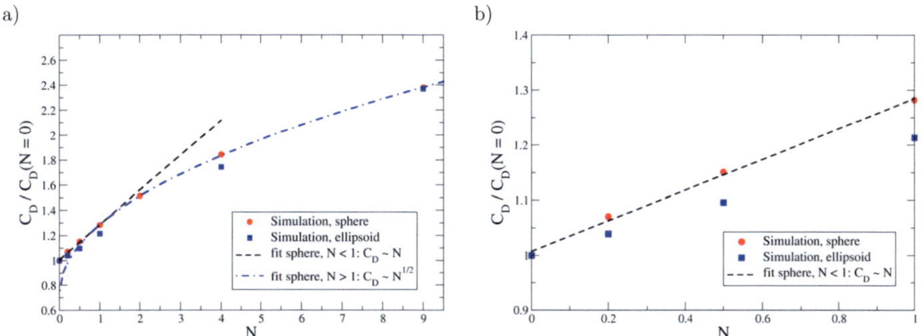

Figure 3.33 Influence of an aligned magnetic field on the drag for $Re_{eq} = 200$ for a fixed sphere and an oblate ellipsoid with $X = 2$, $\phi_z = 90°$. a) Relative change in drag versus magnetic interaction parameter N. b) Close-up for for $N \in [0,\ 1]$.

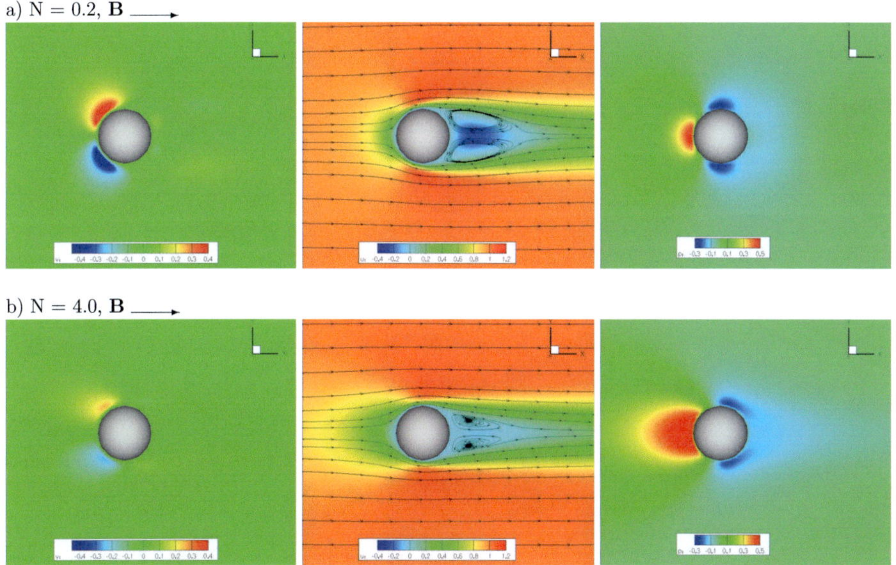

Figure 3.34 Influence of an aligned magnetic field on the flow past a fixed sphere for $Re = 200$. The arrow indicates the direction of the field. a) $N = 0.2$, b) $N = 4.0$. From left to right: Contour through sphere center of transverse velocity v, streamwise velocity u with streamlines and pressure p.

axisymmetric wake for $N = 1.0$ are observed. The relative increase in drag at low N is moderate compared to the sphere, since vortex shedding is suppressed for $N < 1$. Similar scaling of the drag augmentation with N is observed as for the sphere, which was also stated in [314] for an ellipsoid at high Reynolds numbers and interaction parameters.

Moving spheroids

The impact of an aligned magnetic field on the dynamics of moving spheroids is studied. In [245], both oscillating spheroids and freely moving particles were considered. Here, an oscillating ellipsoid under the influence of a homogeneous magnetic field parallel to u_∞ is briefly recapitulated where the setup is otherwise analogous to the one in Section 3.4.5. A more in-depth analysis of the effect of such a magnetic field on the dynamics of a rising bubble in liquid metal is provided in Chapter 5. The ellipsoid has an aspect ratio of $X = 1.25$, a density ratio of $\pi_\rho = 0.001$ and is studied at a Reynolds number of $Re_{eq} = 300$. In the purely hydrodynamic case, the particle undergoes a distinct zig-zag motion and the wake describes a standing wave. The magnetic interaction parameter, N, is now increased progressively taking values of $N = 0$, 0.05, 0.1, 0.2, 0.5, i.e. the focus of this study is on rather weak to moderate field strengths. Statistics were collected for 128 time-units tu_∞/d_{eq} for each value of N. With increasing N, the oscillation in ϕ_z and v_p remains regular, but the amplitudes reduces and also the characteristic frequency decreases. Figure 3.35 provides quantitative access to this data. As the lateral dynamics are damped, the drag on the particle decreases slightly. This is an adverse effect to the monotonous increase in drag with N for the fixed spheroids. An additional simulation was conducted for $N = 4$ and then an overall increase of the average drag coefficient was measured compared to case without magnetic field.

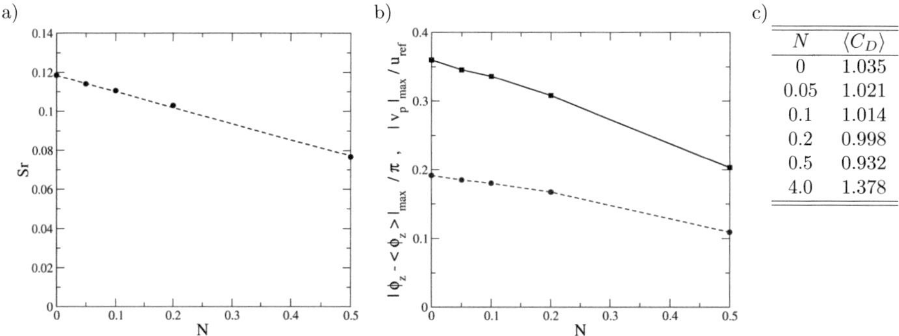

N	$\langle C_D \rangle$
0	1.035
0.05	1.021
0.1	1.014
0.2	0.998
0.5	0.932
4.0	1.378

Figure 3.35 Influence of an aligned magnetic field on the dynamics of an oscillating ellipsoid $X = 1.25$, $Re_{eq} = 300$, $\pi_\rho = 0.001$. a) Strouhal number, Sr, versus magnetic interaction parameter, N. b) Amplitude in oscillation of ϕ_z and y_p versus N. c) Mean drag coefficient versus N.

3.5 Concluding remarks

The immersed boundary method was extended towards particles of more general shape and very light particles. Immersed spheroids were then studied in uniform cross-flow starting with a fixed sphere and gradually increasing the complexity by adding geometrical degrees of freedom, a freely evolving rotational motion and finally the option for the particle to translate in one direction. It has been shown that the particle dynamics are strongly coupled to the wake modes and the vortex shedding. For very low particle densities, a bubble like behavior was found characterized by a zig-zag trajectory. Further, it was discussed how a transverse oscillation can lead to an increase in drag. At last, the influence of an aligned magnetic field was addressed. For fixed spheroids, a monotonous increase in drag was found as the magnetic interaction parameter is increased. On the other hand, for freely oscillating particles the magnetic field damps the transverse motion, which is accompanied by a mild decrease in drag. For higher interaction parameters, the drag again increases compared to the case without magnetic field. The present configuration provides a suitable setup to study specific effects of particle-fluid-interaction at moderate computational costs. With the immersed boundary method, selected degrees of freedom can be blocked or activated. Further research should address helical motion, consider deformable particles and screen the parameter space (Re, π_ρ, X) in more detail.

4 Representation of Bubble Shapes

4.1 Bubble shape regimes

An ascending bubble is deformed by the loads exerted by the liquid while surface tension drives it towards a spherical shape. Gas bubbles rising in liquids therefore adopt shapes which vary from spherical, ellipsoidal, cap-like to largely distorted forms [165, 37]. The well-known regime map of Clift et al. [37] categorizes these bubble shapes and is shown in Figure 4.1. The figure also features bubble shapes that were obtained from present simulations addressed later on.
The shape of an individual bubble rising in quiescent fluid in an unbounded domain is characterized by the terminal Reynolds number and the Eötvös number,

$$Re_t = \frac{\langle v_p \rangle d_{eq}}{\nu}, \qquad Eo = \frac{\Delta\rho\, g\, d_{eq}^2}{\sigma}, \tag{4.1}$$

where $\langle v_p \rangle$ is the average rise velocity, d_{eq} the sphere volume-equivalent diameter, $\Delta\rho = \rho_p - \rho_f$ the density difference between the gas and the fluid, g gravity, ν the liquid's kinematic viscosity and σ is the surface tension. While the Reynolds number is the ratio of inertial to viscous forces, the Eötvös number measures the magnitude of buoyancy forces compared to surface tension forces. In a similar fashion, one could use the Galilei number, $G = \sqrt{|\pi_\rho - 1|\, g\, d_{eq}^3}/\nu$, and the Weber number, $We = \rho_f u_{ref}^2 d_{eq}/\sigma$, to span a bubble shape regime map (where $\pi_\rho = \rho_p/\rho_f$).
In the shape diagram of Figure 4.1, spherical bubbles are found if surface tension is dominant, i.e. at low Eo and also independent of Re_t. As the influence of surface tension decreases, the bubble shapes become ellipsoidal, and largely distorted bubble shapes are observed for high Eo. These comprise cap-shaped bubbles which also can be dimpled or skirted. When moving towards higher Re_t, the bubble shape becomes time-dependent. Vortex shedding leads to path instability and non-rectilinear bubble trajectories including zig-zag, spiraling or rocking bubble motion. Hence, the hydrodynamic forces acting on the bubble also vary in time and so does the bubble shape. This is indicated as wobbling in the shape diagram, where this originally chosen term might be a bit misleading. At the indicated position in the regime map of [37], the pairing of Re_t and Eo corresponds to an ellipsoidal bubble with rather regularly oscillating aspect ratio [84, 81, 167, 244]. Actual wobbling, which might be understood as irregular shape oscillations, is present at higher Eo.
To illustrate the parameter range considered in the present work dealing with liquid metal and MHD systems, Table 4.1 lists material properties of the eutectic alloy GaInSn and compares them to those of water. The liquid metal GaInSn has been selected here as the simulations reported in Section 5.2.1 are conducted for a configuration with argon bubbles in GaInSn. This specific alloy is liquid at room temperature which is an attractive property

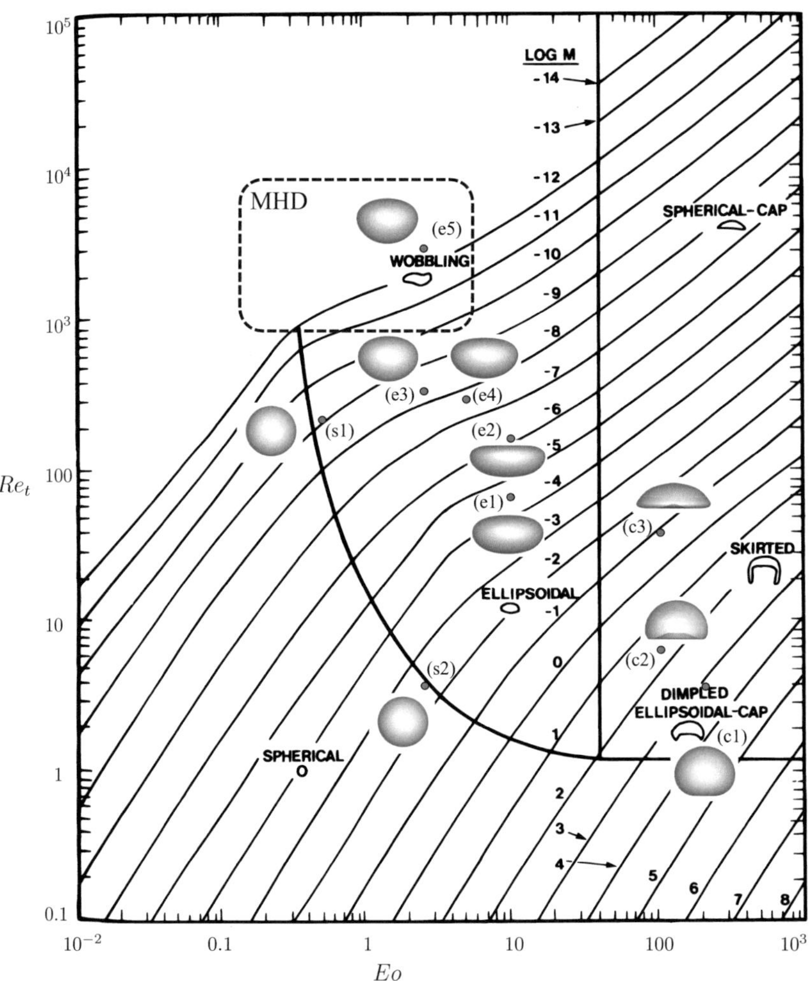

Figure 4.1 Shape regimes from [37] for single bubbles and drops in quiescent fluid and present results from Section 4.4. Shapes as function of terminal Reynolds number, Re_t, and Eötvös number, Eo, (additional parameter Morton number, $M = \left(g\,\mu_f^4\,\Delta\rho\right) / \left(\rho_f^2\sigma^3\right)$, characterizing one specific gas-liquid system, e.g. air-water $M \approx 3\cdot 10^{-11}$).

for its use in experiments. Its density and surface tension are markedly higher than those of water, while the kinematic viscosity is smaller. As a consequence, the Galilei number, which relates buoyancy forces to viscous forces, is higher for an argon bubble in GaInSn than for an air bubble of equal size in water, hence resulting in a higher bubble Reynolds number. The high density ratio and high surface tension are difficult to deal with in many multiphase methods, e.g. the volume of fluid method where spurious currents may occur as numerical artifacts and low time step sizes become necessary. Liquid metals are characterized by a very low Prandtl number due to the very high thermal conductivity making them very attractive for modern energy concepts. The most significant contrast with water is the difference in electrical conductivity by about eight orders of magnitude enabling electromagnetic flow control. An approximate value for tab water is listed for comparison. The Eötvös number is almost the same so that in GaInSn similar bubble shapes can be expected as in water, but at a higher Reynolds number. The MHD parameter region is also sketched in Figure 4.1.

Table 4.1 Material properties of GaInSn and water at a temperature of $20°C$ and ambient pressure of 1 bar. The non-dimensional numbers are calculated for an argon bubble in GaInSn and an air bubble in water, both with an equivalent diameter of $d_{eq} = 4.6$ mm.

	GaInSn	Water
Density ρ_f $[kg\,m^{-3}]$	6361	998
Surface tension σ $[N\,m^{-1}]$	0.533	0.073
Kinematic viscosity ν $[m^2\,s^{-1}]$	$3.46 \cdot 10^{-7}$	$9.82 \cdot 10^{-7}$
Electrical conductivity σ_e $[S\,m^{-1}]$	$3.27 \cdot 10^{6}$	$\approx 5.0 \cdot 10^{-2}$
Galilei number G	2825	995
Eötvös number Eo	2.5	2.8
Prandtl number Pr	$\mathcal{O}(10^{-2})$	7.0

4.1.1 Numerical description of bubble shapes

The most common quantitative measures describing the shape of a bubble or drop are ellipsoidal aspect ratios in both experiments, e.g. [309, 167, 303], and numerical studies, e.g. [187, 169, 81]. Approaches of higher order to evaluate the results for the bubble shape quantitatively are rather rare, e.g. spherical harmonics are used the experiments of [22, 289] and simulations of [133, 98].

In accordance with the occurrence of bubble shapes apparent from the shape diagram, different numerical descriptions of bubble shapes were successively implemented in PRIME. Three major bubble shape representations can be distinguished: Sphere, Ellipsoid, and Spherical Harmonics (SH).

Figure 4.2 displays a spherical bubble (a) with an equidistant forcing point distribution [157], an ellipsoidal bubble (b) with additional surface triangulation and a slice through a cap-shaped bubble represented by spherical harmonics (c) with an additional spherical coordinate surface grid. Depending on the prerequisites of the simulation to be undertaken, a certain representation is chosen. This choice can be interpreted as a regularization in terms of bubble shape and reaches from *a priori* determined, constant, simple bubble shapes to temporally changing, flow-determined, irregular shapes. If for instance a large number of small bubbles shall be considered in a turbulent channel flow [236] or the formation of regular packings in bubble foams is of interest [102, 101], the choice of a representation by spherical bubbles eases the physical interpretation of the results, increases reproducibility and simplifies the numerical modeling.

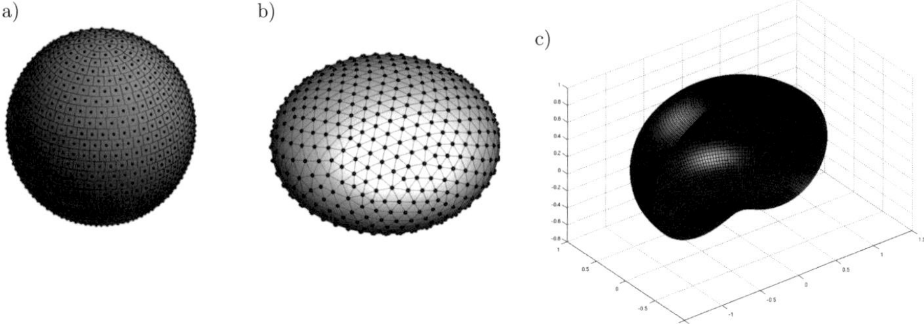

Figure 4.2 Particle shape representations in PRIME. a) Sphere [157], b) Ellipsoid, c) Spherical Harmonics, slice through cap-shaped bubble.

A *temporal variation* in bubble shape or a shape adaptation can be incorporated by:

1) a prescribed shape evolution,

2) a change in shape based on a correlation, e.g. from experimental data,

3) coupling to the hydrodynamic forces acting on the bubble.

The Lagrangian surface mesh is adapted to the instantaneous, analytically described shape in each time step.
Variant 1) constitutes a user-prescribed temporal evolution of the bubble shape. One example for this case is a physically sound description of the merge of two touching bubbles forming one larger bubble which is dealt with in detail in Section 7.5.2.
In variant 2), the bubble shape and its temporal change are obtained from a correlation to a solution variable. For the parameters of an argon bubble rising in the liquid metal GaInSn [244] (Section 5.2.1), the bubble shape is expected to be 'ellipsoidally wobbling' [37]. Hence, the shape parameter X of the ellipsoidal particle has to be varied in time in a physically meaningful way. As suggested in [165], an empirical correlation between the instantaneous bubble Weber number $We(t) = \rho_f \|\mathbf{u}_p(t)\|^2 d_{eq}/\sigma$ and the instantaneous aspect ratio $X(t)$ is employed. Loth [165] presented experimental and theoretical data for the correlation $X(We)$ and provided a fit for moderate to high bubble Reynolds numbers, $Re > 100$. The correlation was obtained using data for the mean rise velocity and mean aspect ratio. It is then shown in [165] that the correlation is also a good fit for instantaneous data $X(We(t))$ by comparison to the time-resolved experiments in [274]. The validity of the approach was also scrutinized by numerical simulations in [2] comparing the instantaneous bubble shape to predictions from a phase-field model and good agreement was found. The functional relation from [165] reads

$$X^{-1}(t) = 1 - 0.75 \tanh(0.165\, We(t)) \,. \tag{4.2}$$

Figure 4.6 shows this curve together with the data given in the review of Loth [165] and present results. With the material properties being constant, a bubble moving with a high

velocity $\|\mathbf{u}_p(t)\|$ will adopt a flat shape while the bubble shape remains spherical at low velocities. The correlation is valid for single bubbles, rising in quiescent fluid in an unbounded domain with $Re > 100$.
Variant 3) provides a more general approach in which the bubble shape is coupled to the loads imposed by the surrounding fluid at the phase interface. Imagine two bubbles rising in quiescent fluid where one bubble is trailing the other. The latter will be ascending faster as the leading one once it is entrained in the wake. However, in contrast to the correlation above, it will be more spherical. For bubbles rising in chains or swarms, or in the vicinity of walls, or for bubbles immersed in a turbulent flow, the bubble shape cannot be easily correlated. Instead the deformation of the bubble by hydrodynamic forces needs to be directly resolved.

4.2 Deformable bubbles

4.2.1 Physical and numerical model

The representation of deformable bubbles is described in this section and was partially published in [246]. The potential displacement energy W is given by

$$W = \int_S [(p_{in} - p_o) - \sigma\,\kappa]\,\delta n\;dA, \tag{4.3}$$

where p_o is the outer pressure, $\Delta p(\mathbf{x}_S) = p_{in} - p_o(\mathbf{x}_S)$ is the local pressure difference over the interface, and $\mathbf{x}_S$ a point on the surface. Viscous stresses are neglected here, as justified in [46, 61]. The bubble volume is imposed to be constant in time and the pressure inside the bubble, p_{in}, is assumed to be constant over the bubble at each instant and obtained as a surface average $p_{in} = \langle\sigma\kappa + p_o\rangle_S$ which gives $p_{in} = 2\sigma/r_{eq}$ for a small bubble at rest in quiescent fluid. The pressure on the outside is $p_o(\mathbf{x}_S) = p_{dyn} + p_{stat}$, where p_{dyn} is interpolated from the pressure field on the Cartesian grid, and p_{stat} is the hydrostatic pressure component. The hydrostatic component was eliminated in the continuous pressure field p, e.g. to better cope with periodic boundary conditions, and it therefore occurs as an additional term. The displacement of the interface normal to the surface is given by δn. Twice the local mean curvature of the bubble surface is denoted by $\kappa(\mathbf{x}_S)$.
Equation (4.3) considers the work that is performed due to the displacement of the interface by pressure forces and the work that is performed due to deformation of the surface by surface tension. The potential displacement energy vanishes in equilibrium, $W = 0$, i.e. the bubble shape adapts so as to satisfy the local force balance. The local residuum is

$$\varepsilon(\mathbf{x}_S) = (p_{in} - p_o) - \sigma\,\kappa\,. \tag{4.4}$$

In case the bubble shape is restricted by some parametrization, (4.4) leads to the requirement $W \Rightarrow min!$ instead of $W = 0$ which holds for a completely unrestricted shape. To find the optimum shape the test function for the deformation is chosen to be $\delta n(\mathbf{x}_S) = \varepsilon(\mathbf{x}_S)$ [61], which yields the quadratic form to be minimized

$$W = \int_S (\varepsilon(\mathbf{x}_S))^2\,dA \Rightarrow min!\;, \tag{4.5}$$

with the constraint of constant bubble volume $V_p = const$ and a constant center of mass. The mass of the dispersed phase hence is conserved exactly, in contrast to other methods

without prior shape assumptions such as volume of fluid methods (VOF).
The present approach constitutes an equilibrium model. In case of non-equilibrium conditions, e.g. the initial conditions for the coalescence process described in Section 7.5.1 and in [250], or for very intense or fast changes in bubble shape, the integral equation (4.3) can be re-written as a local force balance. An additional term is included which is proportional to a specified fluid mass attached to the interface times its acceleration and which thus attenuates the dynamics towards equilibrium also stabilizing the numerics within the explicit coupling scheme. The force balance then reads $m_f\, d\mathbf{u}_{sh}/dt = F_{\Delta p} + F_\sigma$ with $F_{\Delta p}$ and F_σ being the local pressure and surface tension forces, and $\mathbf{u}_{sh}$ the velocity of a point on the bubble surface related to a change in shape with an associated mass of m_f (e.g. $m_f \simeq \rho_f V_E$ chosen here). It is solved again with the constraints $V_p = const$ and a constant center of mass. The non-equilibrium extension is used in the detailed study of a single bubble coalescence event in Section 7.5.1 and good quantitative agreement is found for the temporal evolution of the bubble shape. The equilibrium model is employed for the studies presented below, the examination of single bubble ascent (Section 5.6), as well as for the simulations considering bubble chains (Chapter 6 and Section 7.6).

To sum up, the outer pressure field $p_o(\mathbf{x}_S)$ on the bubble surface and twice the local mean curvature $\kappa(\mathbf{x}_S)$ need to be determined to solve (4.5) for an optimum shape. Two parametrization of the deformable bubble shape are considered: An oblate ellipsoidal shape (Section 4.3) and bubbles represented by spherical harmonics (Section 4.4).

4.2.2 Pressure interpolation

The pressure field of the continuous fluid phase needs to be interpolated from the Cartesian grid to the bubble surface. Two different interpolation strategies were considered. First, the regularized δ-function of Roma [225], which is used in the IBM for the interpolation of the velocity field (Section 3.3.4), is adopted also for the interpolation of the pressure. Second, the more general framework of the Moving Least Squares interpolation (MLS) is employed which allows for a larger variety of basis functions, as well as additional constraints [151, 158, 285]. A brief outline is given in the Appendix K. The accuracy and convergence properties of the latter approach were studied in detail within the diploma thesis of Beetz [12] for the pressure interpolation to the Lagrangian forcing point distribution of an ellipsoid. The implementation was then amended to a general set of points, e.g. the structured SH mesh which enables more efficient calculations on the surface.
An example pressure field is provided in the contour of Figure 4.6 for an oblate ellipsoid in cross flow. It features the stagnation pressure region at the bubble front, low pressure regions on each side where the flow is accelerated, and again moderately increased pressure at the rear of the bubble where the flow recirculates. The figure also allows for a view on the pressure field inside the bubble where an artificial, weak flow is driven by the IBM. The pressure inside the bubble is very close to the reference pressure corresponding to the zero level here. This results in a marked pressure difference over the interface for instance at the front stagnation point. However, the pressure field is substantially smoother then in, e.g., VOF or front-tracking methods where a substantial pressure jump, proportional to the density difference between the phases, needs to be resolved [275]. This usually degrades the performance of the pressure solver and can lead to time step restrictions and numerical difficulties.

The regularized δ-function uses a symmetric stencil with three points in each direction. In order to reduce the influence of points inside the bubble on the interpolation result, the location of the interpolation is slightly shifted towards the outside by $1.5\Delta x$ in radial direction. A more rigorous approach consisting of an extrapolation towards the bubble surface using only outside points should be considered in the future, e.g., in the MLS framework. The parallel implementation comprises simultaneous interpolation on all processes containing a fraction of the bubble surface and gathering of the results on the master process by MPI communication.

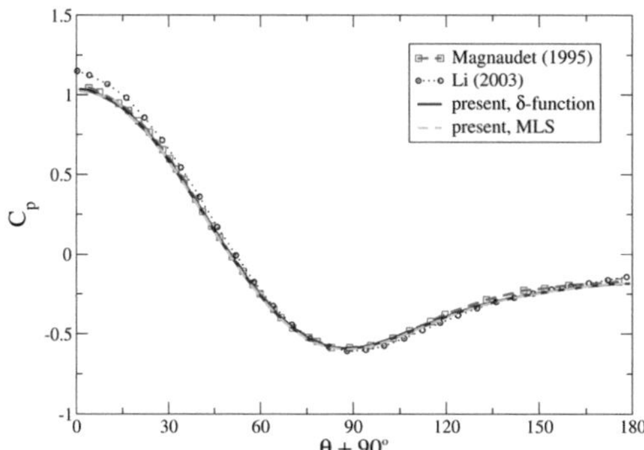

Figure 4.3 Surface pressure distribution for the flow around a rigid sphere for $Re = 100$. Pressure coefficient $C_p = (p - p_{ref}) / \left(\rho\, u_{ref}^2/2\right)$ over angular coordinate measured from front stagnation point. Comparison of present data to simulation data of [159] and [170].

Figure 4.3 shows the pressure profile on a sphere in cross-flow for $Re = 100$. For this Reynolds number, the flow is axisymmetric with respect to the direction of the incoming flow and a steady recirculation zone forms behind the sphere. The simulation was conducted with a spatial resolution of $d_p/\Delta x = 40$ in a sufficiently large domain. Comparing the results obtained with the regularized δ-function and the MLS approach with quadratic basis functions, very similar accuracy is observed. The δ-function provides slightly better smoothing properties, especially on coarse grids (not shown here), which is why it is used in the simulations presented in this work. The graph also contains reference data from [159] and [170] obtained with finite-volume methods employing body fitted grids. Very good agreement is found whereas a minor underprediction of the stagnation pressure is observed in the present case due to the reasons mentioned above. In terms of bubble shape, this would result in slightly more spherical bubbles.

4.2.3 Interface velocity due to shape oscillations

For a deformable particle and a no-slip condition at the interface, the velocity at the bubble surface S consists of three parts $\mathbf{u}_S(\mathbf{r},\, t) = \mathbf{u}_p + \boldsymbol{\omega}_p \times \mathbf{r} + \mathbf{u}_{sh}$, where $\mathbf{u}_p$ is the translational velocity of the particle center, the second term denotes the part due to rotation and $\mathbf{u}_{sh}$ is the velocity induced due to changes in shape. This latter velocity needs to be determined from the computed temporal change in bubble shape.
Figure 4.4 schematically sketches an ellipsoidal shape oscillation and the absolute value of

the fluid velocity that is created in the vicinity of the bubble. In this thought experiment, the bubble is fixed with respect to translation and does not rotate. Hence, only $\mathbf{u}_{sh}$ contributes to the interface velocity. An oblate ellipsoid of temporally constant volume and regularly oscillating aspect ratio $X(t)$ was considered. The scenario is similar to the late stage of shape oscillations subsequent to a coalescence event studied in detail in Section 7.5.1. Note, that in this case the velocity along the axis of rotation $\dot{b}(t) = 2/3X(t)^{-1/3}\,\dot{X}(t)\,r_{eq}$ takes on larger peak values than its counterpart $\dot{a}(t) = 1/3X(t)^{-2/3}\,\dot{X}(t)\,r_{eq}$ directed orthogonally to $\dot{b}$.

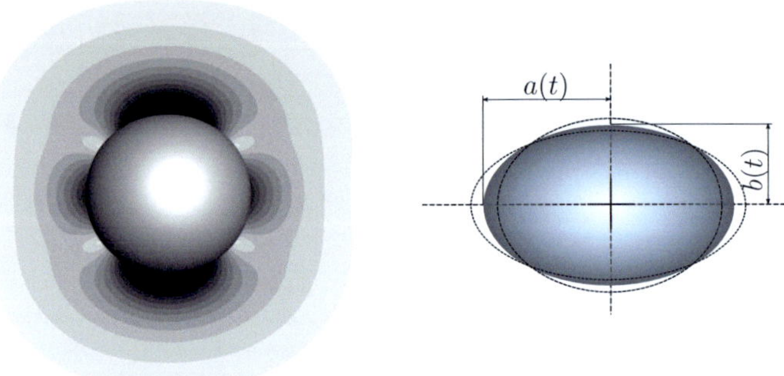

Figure 4.4 Schematic sketch of interface velocity due to shape oscillations. Contour of the absolute value of the induced fluid velocity.

Two variants are implemented for the determination of $\mathbf{u}_{sh}$ depending on the chosen shape representation. The first variant is based on the discrete representation and employs the Lagrangian forcing points $\mathbf{x}_{fp}$. The total temporal change in the location of a forcing point is governed by

$$\frac{\mathrm{d}\mathbf{x}_{fp}}{\mathrm{d}t} = \mathbf{u}_{S,fp} = \mathbf{u}_p + \boldsymbol{\omega}_p \times \mathbf{r}_{fp} + \mathbf{u}_{sh,fp} \tag{4.6}$$

in the laboratory coordinate system. The two contributions due to translation and rotation with respect to the particle center are known from the solution of the respective particle momentum equations. The position of the interface, $\mathbf{x}_{fp}(t)$, is obtained from the shape adaptation. One thus can solve for $\mathbf{u}_{sh,fp}$, e.g. employing backward finite differences (outlined in Section 3.3) for $\mathrm{d}\mathbf{x}_{fp}/\mathrm{d}t$. The latter requires to save old forcing point positions and additional communication in parallel runs.

The second variant is valid for bubbles represented by spherical harmonics discussed in Section 4.4 below. In this case, $\mathbf{u}_{sh}$ can be computed from the recent history of the SH shape coefficients. The temporal evolution of the interface due to a change in the axisymmetric bubble shape can be expressed in the local spherical coordinate system by

$$\frac{\mathrm{d}r(\theta,t)}{\mathrm{d}t} = \sum_{n=0}^{N_{SH}} \frac{\mathrm{d}\,q_n(t)}{\mathrm{d}t} P_n(\theta), \tag{4.7}$$

again employing backward finite differences for $\mathrm{d}q_n/\mathrm{d}t$. A transformation to Cartesian coordinates yields $\mathbf{u}_{sh}^{Loc} = \left(u_{sh}^{loc}, v_{sh}^{loc}, w_{sh}^{loc}\right)$. Finally, a transformation from the local coordinate system of the spherical harmonics to the global, laboratory system using the rotation matrix $\mathbf{A}$ introduced in Section 3.2 yields

$$\mathbf{u}_{sh} = \mathbf{u}_{sh}^{lab} = \mathbf{A}^T \cdot \mathbf{u}_{sh}^{loc} , \tag{4.8}$$

which can then be incorporated into the IBM and evaluated at discrete locations to obtain $\mathbf{u}_{sh,fp}$. The shape history $q_n(t)$, $n = 0, \ldots, N_{SH}$ needs to be saved for a number of recent time levels to enable the evaluation of the backward finite differences and it has to be communicated in parallel simulations. A rigorous validation was conducted by a prescribed shape oscillation of an ellipsoid, $X(t)$, and comparison of the computed interface velocity to the analytical solution (not shown here). Further validation is apparent from Section 7.5.1 where the computed interface velocity is compared to experimental data for two bubbles undergoing coalescence.

4.3 Ellipsoidal bubbles

4.3.1 Algorithm for shape adaptation

A general ellipsoid is defined by its semi-axes a, b and c. The surface $\mathbf{x}_S = (x_S, y_S, z_S)$ is parametrized as

$$x_S = a \cos\eta \sin\xi, \qquad y_S = b \sin\eta \sin\xi, \qquad z_S = c\cos\xi \tag{4.9}$$

with $\eta \in [0; 2\pi)$ and $\xi \in [0; \pi]$. Note, the surface parameters η and ξ are not angles of a spherical coordinate system.
The local mean curvature $\kappa(\mathbf{x}_S)$ of this ellipsoid can be determined analytically:

$$\kappa = 2\frac{abc[3(a^2+b^2) + 2c^2 + (a^2+b^2-2c^2)\cos(2\xi) - 2(a^2-b^2)\cos(2\eta)\sin^2\xi]}{8[a^2b^2\cos^2\xi + c^2(b^2\cos^2\eta + a^2\sin^2\eta)\sin^2\xi]^{3/2}} . \tag{4.10}$$

Physical observations suggest to model the bubble shape by an oblate ellipsoid with semi-axes $a = c > b$, valid in a wide range of the regime map of [37, 165]. The present implementation is similar to the approach outlined in [60, 61].
For the ellipsoidal bubble representation, the Lagrangian forcing points (index fp) serve as the basis for the algorithm. The following steps are undertaken for the evaluation of the potential displacement energy:

- Interpolation of the fluid pressure onto the Lagrangian forcing points, calculation of local pressure difference $\Delta p_{fp}^{(l)}$ and transformation of the pressure field to a coordinate system aligned with the major axes of the ellipsoid

- Calculation of the parametrization $(\eta, \xi)_{fp}^{(l)}$ of the forcing point $\mathbf{x}_{fp}^{(l)}$ from (4.9).

- Evaluation of curvature $\kappa_{fp}^{(l)}$ and local residuum $\varepsilon_{fp}^{(l)}$ from (4.10) and (4.4).

- Integration over forcing points (or a specific subset [60]) and calculation of the total potential W, equation (4.5) becomes $W = \sum_{l=1}^{N_L} \left(\varepsilon_{fp}^{(l)}\right)^2 A_{fp}^{(l)} \Rightarrow min!$

A single parameter has to be determined by the minimization problem (4.5), the ellipsoid aspect ratio $X = a/b$. The following *optimization strategy* is pursued:
Starting from the initial value $X_{j=1} = X_n = X(t_n)$ a Newton algorithm (iteration index j) is used to compute the optimum aspect ratio X_{opt} at time t_n. This optimum has to fulfill

$$W' = \frac{\mathrm{d}W}{\mathrm{d}X} = 0, \qquad W'' = \frac{\mathrm{d}^2(W)}{\mathrm{d}X^2} > 0. \tag{4.11}$$

In order to determine these derivatives, W is calculated for X_j and two neighbor states $W^{(-)}$ and $W^{(+)}$ with a 'frozen' pressure field $p_o^n(\mathbf{x}_S)$ under the constraint $V_p = const.$

$$\begin{aligned} W^{(-)} &= W(X_j - \Delta X) \\ W_j &= W(X_j) \\ W^{(+)} &= W(X_j + \Delta X) \end{aligned}$$

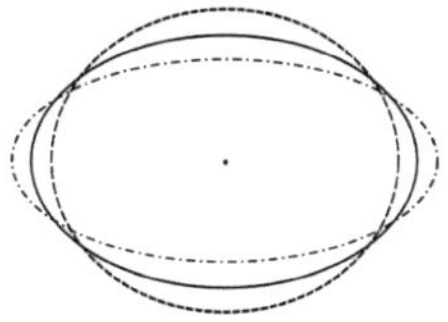

The derivatives are calculated by second order divided differences with

$$W'(X_j) = \frac{W^{(+)} - W^{(-)}}{2\Delta X}, \qquad W''(X_j) = \frac{W^{(+)} - 2W_j + W^{(-)}}{(\Delta X)^2}. \tag{4.12}$$

The next step in the Newton algorithm is obtained by

$$X_{j+1} = X_j - \frac{W'(X_j)}{W''(X_j)}. \tag{4.13}$$

The solution of the minimum search is called X_{opt}.
Under-relaxation of the change in aspect ratio is necessary for stability due to the CFL constraint and the 'frozen' pressure field. The new aspect ratio is thus computed from $X_{new} = X_n + w\,(X_{opt} - X_n)$ with the under-relaxation factor $w < 1$. See for instance [18] for a discussion of weak versus strong (explicit versus implicit) coupling in fluid-structure interaction problems. The new value X_{new} is applied at $t_n + \Delta t_{sh}$ (where $\Delta t_{sh} \approx 10\Delta t$ is chosen here) and the intermediate shapes are determined by linear interpolation between X_n and X_{new}. After the time interval Δt_{sh} the shape optimization is repeated with an updated pressure field. In future developments of the algorithm, information from within the time-interval between two adaptation steps should be used to improve the shape coupling. The algorithm is shown schematically in Figure 4.5a) for the adaptation towards a terminal shape X_t.

For parallel runs, the solution of the minimization problem is conducted on the master process only. The current ellipsoid aspect ratio is communicated to the slave processes to enable the distribution of forcing points. For moving particles, additionally X_n and X_{new} need to be exchanged if a particle leaves its local domain.

4.3.2 Results and comparison

The approach was validated by computing the *quasi-stationary shape of a fixed ellipsoidal bubble in cross flow.* A box of extent $\mathbf{L} = (14,\ 7,\ 7)d_{eq}$ is discretized equidistantly with

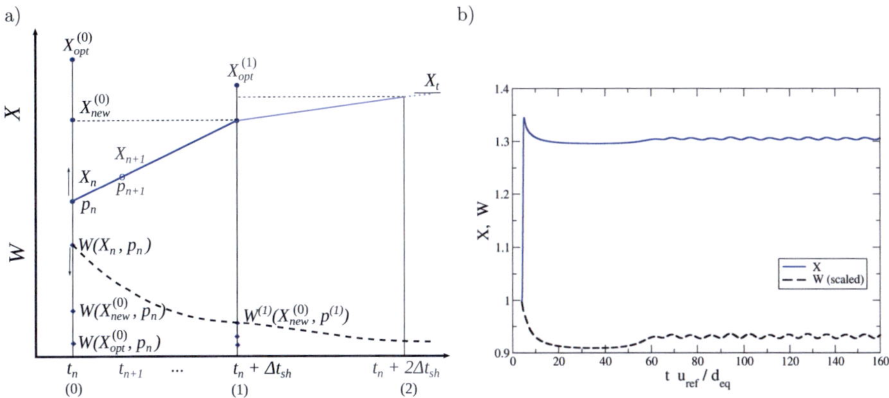

Figure 4.5 Adaption of bubble shape to hydrodynamic forces. a) Schematic sketch of shape adaptation algorithm. b) Temporal evolution of ellipsoid aspect ratio X and scaled potential displacement energy W for $We = 2$ and $Re = 250$.

$\mathbf{N} = (256, 128, 128)$ cells, corresponding to a spatial resolution of $d_{eq}/\Delta x \approx 18$. The bubble surface is represented with $N_L = 2122$ Lagrangian forcing points. The time step is chosen to yield $CFL = 0.5$. The particle position is $\mathbf{x}_p = (0.5, 0.5, 0.5)\, L_y$, and the particle has an initial aspect ratio of $X_0 = 1.0$. A uniform inflow is prescribed with $u_\infty = u_{ref}$ and a convective outflow condition is applied at the opposite boundary as well as periodic conditions at the lateral boundaries.

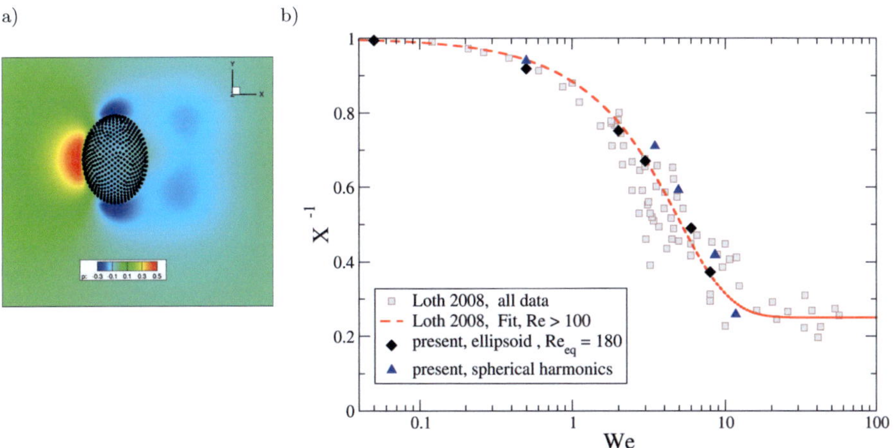

Figure 4.6 Bubble shape coupled to fluid loads. a) Instantaneous pressure field (close-up) for the final ellipsoidal shape with $Re_{eq} = 180$, $We = 2$. b) Bubble aspect ratio versus Weber number. Comparison of present simulation results for ellipsoidal and SH bubbles with data and fit from [165].

To check whether a distinct global minimum in W and therefore an optimum shape exists,

a preliminary simulation is performed stepping through the parameter range $X \in [1.0, 4.0]$ with $\Delta X = 0.1$ and monitoring W (not shown here). Indeed, the potential displacement energy takes a global minimum at an aspect ratio close to the value given in Figure 4.6 for a prescribed Weber number. It now needs to be studied how fast the optimization strategy finds this optimum. The temporal evolution towards the final shape is displayed in Figure 4.5b) for $We = 2$ and $Re_{eq} = 250$. The adaptation process is started $4\, d_{eq}/u_{ref}$ after initialization. Starting from the initial aspect ratio of $X = 1.0$ the shape adapts towards the equilibrium shape of $X = 1.32$ within approximately $2\, d_{eq}/u_{ref}$ time units also obeying the CFL-constraint. A moderate overshoot is observed while likewise the local flow field adjusts to the new particle shape. The aspect ratio decreases very slightly as the global flow field develops. At $t \approx 50\, d_{eq}/u_{ref}$ vortex shedding sets in as to be expected for the chosen Reynolds number. The residual in the surface potential W again somewhat increases since the chosen shape parametrization is axisymmetric, but the flow and thus the fluid loads just lost this property. The ellipsoid aspect ratio then oscillates periodically with a frequency strongly correlated to the vortex shedding.
Results for different values of We and a constant Reynolds number $Re = 180$, as well as a comparison with data of single, clean bubbles from the literature [165] is given in Figure 4.6b). Overall good agreement is found for all We considered. Note that for large Weber numbers, the bubble shape deviates from ellipsoidal and the bubble obtains, e.g., a cap shape. In this case a more sophisticated parametrization of the bubble shape is needed allowing more degrees of freedom which is introduced in the next section.

4.4 Bubbles represented by spherical harmonics

4.4.1 Algorithm for shape adaptation

The bubble shape is described analytically by a series expansion in spherical harmonics (SH). In the current framework, there are basically two options to increase the degree of freedom for the bubble shape. On the one hand, one could use the triangulated discrete forcing point distribution (Figure 4.2b)) and move each single marker point to satisfy the equilibrium of the potential displacement energy. One major drawback of this approach is that the calculation of surface curvature is rather inaccurate for shapes given by discrete points. More details on this matter are provided in Section 3.1.
It seems therefore reasonable to use a continuous, analytical description of the bubble shape. Here, spherical harmonics shall be used for this purpose. The bubble is described in a local spherical coordinate system by

$$r(\theta, \phi) = \sum_{n=0}^{\infty} \sum_{m=-n}^{n} a_{nm} Y_n^m(\theta, \phi)\,, \tag{4.14}$$

with the spherical harmonic functions $Y_n^m(\theta, \phi)$ [77]. The Appendix J provides the quite extensive mathematical formulae related to the SH, e.g. for the determination of pointwise quantities like surface normal vectors or integral quantities like bubble volume. Although all equations are given in a general form and the implementation was performed in a similar fashion, the discussion of bubble shapes shall be restricted to *axisymmetric* shapes from here on. Note that the axis of rotation may have an arbitrary orientation in the laboratory system. Three-dimensional SH shapes are considered for bubble coalescence in Section 7.5.2. We use a local spherical coordinate system $(r,\, \phi,\, \theta)$ centered in $\mathbf{x}_p$ where the coordinate axes are aligned with the principal axes of inertia of the bubble. The ansatz simplifies for axisymmetric shapes in ϕ to

$$r(\theta) = \sum_{n=0}^{\infty} q_n P_n(\cos(\theta)) \approx \sum_{n=0}^{N_{SH}} q_n P_n(\cos(\theta)) \, . \tag{4.15}$$

with q_n designating the coefficients of the series expansion and P_n the Legendre polynomials of order n. The surface vector $\mathbf{x}_S$ reads

$$x_S = r(\theta,\phi)\sin(\theta)\cos(\phi), \quad y_S = r(\theta,\phi)\sin(\theta)\sin(\phi), \quad z_S = r(\theta,\phi)\cos(\theta). \tag{4.16}$$

The local mean curvature $\kappa(\mathbf{x}_S)$ is determined analytically as in [77] (Appendix J) with

$$\kappa = f\left(r;\ r_{,\theta};\ r_{,\theta\theta};\ \mathbf{x}_{S,\phi};\ \mathbf{x}_{S,\theta};\ \mathbf{n}_{S,\phi};\ \mathbf{n}_{S,\theta}\right) \tag{4.17}$$

involving first derivatives of the surface vector $\mathbf{x}_S$ and the surface normal vector $\mathbf{n}_S$ with respect to ϕ and θ, as well as first and second derivatives of the local radius $r(\theta)$ which is only a function of θ in the axisymmetric case. The calculation of surface curvature was conscientiously validated against the solution for ellipsoids from equation (4.10). Note that an ellipsoid is not a direct subset of the SH representation and therefore the test scrutinizes all equations involved. Using $N_{SH} = 12$ and an appropriate Gaussian quadrature, the maximum error in the surface radius is smaller than 0.01% and the maximum error in surface curvature is determined to be 0.43%, where the maximum deviations occur at the poles. The description is hence very accurate and the error in curvature is substantially smaller compared to a discrete representation of the surface. The results are plotted in Figure 4.7 and the corresponding graphs of the analytically described ellipsoid and the SH representation collapse nicely.

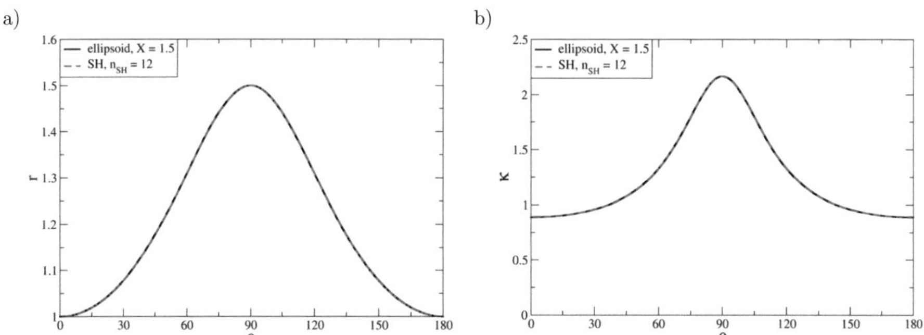

Figure 4.7 Representation of an oblate ellipsoid with $X = 1.5$ by spherical harmonics with $N_{SH} = 12$. a) Surface radius. b) Surface curvature.

The algorithm determining the bubble shape and the involved *optimization strategy* differ slightly from those pursued for the ellipsoidal shape. The SH shape adaptation is performed in the local spherical coordinate system wherein the angles θ_i of the discrete, structured representation are chosen according to a Gaussian quadrature of n_θ points. First, the pressure is interpolated to the SH surface and averaged in circumferential direction ϕ. To accurately capture the pressure field, a sufficiently dense distribution of n_θ points is used, exceeding the necessity of the quadrature determined by N_{SH}. Then a derivative-free, iterative procedure to determine the optimum shape of $N_{SH}+1$ degrees of freedom is employed instead of the Newton algorithm for the single parameter X in the ellipsoidal case. It consists of the following steps: A local residuum $\varepsilon(\theta_i)$ can be calculated from (4.4) using the local pressure difference $\Delta p(\theta_i)$ and the local curvature $\kappa(\theta_i)$. The surface is then shifted, locally weighted by ε. The shape coefficients q_n and the curvature $\kappa(\mathbf{x}_S)$ of the adapted shape are determined under the constraints of constant bubble volume and constant center of mass. The displacement energy W is re-calculated until an optimum shape is found for the given pressure field. A damped update of the bubble shape $q_n(t)$ is linearly distributed over the interval Δt_{sh}. The change in shape takes into account the CFL-criterion.

4.4.2 Results for a prescribed pressure field

As a first and rigorous validation, the algorithm is shown to compute the correct bubble shape for a prescribed, constant pressure field. The latter together with data for the bubble shape was received from private communication with J. Degroote, the author of [46, 45]. In these references, bubble dynamics are studied using front-tracking based on a partitioned fluid-structure interaction algorithm. The coupling is implicit and body fitted grids are employed. Those numerical results are also in very good agreement with corresponding experiments from [164]. The studied case corresponds to an air bubble in water which had just detached from an injection needle. An bubble equivalent radius of $r_{eq} = 2.165$ mm and thus an Eötvös number of $Eo = 2.6$ were obtained from the data. In the present simulation, the received pressure field is first linearly interpolated to the $n_\theta = 12$ grid points of the Gaussian quadrature. Starting from an initial ellipsoid with $X = 1.2$ the shape is then adapted minimizing the potential surface energy as described above. Figure 4.8 shows that the shape computed with the present SH algorithm is in very good agreement with the reference data.

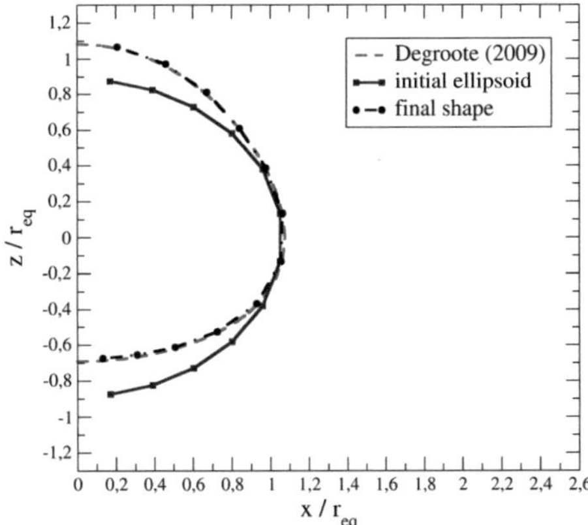

Figure 4.8 Spherical harmonic bubble shape coupled to fluid loads. Comparison of present simulation results with simulation data from [46].

4.4.3 Results for freely rising bubbles

The bubble shape of single, freely rising bubbles in quiescent fluid is studied. The major aim of this section is to illustrate the capability of the SH approach to represent a wide range of bubble shapes apparent from the shape diagram in Figure 4.1, but also to address the limits of the present approach. At first, a comparison of bubble shapes is presented for low Re and high Eo with predictions by a hybrid particle level-set method (HPLS) in [82, 81, 83] and experiments documented in [13]. The expected bubble shapes are in the cap regime.

Two simulations are conducted where the Eötvös number is fixed with $Eo = 116$ and the Galilei number is varied to be $G = 13.9$ (c1) and $G = 62.0$ (c2), respectively. The indications c1 and c2 relate to the cases cap 1 and 2 in Figure 4.1, respectively.

The setup comprises an individual bubble which is represented by spherical harmonics with $N_{SH} = 24$, $n_\theta = 120$ and $n_\phi = 12$. It is initially spherical at rest and rises under the action of gravity in stagnant liquid. The size of the computational domain and the discretization are adopted from [82] with $\mathbf{L} = (4, 8, 4)\, d_{eq}$ and $\mathbf{N} = (240, 480, 240)$ corresponding to $d_{eq}/\Delta x = 60$. The time step is adapted to yield $CFL = 0.5$. Periodic boundary conditions are applied in all three directions. A fringe zone is introduced [81] traveling at some distance in front of the bubble in which the fluid velocity is driven to zero to compensate for the rather small domain extent in the direction of gravity. The distance of this fringe zone is chosen to be $3d_{eq}$ with respect to the bubble center and it has a height of $0.25d_{eq}$. In this zone, a forcing term is introduced to the Navier-Stokes equations proportional to the difference of the current fluid velocity and the desired zero velocity very similar to the IBM discussed above.

The results of the two simulations are gathered in Figure 4.9 alongside the numerical and experimental reference data.

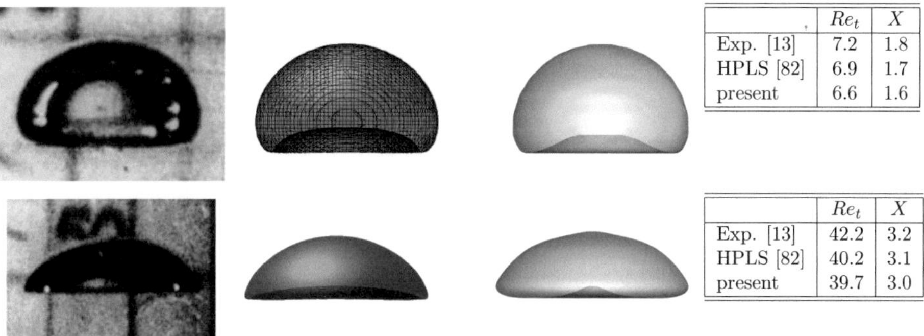

	Re_t	X
Exp. [13]	7.2	1.8
HPLS [82]	6.9	1.7
present	6.6	1.6

	Re_t	X
Exp. [13]	42.2	3.2
HPLS [82]	40.2	3.1
present	39.7	3.0

Figure 4.9 Cap bubble shapes from experiment of [13], HPLS simulation of [82] and present results with $Eo = 116$, $G = 13.9$ (top row, c2) and $Eo = 116$, $G = 62.0$ (bottom row, c3). Comparison of rise Reynolds number Re_t and bubble aspect ratio X .

A generally good agreement is found in both simulations. The bubble shapes agree well with the shapes from the experiment and the simulation with the HPLS-method. The rise Reynolds number as well as the aspect ratio are somewhat underpredicted in both cases by the current simulations where the error is less apparent at higher rise Reynolds numbers. The aspect ratio is here defined as $X = 2\max(x_S)/(\max(z_s) - \min(z_s))$. The main reasons for the deviation are the no-slip boundary condition that is enforced by the IBM and neglecting viscous stresses for low Re. A comparison between a rigid ellipsoid with $X = 2$ and an geometrically identical bubble is conducted in [159] showing the differences of a no-slip to a free-slip condition at the interface. The drag coefficient is larger for rigid particles and the difference is larger at low Re where viscous contributions play a marked role. A larger drag yields a lower rise velocity and thus a smaller rise Weber number and therefore smaller deformation. The cap shape is predicted correctly by the spherical harmonics algorithm including the formation of a dimple or under cut at the rear side of the bubble. The shape of the dimple, however, deviates from the predictions by the HPLS simulation. Also the representation of the quite sharp corner for the very flat cap-shape bubble for $G = 62.0$ is problematic with the SH-series. One should further keep in mind, that the SH-approach allows only for a unique mapping of r onto the (ϕ, θ)-space, i.e. the predictions are limited in terms of representing dimples and undercuts or skirted bubbles. This regime of very high Eo-bubbles is beyond the scope of this study which is focused on spheroidal bubbles at high Reynolds numbers.

Table 4.2 Overview of simulations regarding SH bubble shapes.

Run	(s1)	(s2)	(e1)	(e2)	(e3)	(e4)	(e5)	(c1)	(c2)	(c3)
	spherical			ellipsoidal					cap	
Eo	0.5	2.5	10	10	2.5	5.0	2.5	200	116	116
G	170	10	80	170	280	280	2825	10	13.9	62.0
Re_t	224	3.8	66.4	159	336	298	3031	3.6	6.6	39.7

Further simulations were conducted to scan the shape diagram. An overview of all simulations regarding the SH bubble shapes is provided in Table 4.2. The setup differs slightly from the one described above with now $\mathbf{L} = (6.4, 12.8, 6.4)\, d_{eq}$ and $\mathbf{N} = (256, 512, 256)$. The results are displayed in the shape regime map of Figure 4.1 and selected resulting bubble aspect ratios are plotted over the terminal Weber number in Figure 4.6. The latter shows good quantitative agreement with the correlation of equation (4.2) with a slight underprediction in X which probably also stems from the chosen definition of the aspect ratio for the SH bubbles. Three simulations were performed alongside the virtual boarder to the spherical regime. The parameters $Eo = 0.5$, $G = 170$ (s1) and $Eo = 2.5$, $G = 10$ (s2) led to almost completely spherical bubbles with the first result also being confirmed in [2] by a phase field approach. The pairing $Eo = 200$, $G = 10$ (c1) yields a modest cap shape with a very mild dimple. Additional simulations were conducted in the ellipsoidal regime. The two runs $Eo = 10.0$, $G = 80$ (e1) and $Eo = 10.0$, $G = 170$ (e2) have been chosen to illustrate front-to-back asymmetry. While the hydrostatic pressure difference from top to bottom leads to a flatter shape at the rear, an increase in the hydrodynamic pressure component at the front compared to the back leads to symmetric ellipsoidal shapes and front-flattened shapes. The two simulations $Eo = 2.5$, $G = 280$ (e3) and $Eo = 5.0$, $G = 280$ (e4) show shape oscillations around the displayed average shapes related to a zig-zag trajectory and corresponding velocity oscillations. The rise of a single bubble in liquid metal with $Eo = 2.5$, $G = 2825$ (e5) is discussed in detail in Section 5.6 and published in [246]. Distinct path and shape oscillations occur. However, the average shape corresponds closely to the one at $Eo = 2.5$, $G = 280$, i.e. an increase of the rise Reynolds number by an order of magnitude shows a negligible

influence on the average bubble shape.
The focus of this section lied on the prediction of steady shapes and time-averaged bubbles shapes. The computed bubble shapes are also in good agreement with the observations in [17]. Nonetheless, the SH algorithm also works for temporally varying shapes. Apart from the aforementioned numerical experiments, a simulation of bubble-wall collision and a comparison to experimental data of [318] is conducted in Section 7.4.3 and excellent agreement is found for strong, time-dependent bubble deformations.

4.5 Concluding remarks on bubble shapes

This section outlined the representation of particle shapes in the PRIME code reaching from simple spherical particles to complex, flow-determined bubble shapes. The shortcomings and the advantages of the present approach are now summarized: A large variety of shapes can be described. Basically the full shape diagram is covered, but a certain limitation remains due to the chosen parametrization. Very strong deformations of individual bubbles are difficult to capture, also the approach is limited in terms of particle topology, e.g. splashing of drops or shearing-off of smaller bubbles from the skirt of a bigger one would be challenging. Nevertheless, Section 7.5.2 describes the modeling of bubble coalescence, i.e. the merge of two bubbles, within the framework of spherical harmonics. The present weak, fully explicit coupling between bubble shape and fluid poses further limitations within the temporal evolution of the bubble shape. Explicit coupling can become numerically unstable and requires low time steps especially for strong or fast bubble deformations. As an outlook, implicit coupling should be considered. Lower order methods for the pressure field [45] could mimic a fully implicit treatment and would help to avoid substantial performance losses. If bubbles at low Reynolds numbers are of interest, viscous stresses should be incorporated into the fluid loads deforming the bubble.
Regarding the advantages, the shape of each individual particle is analytically described and therefore the constraint of constant volume can be implemented easily. In this way the mass of the disperse phase can be conserved exactly. The proposed algorithm for the simulation of shape-varying bubbles employs two different shape representations. One is the highly regular definition of the shape by analytical functions which allows to determine curvature terms in closed manner, simultaneously performing regularization. The other is a discrete one, defining Lagrangian markers on the surface, in order to accomplish the IBM-coupling between fluid and bubble surface. It is a useful feature of the employed method that the particle shape can be modeled directly. Experimental information on the bubble shape, if available, can be introduced. The delicate introduction of surface tension forces are not needed in the present model. This yields high robustness and avoids spurious currents [275]. Direct modeling of the bubble shape is very efficient and allows time steps of $CFL \approx 1$ for moderate shape oscillations which is usually not achieved with typical VOF methods resolving the density jump at the interface [149]. An exactly defined phase boundary, in contrast to a diffuse interface, is also advantageous for the modeling of particle-particle and particle-wall interactions [136] which will be discussed in Chapter 7.

5 Single Bubble Ascent Influenced by a Magnetic Field

5.1 On the rise of a single bubble

The ascent of a single bubble in a quiescent liquid is a fascinating phenomenon, for the layman as well as for the scientist. The trajectory of the bubble which can exhibit forms ranging from straight vertical ascent to chaotic irregular motion, and regimes of shape ranging from strictly spherical to irregularly wobbling still challenge physicists and engineers. An interesting review assembling the early views on rising bubbles is given in [214]. In this reference, the term Leonardo's paradox is suggested for the tendency of sufficiently large bubbles to rise along a zig-zag or spiraling path rather than along a rectilinear one. The reason for the latter is attributed to the structure of the bubble wake. Two-threaded vortices of opposite circulation induce a lift force on the bubble deflecting it from a strictly vertical trajectory. A review on the hydrodynamic forces acting on isolated, spheroidal high-Reynolds-number bubbles and the associated motion is provided in [168]. The vortical structures in the wake of air bubbles in water have been analyzed by modern optical experimental techniques like Schlieren optics [44], digital particle image velocimetry [27] or dye visualization [234]. Alternately shed vortex filaments are observed for a bubble rising in zig-zag, while a spiral trajectory is characterized by a continuous pair of parallel vortices wrapped around the axis of the helix. In experiments, it has been observed frequently that the path first follows a zig-zag and later on changes to a helical shape [230, 274], whereas a transition in the opposite direction has not been reported so far. Not surprisingly, the structures in the wake behind bubbles rising in zig-zag are similar to those observed behind rising solid spheres following a zig-zag trajectory [113]. It has been shown experimentally [59] as well as numerically [186, 187, 169] that path oscillations can appear in the absence of shape oscillations which proves that indeed the vortex structures in the wake are responsible for the former. This is extensively discussed in the review of Ern et al. [63] which assembles current knowledge about the wake of fixed bodies and its relation to the onset and development of path instabilities of both bubbles and rigid objects.
Most experimental and numerical work on bubbles so far has been conducted for the air-water system, often using hyper-clean water which is almost free of contaminants and therefore justifies the application of a shear-free boundary condition at the gas-liquid interface [168, 169]. Nevertheless, there is a variety of industrial applications where gas bubbles play an important role and where these conditions are not met. The continuous casting process in metallurgy is one example [272, 271]. Here, gas bubbles are injected into the melt to clean the liquid metal from contaminants and to stir and homogenize the liquid phase [320]. Magnetic fields are used in liquid metal processes to starry [262] and to stabilize the flow regimes [300]. Liquid

metals are prone to oxidation, and in general a melt is never free of contaminants so that an oxide layer forms at the gas-liquid interface. Furthermore, contaminants and inclusions agglomerate at the bubble surface. The appropriate condition for the velocity at the bubble surface hence is the no-slip condition. This is backed by the observation that the drag of a fully contaminated spherical bubble corresponds to that of a solid sphere [168, 66].

Liquid metals are opaque and therefore experimental data are difficult to obtain and rare. The optical measurement techniques specified above hence cannot be used to get detailed insight into liquid metal multiphase flows. Ultrasound Doppler velocimetry is an alternative approach in this case and has been used to study the motion of a single bubble [319] and a bubble-driven liquid metal jet [320] under the influence of magnetic fields. Local conductivity probes have also been used to measure the rise velocity of bubbles in mercury [183] as well as the behavior of gas bubbles in turbulent liquid metal magnetohydrodynamic flows [55, 56].

Direct numerical simulation of bubbles in liquid metals is challenging due to the large differences of density and viscosity between the phases and the high bubble Reynolds number typically encountered. As a result, there are only very few phase-resolving simulations of bubbles in liquid metal under the influence of a magnetic field. A rising bubble in a small enclosure under a vertical magnetic field was computed in [256] by means of a volume of fluid approach with reduced density and viscosity ratio and very moderate Galilei number. Gaudlitz and Adams [84] simulated the influence of a vertical magnetic field on the rise of a single bubble in electrically conductive liquids with a hybrid particle level set method neglecting the effect of interface contamination. The numerical parameters of this case correspond to a small bubble in mercury, i.e. the Galilei number is smaller by a factor of five compared to the present study.

It is known that homogeneous magnetic fields substantially modify vortical structures in turbulent flows [139, 15] as well as the pressure field around fixed objects [173]. Therefore, a considerable impact of such a field on the bubble dynamics is to be expected which indeed was observed in experiments [319, 320]. Despite these studies the actual influence of a magnetic field on bubbles in liquid metal is still not fully understood. In particular, the impact of a magnetic field on the interaction between bubble wake and bubble dynamics in metallurgical systems is unclear and also the modification of the bubble shape in that case is not fully understood to this date. This is mostly due to the lack of visual data impeded by the opaque liquid metal.

The aim of the present paper is to fill this gap and to provide insight into the influence of a longitudinal magnetic field on bubble wake and bubble dynamics. Phase-resolving direct numerical simulations of an argon bubble in the liquid metal GaInSn have been conducted for different values of magnetic interaction. The three-dimensional data of high spatial and temporal resolution obtained from the simulations are evaluated, visualized and compared against experimental data.

The chapter is structured as follows: Section 2 gives a short description of the equations to be solved and the numerical approach employed, as well as a refinement study quantifying the numerical error. Section 3 contains the numerical results for the ascent of a single bubble with and without a magnetic field. Visualizations are presented to highlight conspicuous flow features in the bubble wake. Furthermore, the numerical results are compared against available experimental findings and other simulation data. The last section summarizes the results of the present study and outlines future research directions. Part of this chapter was published in [244, 248].

5.2 Physical model and numerical method

5.2.1 Parameters of single bubble ascent

The problem of a single particle rising or falling in a pool of quiescent fluid due to the effect of buoyancy is governed by three parameters [63]: The particle-to-fluid density ratio, $\pi_\rho = \rho_p/\rho_f$, the Galileo number, $G = \sqrt{|\pi_\rho - 1|\, g\, d_{eq}^3}/\nu$, and a geometrical parameter relating to the shape of the particle, such as the ratio of diameter to height for a cylinder or the aspect ratio for an ellipsoid of rotation for example. Here, g is gravity, d_{eq} is the diameter of a volume-equivalent sphere and ν is the kinematic viscosity of the liquid. The terms bubble and particle are practically used as synonyms, with index p throughout. Indeed, the term 'particle' in the literature often designates any element of a disperse phase, be it solid, fluid or gaseous [37]. In case of a rising bubble, the density ratio is very small and the motion is predominantly governed by the inertia of the fluid. The Galileo number determines the ratio of the driving buoyancy force to the viscous forces. Inserting the gravitational velocity, $u_{ref} = \sqrt{|\pi_\rho - 1|\, g\, d_{eq}}$, into the definition of G yields a reference Reynolds number, Re_{ref}. The latter velocity scale, u_{ref}, and in a similar fashion the reference time, $t_{ref} = \sqrt{d_{eq}/g}$, are used for scaling here, together with the reference length d_{eq}.
The shape of a single rising bubble is governed by viscous and pressure forces deforming the interface and by the stabilizing effect of surface tension driving the bubble shape towards a spherical one. The Eötvös number $Eo = \Delta\rho\, g\, d_{eq}^2/\sigma$ which is the ratio of buoyancy force to surface tension force therefore can be used to characterize the bubble shape. Here, $\Delta\rho$ denotes the density difference between the phases and σ the surface tension. The three parameters G, π_ρ and Eo characterize the system and are known *a priori*.
The bubble velocity, $\mathbf{u}_p = (u_p, v_p, w_p)^T$, is a result of the simulation. The rise velocity v_p can then be used to determine the bubble Reynolds number $Re = v_p\, d_{eq}/\nu$. The instantaneous, vertical component of the bubble velocity is used here to calculate $Re(t)$ because this component was measured in the corresponding experiments [319].

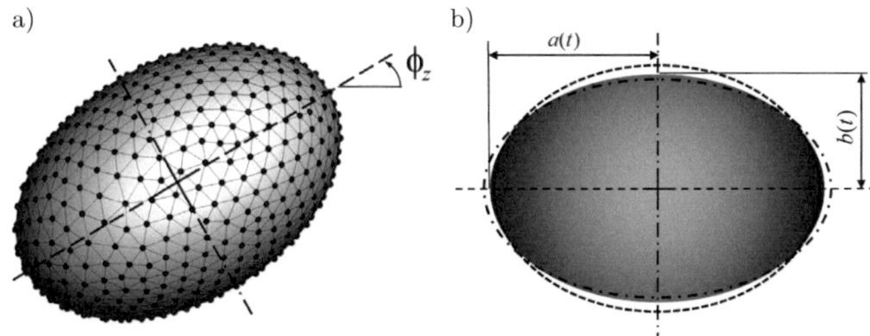

Figure 5.1 (a) Ellipsoidal bubble with $N_L = 664$ Lagrangian forcing points, tilted by $\phi_z = 30°$. (b) Shape oscillation $X(t) = a(t)/b(t)$. The discontinuous lines indicate the states with maximum and minimum aspect ratio observed in the simulation with $N = 0$ reported below.

In the present study, the bubble shape is approximated as an oblate ellipsoid of rotation with aspect ratio $X = a/b$ and semi-axes $a = c > b$ (Figure 5.1b). The surface of each individual bubble is described using a set of Lagrangian marker points interconnected by a triangular mesh (Figure 5.1a). For the parameters of an argon bubble in GaInSn, the

bubble shape is expected to be 'ellipsoidally wobbling' [37]. The Weber number is defined as $We = \rho_f \|\mathbf{u}_p\|^2 d_{eq}/\sigma$ using the absolute value of the bubble velocity. The instantaneous Weber number $We(t)$ is used to characterize the time-dependent bubble shape employing the correlation $X^{-1}(t) = 1 - 0.75 \tanh(0.165\, We(t))$ for the instantaneous aspect ratio of the bubble.
Finally, the magnetic interaction parameter, $N = \sigma_e B^2 d_{eq}/(\rho_f u_{ref})$, is introduced to quantify the relative strength of magnetic forces. It represents the ratio of magnetic forces to inertial forces [319, 139]. The magnetic Reynolds number, $R_m = \mu_0 \sigma_e d_{eq} u_{ref} \approx 4 \cdot 10^{-3}$, is indeed substantially smaller than unity and the quasi-static approximation is justified. The material properties of a 4.6 mm argon bubble in GaInSn were used to calculate R_m and μ_0 designates the magnetic permeability of free space. The electric conductivity for both phases was modeled to be the same for technical reasons at that time. This seems adequate here as the focus is on the influence of a magnetic field on the bubble wake. The effect of a phase-dependent electrical conductivity is then reviewed in the additional simulations of Section 5.6.

5.2.2 Refinement study

A grid refinement study was carried out to estimate the numerical error of the spatial and temporal discretization and to determine the overall order of convergence of the method in the present setup. It was conducted for the initial phase of the ascent during which the bubble accelerates and changes its shape from spherical to ellipsoidal with $X \approx 1.5$. Refinement is performed simultaneously for the spacing of the equidistant Cartesian grid, the Lagrangian surface mesh and the time step. Consequently, the CFL number is kept approximately constant. The number of forcing points N_L on the surface of the oblate ellipsoid is increased as well throughout the grid refinement (Appendix B).
The refinement study was conducted in a cubic domain of extent $L = 6.0\, d_{eq}$ in all three directions, and an equidistant grid of N_x^3 points was used with periodic boundary conditions in all three directions. Gravity acts in negative y-direction. The setup basically corresponds to the one of the simulations presented later on, where a significantly longer extent of the computational domain in vertical direction was used, though. A single bubble is considered with a Galilei number of $G = 2825$, an Eötvös number of $Eo = 2.5$, and a density ratio of $\pi_\rho = 10^{-3}$ corresponding to a 4.6 mm argon bubble in GaInSn. Note that with $\pi_\rho \ll 1$ the results become independent of ρ_p. The particle is initially at rest, $\mathbf{u}_p = 0$, $\boldsymbol{\omega}_p = 0$, in quiescent fluid, i.e. $\mathbf{u} = 0$ in the whole domain. The initial bubble position was chosen to be $\mathbf{x}_{p,0} = (3.0, 0.54, 3.0)\, d_{eq}$. According to the shape correlation (4.2) the bubble has a spherical shape $X_0 = 1.0$ at the beginning of the simulation. A small initial inclination angle of $\boldsymbol{\phi}_0 = (0, 0, 0.05)\, \pi$ was applied which is of no relevance for a sphere, but gives a very small bias towards a zig-zag in the xy-plane once the bubble starts to deform.
We consider the initial acceleration of the bubble for a fixed duration $t_{sim} = 3$ in dimensionless time units, roughly sufficient for the bubble to reach its terminal velocity. The temporal evolution of the bubble Reynolds number (based on v_p) is shown in Figure 5.2a) for different numerical resolutions. At the end of the simulation, $t = t_{sim}$, the bubble has traveled a distance in y of about $3\, d_{eq}$, corresponding to slightly more than half the size of the computational domain (Figure 5.2.) The discretization error is estimated at $t_e = 1.0$ by comparison of the computed instantaneous particle Reynolds number with the value obtained using the finest grid. In the reference case, the Eulerian grid has a spatial resolution

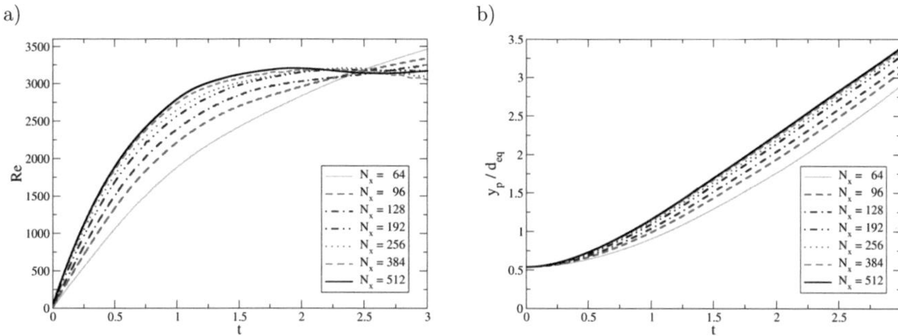

Figure 5.2 Refinement study. a) Bubble Reynolds number over time as a function of grid spacing. b) Vertical position. $N = 0$.

of $N_x = 512$ corresponding to $d_{eq}/\Delta x = 85.3$ gridpoints over the equivalent diameter and a total number $134.2 \cdot 10^6$ cells. A set of $N_L = 24976$ Lagrangian forcing points was used in this case to represent the bubble surface and a non-dimensional time step of $\Delta t = 1.25 \cdot 10^{-3}$ was employed. By means of the fit depicted in Figure 5.3, excluding the two coarsest grids, a convergence order of about 1.7 is obtained for the systematically refined grids employed. The fluid discretization alone is second order accurate for single-phase simulations [134]. The direct forcing scheme utilized with the immersed-boundary method for coupling the dispersed phase to the fluid yields a reduction of the order of convergence. The result for the present configuration is in line with the data in [136].

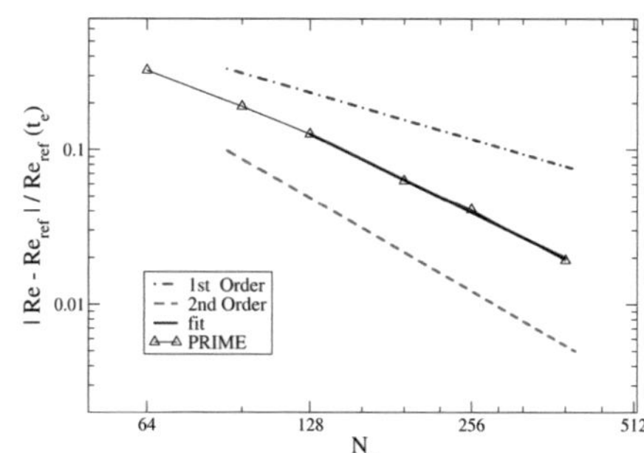

Figure 5.3 Relative error in *Re* at $t = 1.0$ for the simulation in Figure 5.2.

Based on the results of the refinement study, the resolution $N_x = 256$ was chosen for the simulations in the large computational domain. With an error of about 4%, it provides a good compromise between accuracy and computational effort. Further refinement would exceed the available computational resources. The chosen resolution therefore does not correspond to a full DNS, but will be adequate to provide valuable and detailed insight into the physics of this magnetohydrodynamic, multiphase flow. Interpreting the results of the refinement study in physical terms we find that the time scale for the initial acceleration of the bubble is longer on coarser grids. A coarse resolution also yields higher Reynolds numbers at the end of these simulations (see Figure 5.2).

5.3 Results for a single bubble without magnetic field

5.3.1 Setup of simulation without magnetic field

This section presents a simulation of a single bubble in liquid metal without magnetic field and a comparison against the experimental data of [319]. The physical parameters of the bubble correspond to those used for the refinement study above, $G = 2825$, $Eo = 2.5$ and $\pi_\rho = 10^{-3}$, which relate to an argon bubble with $d_{eq} = 4.6$ mm in eutectic GaInSn. As no magnetic field is applied, the magnetic interaction parameter is $N = 0$.
Compared to the refinement study, the computational domain was enlarged in the direction of gravity to resolve as much as possible of the bubble dynamics. The box extends over $\mathbf{L} = (L_x, L_y, L_z) = (6.0, 30.0, 6.0)\,d_{eq}$ and was discretized with a spatial resolution of $\mathbf{N} = (256, 1280, 256)$ points yielding a total of 83.9 Mio cells of the Eulerian grid. The bubble was represented with $N_L = 9093$ Lagrangian forcing points distributed over its surface. The time step is $\Delta t = 2.5 \cdot 10^{-3}$ in dimensionless units. Boundary conditions and initial conditions are the same as in the refinement study of Section 5.2.2, i.e. periodic conditions were applied in all three directions while the fluid as well as the bubble were initially at rest.

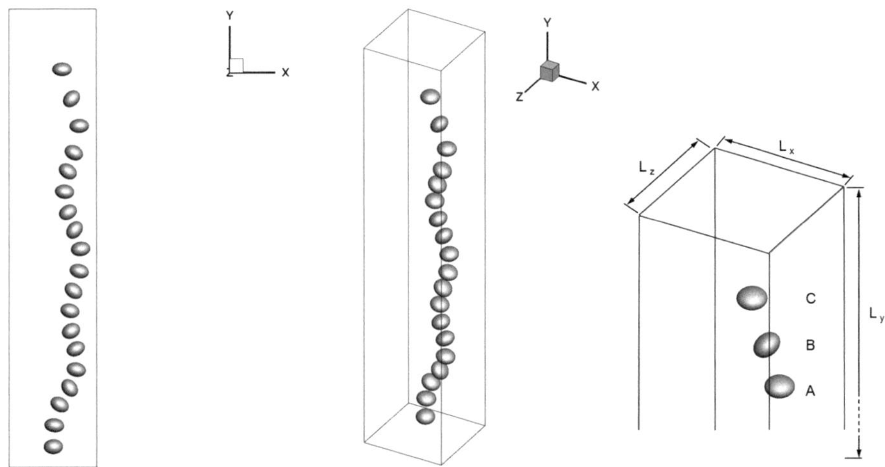

Figure 5.4 Computational domain of size $(6,\ 30,\ 6)\,d_{eq}$ and events A, B, C of the bubble trajectory as indicated in Figure 5.10 below.

In the experiments by Zhang et al. [319], an open cylindrical container with a diameter of $D = 100$ mm and a height of $H = 220$ mm was used corresponding to $D \times H \approx (22 \times 48) d_{eq}$ for $d_{eq} = 4.6$ mm. The bubble was injected at the bottom center. A box with a quadratic cross section $(L_x \times L_z)$ is used in the present study for technical reasons with periodic boundary conditions which mimic a somewhat larger domain. Due to the high computational cost, especially the horizontal extension had to be reduced whereas a moderate reduction was chosen concerning the height. The areal blockage $\pi d_{eq}^2 / (4 L_x\, L_z)$ is about 2%. Gaudlitz [81] used a lateral extent of only $4 d_{eq}$, also with periodic boundary conditions, for a simulation of single bubble ascent at lower Re. It has been shown in [257] that the added mass coefficient of a spherical bubble horizontally aligned with a second bubble equals the one of a single bubble if the distance exceeds $3d_{eq}$. The wake of two spheres placed side by side in uniform flow is only very weakly coupled if the spacing is larger than $3.5d_{eq}$ [242]. A sphere next to a solid wall was studied in [317]. In [124] it is shown that for the largest Reynolds number considered, $Re = 300$, and a wall distance of $4d_{eq}$, drag and lift as well as the Strouhal number deviate only slightly from the values obtained in an unbounded fluid. For these reasons the horizontal extent of the computational domain selected for the present study in combination with periodic boundary conditions is adequate to represent the conditions of the experiments in the wide cylinder.

5.3.2 Results of the simulation and comparison with experimental data

A runtime of 60×61.5 CPU hours on 60 cores of an SGI Altix 4700 is needed for one crossing of the above box taking about 30 dimensionless units in time. The bubble Reynolds number based on the vertical velocity v_p is plotted over time in Figure 5.5. After an initial acceleration the bubble rise velocity starts to oscillate quasi-periodically. The experimental data of [319] is displayed in the same graph for comparison. These data were obtained using single-sensor ultrasound Doppler velocimetry which allows to measure the velocity component along a line.
An overview of the characteristic figures calculated from the instantaneous Reynolds number $Re(t)$ is given in Table 5.1, where $Re_t = \langle Re \rangle_t$ denotes the average rise Reynolds number obtained from a time average over the interval $t \in (6, 29.2)$ and σ_{Re} the corresponding standard deviation. The average rise Reynolds number is in excellent agreement with the data from the measurements. Concerning the oscillation in $Re(t)$, an underestimation of the amplitude characterized by σ_{Re} is recognized. Asymmetric bubble deformation, a deviation from an ellipsoidal shape, and partial slip at the bubble surface might be the reasons for the deviation, besides the remaining discretization error discussed in Section 5.2.2. The frequency on the other hand agrees well with the value reported in [319]. The dominant frequency f_{Re} of the oscillation in $Re(t)$ was obtained from the Fourier spectrum by means of a discrete Fourier transform (DFT) of Re computed with a Hanning window function to account for the non-periodic time signal. In addition, the frequency was determined from the roots in $Re(t) - Re_t$ and in the original experimental work of [319] by a least square curve fit to a sine function. Comparing the results to some extent assesses the uncertainty in the determination of f_{Re} due to the irregular oscillation and the limited period of time.

Table 5.1 Results for a single bubble without magnetic field compared to experimental data of [319]. $Re_t = \langle Re \rangle_t$ is the temporally averaged Reynolds number, σ_{Re} the corresponding standard deviation, $f_{Re} = f^*/f_{ref}$ with f^* being the dominant frequency in Hz and $f_{ref} = \sqrt{g/d_{eq}}$.

	Re_t	σ_{Re}	f_{Re} (DFT)	f_{Re} (roots)	f_{Re} (sine-fit [319])
Simulation	2871	245	0.276	0.270	—
Experiment [319]	2879	369	0.297	0.289	0.280

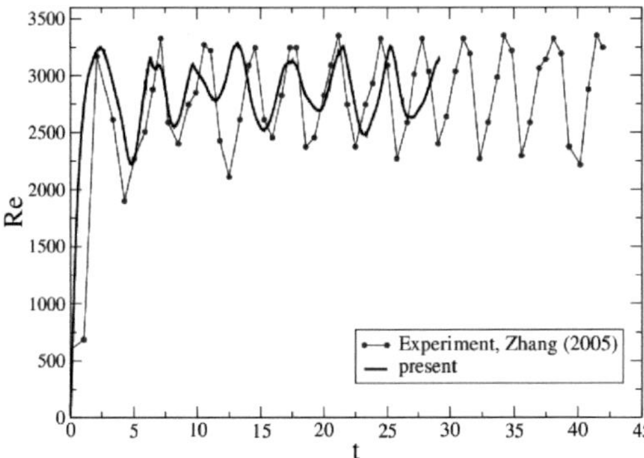

Figure 5.5 Bubble Reynolds number over time for $N = 0$ and comparison to experimental data of [319].

Additional data, interpretation and comparison with literature

Only the vertical component of the bubble velocity over time could be determined in the experiments [319] due to the measurement technique employed. The present simulations now offer full access to all velocity and pressure data for the continuous liquid metal phase as well as 3D data of the bubble trajectory. Therefore the simulations can deliver valuable complementary information on the bubble dynamics. This is reported in Figure 5.6. Indeed, a zig-zag trajectory with lateral drift is observed in Figure 5.6a) as conjectured by the experimentalists [319]. The maximum in $Re(t)$ occurs at extreme points of the bubble path $x_p(t)$. In these points, the bubble is oriented with its small semi-axis parallel to the gravity vector, i.e. the inclination angle ϕ_z is approximately zero. The amplitude of the zig-zag, measured between two extreme points of the path, is approximately $\Delta_{xz} = 1.15\, d_{eq}$. An oscillation in bubble inclination is found as well and plotted in Figure 5.6b). The temporal change in the orientation ϕ_z around the z-axis of the laboratory system is clearly associated with the zig-zag along x. Maximum tilting of the bubble is found closely after a local minimum in $Re(t)$ and approximately half way between the turning points of the zig-zag trajectory where the lateral velocity is largest. Towards the end of the simulation and with the onset of the lateral drift the other two rotation angles ϕ_x, ϕ_y also deviate from zero and oscillate with a higher frequency. The maximum inclination angle is found to be $|\phi_z|_{max} \approx 36°$.

These values can be compared to data from the literature. Lateral distances between two extreme points in a zig-zag trajectory of $1.0 \ldots 1.3\, d_{eq}$ and a maximum tilting of $27 \ldots 30°$ are reported for air-water experiments [167, 27] and for simulations of air bubbles in water [81, 187] at lower Reynolds numbers.

Due to high contamination and oxidation of the gas-liquid metal system a no-slip boundary condition is used here on the bubble surface as justified above. Therefore, besides the higher Reynolds number of the present simulation also the boundary condition at the bubble surface differs from the aforementioned simulations for air bubbles in water. Markedly larger inclination angles are reported for rigid spheroids compared to bubbles in clean water [168], and it is found in [63] that oblate bodies may follow highly non-linear trajectories with large rotation rates if the Reynolds number is high enough. After the initial transient in the present simulations the aspect ratio of the oblate ellipsoid determined according to equation (4.2) oscillates in the interval $X \in [1.35;\ 1.57]$. The bubble shapes for the mean, the minimum and the maximum aspect ratio are displayed in Figure 5.1b) to convey an impression of the amount of shape modification during the presented simulation.

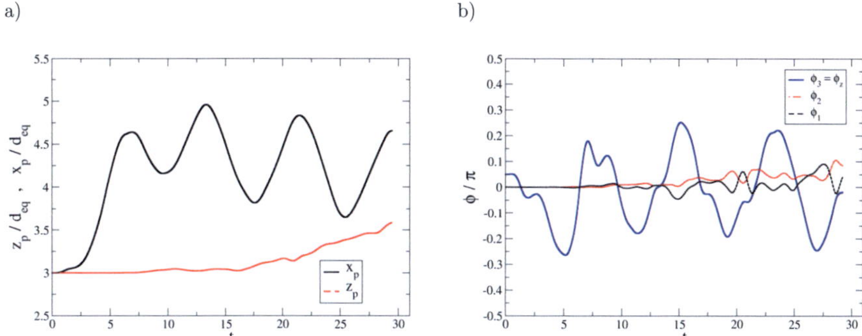

Figure 5.6 Zig-zag trajectory for $N = 0$. a) History of lateral bubble center coordinates x_p and z_p, non-dimensionalized with d_{eq}. b) Bubble orientation over time described by the angles of orientation.

Simulation of a smaller bubble

A smaller bubble with $G = 1488$, $Eo = 1.05$ corresponding to an argon bubble of $d_{eq} = 3.0$ mm in GaInSn was studied as well. The parameters were adjusted in the simulation by changing viscosity and surface tension with all other parameters unchanged. For this smaller bubble, the time-averaged Reynolds number is $Re_t = 1822$. The shape of this bubble remains almost spherical. The rise velocity again oscillates around its mean and a zig-zag path is observed. The characteristic frequency, calculated from a Fourier spectrum of $Re(t)$, is $f_{Re} = 0.222$ which is in good agreement with the data of [163] for an air bubble in water at similar Re_t. In [163], the non-dimensional frequency is determined from vortex shedding visualized by optical measurements. If Re is high enough, a lock-in occurs between vortex shedding and oscillations in the rise velocity of bubbles. No experiments in liquid metal were conducted for this small bubble.

5.4 Results for a single bubble with magnetic field

5.4.1 Setup of simulation with magnetic field

A longitudinal, homogeneous magnetic field in the direction of gravity is now applied with the magnetic interaction parameter being $N = 0.5$ and $N = 1.0$, respectively. All other parameters remain the same as for the simulation with $N = 0$ reported in the previous section. Only the domain size was increased in y-direction to $\mathbf{L} = (6.0,\, 48.0,\, 6.0)\, d_{eq}$ with a mesh of $\mathbf{N} = (256,\, 2048,\, 256)$, i.e. 134.2 Mio grid points. A longer box size in the direction of ascent is necessary because the zig-zag becomes stretched out and the characteristic frequency decreases under the impact of a magnetic field as will be shown later. This box size now corresponds to the height of the experimental container of 220 mm for the $d_{eq} = 4.6$ mm bubble.

5.4.2 Overview of results

A longitudinal magnetic field has significant impact on the bubble dynamics. The influence of the magnetic field is discussed for a single ascending bubble with the parameters $G = 2825$ and $Eo = 2.5$. Quantitative results of the simulations are summarized in Table 5.2. When applying a longitudinal magnetic field this bubble rises faster and the oscillations in $Re(t)$ are damped as shown in Figure 5.7a). The maximum inclination of the bubble decreases from $|\phi_z|_{max}(N = 0) \approx 36°$ in the absence of a magnetic field to $|\phi_z|_{max}(N = 1) \approx 17°$ for the strongest longitudinal field considered which is a reduction by more than 50% (Figure 5.7b)). The corresponding path deviates less from the vertical as shown in Figure 5.8. A zig-zag trajectory is found for all values of N considered with the transverse distance between two extreme points being reduced by the magnetic field. The time scale for one zig-zag increases with increasing magnetic field, i.e. the path oscillation is stretched in the direction of gravity. At the same time the amplitude of the oscillations is somewhat smaller, as shown in Figure 5.8 and Table 5.2. The resulting 3D bubble trajectory is therefore more rectilinear. With less energy being periodically transferred to transverse motion and rotation, the amplitude of the oscillation in $Re(t)$ decreases with N and the bubble rises faster (Figure 5.7a)). The time-averaged rise Reynolds number increases by about 8% for the largest field. The oscillation in $Re(t)$ appears more regular in the case with $N = 0.5$ and even more for $N = 1$ compared to the case without magnetic field.

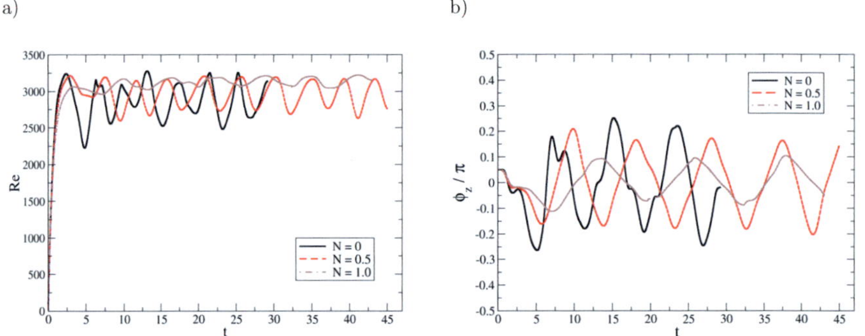

Figure 5.7 Bubble dynamics for the three cases $N = 0$, 0.5, 1.0 and $G = 2825$, $Eo = 2.5$. a) History of bubble Reynolds number. b) History of inclination angle ϕ_z.

Table 5.2 Summary of simulation results.

$Eo = 2.5$, $G = 2825$, fine	Re_t	f_{Re} (DFT)	$\lvert\phi_z\rvert_{max}$	Δ_{xz}(Zig-Zag)
$N = 0$	2871	0.276	36°	$1.15\,d_{eq}$
$N = 0.5$	2957	0.233	31°	$1.08\,d_{eq}$
$N = 1.0$	3132	0.181	17°	$0.78\,d_{eq}$
$Eo = 2.5$, $G = 2825$, coarse				
$N = 0$	3029	0.297	35°	$0.96\,d_{eq}$
$N = 0.5$	3054	0.246	29°	$0.92\,d_{eq}$
$N = 1.0$	3202	0.185	15°	$0.73\,d_{eq}$

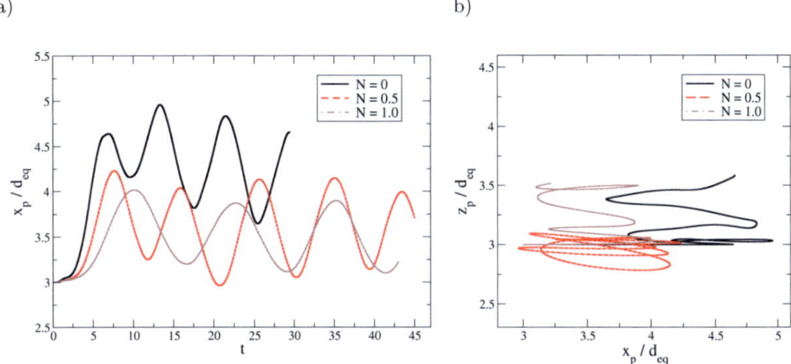

Figure 5.8 Assessment of lateral motion with and without magnetic field. $N = 0$, 0.5, 1.0 and $G = 2825$, $Eo = 2.5$. a) History of lateral bubble center coordinates x_p non-dimensionalized with d_{eq} illustrating the zig-zag trajectory of the bubble. b) Top view on domain, x_p versus z_p.

5.4.3 Comparison of results to experimental data and interpretation

A comparison with the experimental data from [319] is provided in Figure 5.9. Here, the average rise Reynolds number and the Strouhal number are normalized with the value in the absence of a magnetic field ($N = 0$) and are plotted over the magnetic interaction parameter, N ,for different Eötvös numbers, Eo, reflecting different bubble sizes.

First, the *time-averaged Reynolds number dependency on N and Eo* is studied. Whether the time-averaged rise velocity decreases or increases with increasing magnetic interaction depends on the bubble size. This complex behavior is found in both, the present simulations and the experiments in the literature [319]. An increase in rise velocity and hence Re_t with increasing magnetic field is found for large bubbles, i.e. large Eo, and a decrease in Re_t for small bubbles, i.e. small Eo. The reason for this phenomenon is a competition between adverse effects generated by the magnetic field: On the one hand, a longitudinal magnetic field increases the drag of an object. In order to pass around the object fluid elements need to move in a direction perpendicular to the magnetic field which gives rise to a Lorentz force so that the resulting pressure force on the particle increases with magnetic interaction. This was shown experimentally for the flow around a fixed sphere and a disc at high Reynolds number and moderate to high magnetic interaction [173, 314], as well as numerically for spheres and ellipsoids at moderate Re in [245] and in Section 3.4.6. On the other hand, the magnetic field suppresses the lateral dynamics and the bubble rises on a more rectilinear trajectory. Already in the inviscid situation this leads to a larger rise velocity, simply because the trajectory is shorter. Considering viscous effects in addition, less energy is transferred towards rotation and towards motion in the transverse direction where it is dissipated further increasing the rise velocity. The amplitude of the changes in $|Re_t(N)/Re_t(N=0)|$ with N lies well within the band spanned by the data obtained from the experiments. The measurements, however, show a different threshold in Eo for the reversion of the trend, i.e. Re_t increases or decreases with N at slightly larger Eötvös numbers. This is visualized in Figure 5.9a).

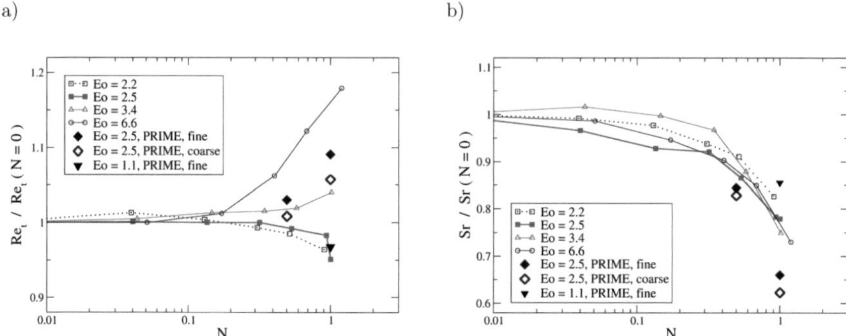

Figure 5.9 Impact of magnetic interaction N. a) Relative change with N in average rise Reynolds number. b) Relative change with N in Strouhal number. Present simulations (bold symbols) with code PRIME compared to experimental data of [319].

Further, the *dependency of the characteristic frequency on* N *and* Eo is studied. Different ways of defining a non-dimensional frequency have been proposed in the literature, so that before reporting the results obtained a few comments on this issue are appropriate. The equivalent bubble diameter d_{eq} is a natural reference length in any case. But one can either choose the *a posteriori* determined time-average rise-velocity $\langle v_p \rangle_t$ or the *a priori* known gravitational velocity scale $u_{ref} = \sqrt{d_{eq} g}$ to determine a reference time scale. As in the experiments [319], the Strouhal number, $Sr = f^* \, d_{eq} / \langle v_p \rangle_t$, is employed here which is based on the dominant frequency, f^*, in the oscillation of the vertical bubble velocity, $v_p(t)$, and on the average rise velocity, $\langle v_p \rangle_t$, of the bubble. The average rise velocity itself is a function of bubble size and magnetic interaction. In contrast, the dimensionless frequency, $f_{Re} = f^* / f_{ref} = f^* / \sqrt{g/d_{eq}} = f^* \, d_{eq} / u_{ref}$, is based on the constant reference velocity given by the gravitational velocity scale. Therefore the Strouhal number measures the combined effect of an additional parameter on both, the average velocity and the frequency in the velocity oscillations.

The determination of Sr and f_{Re} can only be based on a small number of periods of the oscillation in $Re(t)$ due to the size of the computational domain. The number of periods observed in the experiments is the same for $N > 0$. A decrease of Sr and of f_{Re} is found with increasing strength of the magnetic field for all bubble sizes as shown in Figure 5.9b). The relative change in Sr is less pronounced for small bubbles than for large bubbles in the simulation. For the larger bubble with $Eo = 2.5$, the reduction in Sr is over-predicted at large magnetic interaction parameters in the simulations. To quantify the influence of spatial resolution on the result, 'coarse' simulations with an isotropic grid of step size 1.5 times the one of the common grid, i.e. $N_x = 192$, have been conducted for this bubble using the same values of the interaction parameter $N = 0,\ 0.5,\ 1.0$. It appears that the results for the relative change in Sr on the finer grid are closer to the experimental data.

Overall, the agreement of the results of experiments and numerical simulations is promising. All dominant effects of the magnetic field have been captured by the simulation.

5.5 Comparative analysis of wake with and without magnetic field

5.5.1 Coherent structures in the wake

For the bubble with $G = 2825$, $Eo = 2.5$, particular instants in time are chosen for further analysis. These events correspond to characteristic points in the bubble trajectory, extreme points of bubble inclination $\phi_z(t)$ and path $x_p(t)$. They are highlighted as dots in Figures 5.10 and are denoted as A, B, C in chronological order. The inclination $\phi_z(t)$ is zero for event A and C. The time of event B was chosen half way between the time of A and C corresponding to approximately an extremum in inclination $\phi_z(t)$. Events A and C mark approximately the turning points of the zig-zag in $x_p(t)$. We therefore restrict the discussion to half a period of the zig-zag. At instant B, the transverse velocity u_p is close to a maximum and the instantaneous rise Reynolds number $Re = v_p d_{eq}/\nu$ just passed a minimum (Figure 5.10c). As discussed in the introduction, the trajectory of the bubble is closely related to the structure of its wake. In experiments with liquid metal, an analysis of coherent vortex structures is difficult. The presently available experimental techniques only provide selected one- or two-dimensional data with relatively coarse resolution [319, 320, 306, 218]. Here, the present simulations can close this gap and furnish insight into vortical structures of the bubble wake. This is provided in Figures 5.11 to 5.13 based on instantaneous 3D velocity and pressure data available from the simulations. The dominant role of streamwise vorticity has been emphasized by several others, for instance in [27, 63, 168]. Figure 5.11 and 5.12 for this reason show the vertical component of the vorticity, ω_y, for event B and C, respectively. Two iso-surfaces are depicted, one with a positive and one with a negative value, so that counter-rotating vortices can be detected. In these plots, complementary views of the same structure are shown differing by an angle of 90°. The magnetic interaction parameter increases from left to right from $N = 0$ to $N = 0.5$ to $N = 1.0$.

The vorticity in the wake is distributed in an undulating pattern as a consequence of two effects. One is the von Kármán instability of the wake leading to an alternating vortex pattern even if the bubble would rise along a straight path. Additionally, once the path of the bubble oscillates in horizontal direction, vorticity is generated at varying horizontal positions, so that even without the wake instability a zig-zag trajectory yields a zig-zag shape of the vorticity pattern. Also recall that in inviscid smooth fluid flows, vortex lines move with the fluid [171]. In all cases, one can observe that vorticity is shed pair-wise with alternating sign in the zy-plane. These counter-rotating vortex filaments induce a velocity in the x-direction according to the Biot-Savart law yielding a tilting of the bubble and hence the observed zig-zag motion in the xy-plane. Substantial damping of the vortical structures in the bubble wake by the vertical magnetic field is found. Especially small structures vanish with increasing N while the larger vortex filaments are more aligned with the field. The vertical orientation of the vortices in turn is also caused by the more rectilinear trajectory of the bubble.

While iso-contours of the vertical vorticity component give access to coherent structures of smaller scales, iso-contours of pressure can be used to visualize larger scales [72]. As vortex cores are characterized by low pressure regions, iso-contours of the pressure coefficient $C_p = p \,/\, \left(\rho_f u_{ref}^2/2\right)$ are displayed in Figure 5.13 for event C.

Vortex rings form in the wake of the bubble in the absence of a magnetic field triggered by the zig-zag path which have also been visualized in experiments at similar Reynolds

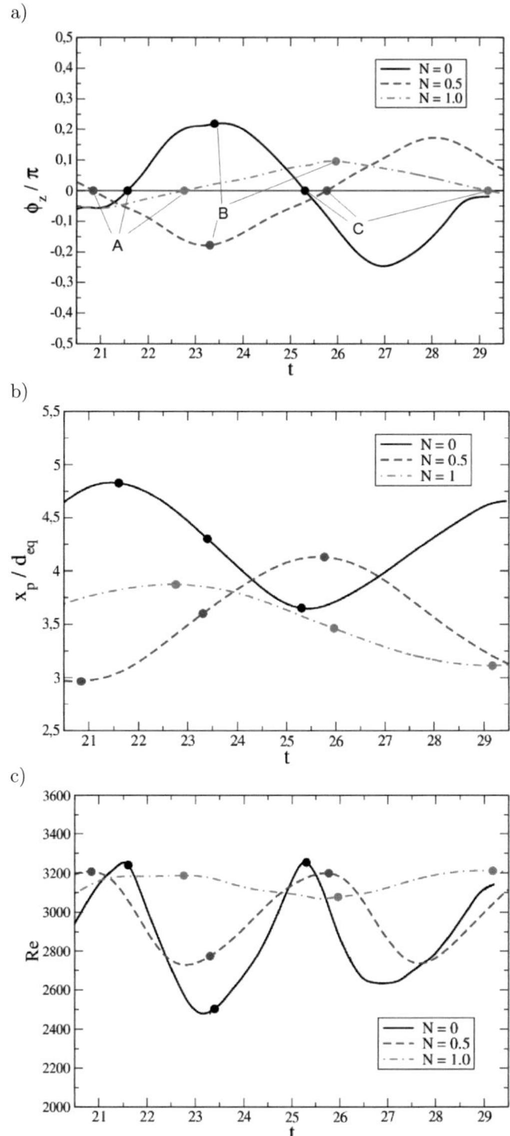

Figure 5.10 Selected instants in time A, B, C characteristic for the bubble trajectory. They are defined in the plot of the inclination angle ϕ_z and marked by dots in the other plots of bubble position x_p and bubble Reynolds number for the three cases $N = 0,\ 0.5,\ 1.0$; $G = 2825$, $Eo = 2.5$ in all cases.

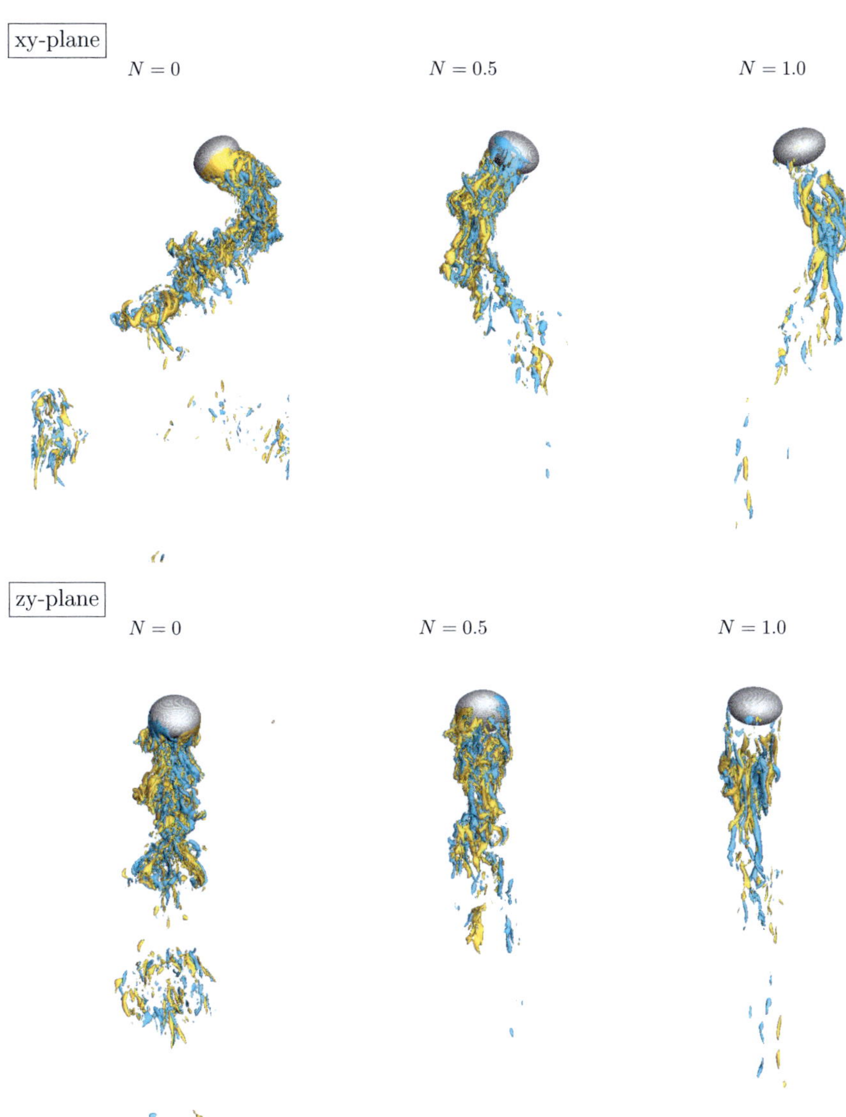

Figure 5.11 Event B: Iso-contours of $\omega_y d_{eq}/u_{ref} = \pm 6$.

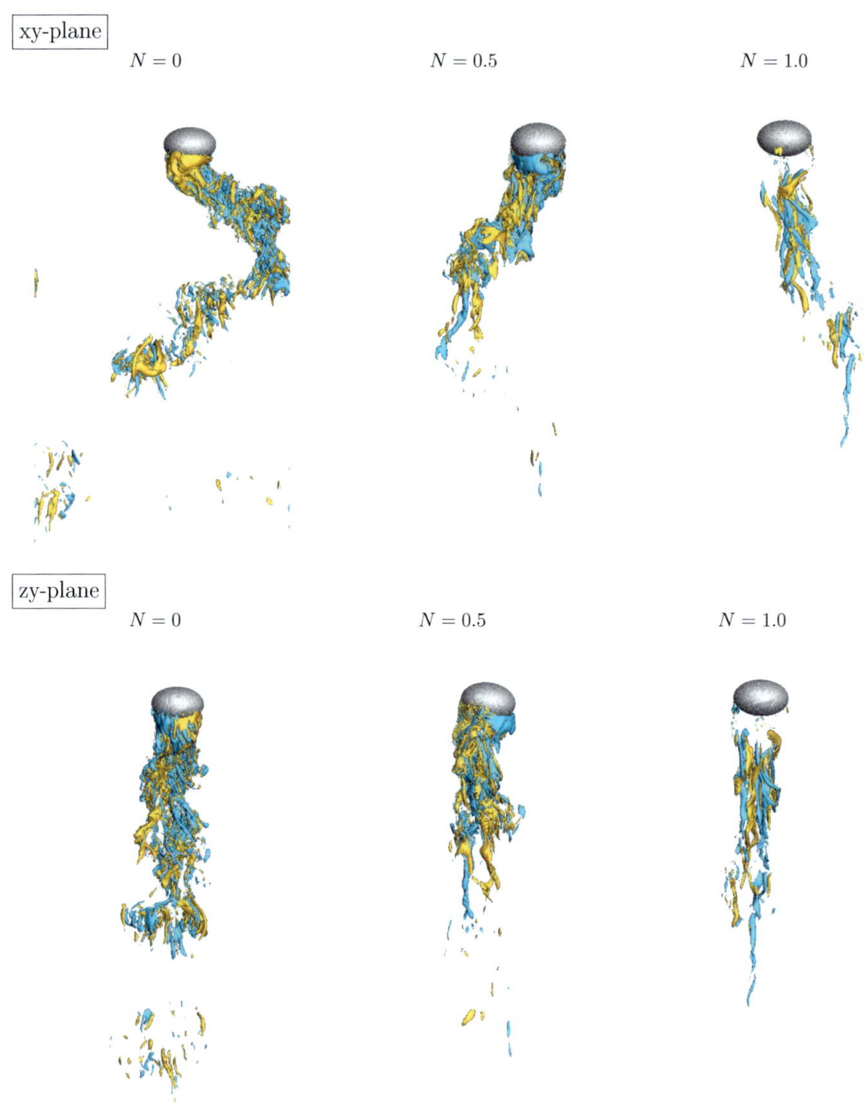

Figure 5.12 Event C: Iso-contours of $\omega_y d_{eq}/u_{ref} = \pm 6$.

number [113]. A $4R$ vortex mode [113] is associated with one zig-zag period consisting of two primary vortex rings at the extreme points of the path and two secondary rings in between at maximum, absolute inclination. These rings are less pronounced in the case $N = 0.5$ and eventually vanish for $N = 1.0$ due to the rather rectilinear path. The 'two-legged' structure of the bubble wake is clearly visible in the snapshot of $N = 0.5$ at the chosen value for C_p.
The pressure iso-contours also show a region of high pressure at the front of the bubble and a low pressure region aside from and behind the bubble. Both regions increase in size with increasing magnetic interaction indicating an augmentation of the pressure drag on the bubble with N as discussed in [173]. The Lorentz force is generated by the transverse velocity components u and w for a vertical magnetic field, so that increasing magnetic interaction leads to a damping of these lateral velocity components and to a straightening and stretching of the path lines of fluid elements around the bubble in vertical direction resulting in the described change in the pressure field. The effect is also visible in the graphs of Figure 5.14 where the extent of non-zero transverse vorticity in front of the bubble increases with stronger magnetic fields.

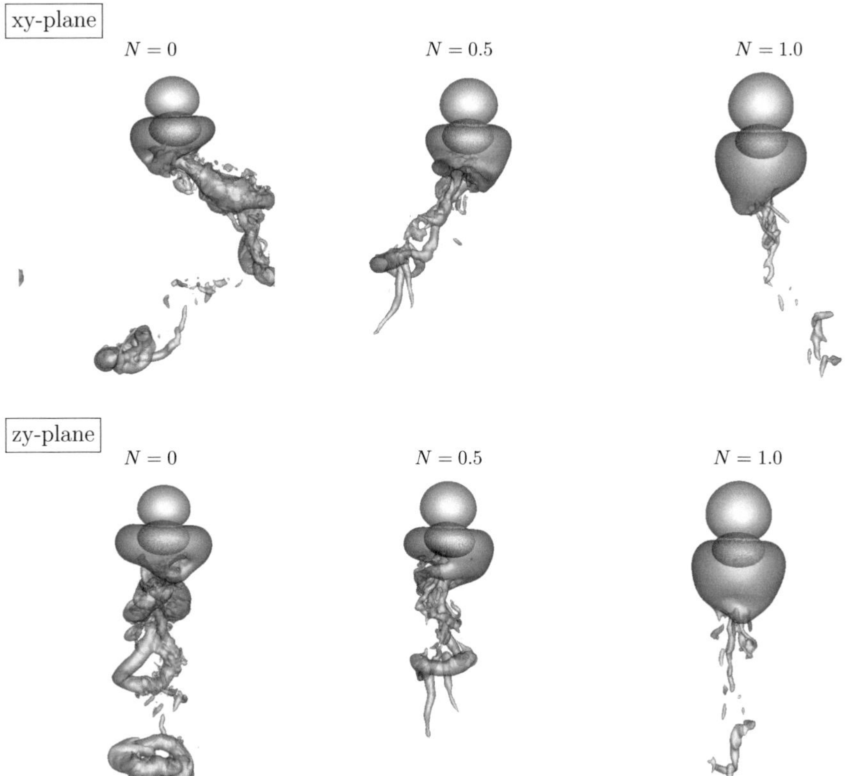

Figure 5.13 Event C: Iso-contours of pressure with $C_p = p \,/\, \left(\rho_f u_{ref}^2/2\right) = \pm 0.24$.

5.5.2 Quantification of the damping effect in the bubble wake

The vorticity in the bubble wake has been found to be the crucial quantity in understanding bubble dynamics and path oscillations [63, 168, 27]. Therefore, the focus is now on the quantification of the damping effect resulting from the applied magnetic field. The absolute value of the normalized vorticity component in y-direction, $\tilde{\omega}_y = \omega_y \, u_{ref}/d_{eq}$, is integrated in xz-planes according to

$$\langle|\omega_y|\rangle_{xz} = \frac{1}{d_{eq}^2} \iint |\tilde{\omega}_y| \, dx \, dz \, . \tag{5.1}$$

The integration in equation (5.1) is conducted over the entire xz-plane and normalized with d_{eq}^2. The transverse components $\langle|\omega_x|\rangle_{xz}$ and $\langle|\omega_z|\rangle_{xz}$ are determined in an analogous way. Sample results for event C are plotted over the vertical distance from the bubble center for increasing magnetic interaction $N = 0,\ 0.5,\ 1.0$ in Figure 5.14. The plots show global maximum values of $\langle|\omega_x|\rangle_{xz}$ and $\langle|\omega_z|\rangle_{xz}$ at the front of the bubble in all cases. The values of the maxima are similar since the bubble has zero tilting at event C and therefore the geometrical configuration is symmetric with respect to x and z at this instant in time. With increasing magnetic interaction the maximum in $\langle|\omega_x|\rangle_{xz}$ and $\langle|\omega_z|\rangle_{xz}$ is reduced and the region of non-zero vorticity extends further upstream. In the bubble wake, considerable damping of all vorticity components is found when a magnetic field is applied. The peaks of $\langle|\omega_y|\rangle_{xz}$ for $N = 0$ in Figure 5.14a vary in amplitude due to asymmetries in the zig-zag and tilting of the bubble as well as due to irregular vortex shedding. With increasing magnetic interaction, the bubble wake contains less vertical vorticity and the values of the extrema in the plot are substantially reduced.

The damping effect of a vertical magnetic field is anisotropic. Joule damping associated with the Lorentz force acts linearly on all scales with a privileged direction [139]. This is now assessed by means of the average weight of $|\omega_y|$ compared to the total vorticity $|\boldsymbol{\omega}|$ using the quantity

$$\Gamma_y = \frac{1}{n_{xz}} \sum_{i=1}^{n_{xz}} \frac{\langle|\omega_y|\rangle_{xz}^{(i)}}{\langle|\boldsymbol{\omega}|\rangle_{xz}^{(i)}} \, . \tag{5.2}$$

With the present data, $n_{xz} = 1280$ equi-distributed xz-planes have been used for the interval $(y_p - y)/d_{eq} \in [-5;\ 25]$. The quantity Γ_y is reported in Table 5.3 for event C. The table also lists an integral measure of vorticity for all three components obtained by summation over all xz-planes. The difference in the damping of ω_x and ω_z is related to the privileged direction of the zig-zag. A roughly linear decrease of Γ_y with N is found, i.e. the vorticity component ω_y is the one which is affected most by the magnetic field. In general, a vertical magnetic field leads to homogenization of the transverse velocities u and w and therefore reduces the gradients of these components with respect to z and x which enter in ω_y.

	$N = 0$	$N = 0.5$	$N = 1.0$
Γ_y	0.471	0.337	0.210
$\sum_{i=1}^{n_y} \langle|\omega_y|\rangle_{xz}^{(i)}$	73.2	36.7	25.8
$\sum_{i=1}^{n_y} \langle|\omega_x|\rangle_{xz}^{(i)}$	75.8	50.4	48.5
$\sum_{i=1}^{n_y} \langle|\omega_z|\rangle_{xz}^{(i)}$	74.5	45.3	39.6

Table 5.3 Event C: Average weight Γ_y of ω_y compared to total vorticity according to (5.2) and integral measure of vorticity for all three components.

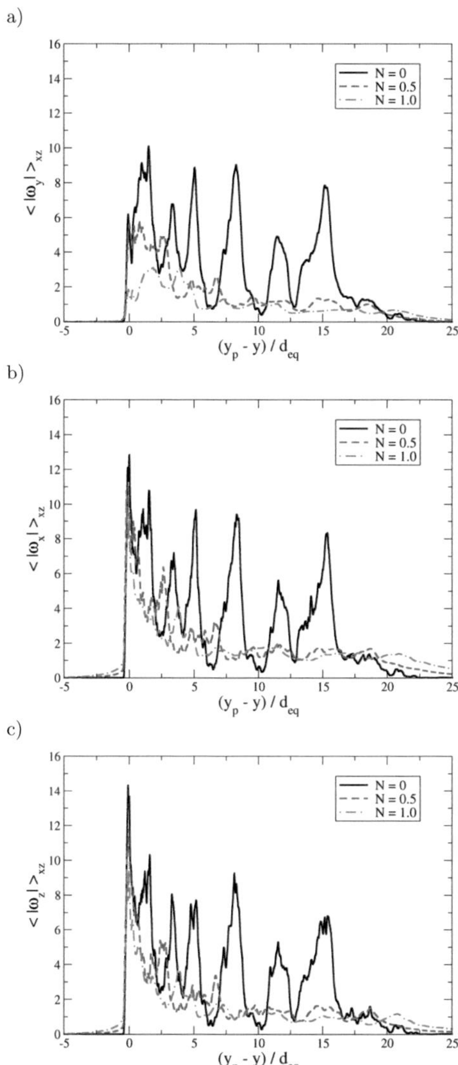

Figure 5.14 a) Absolute vertical vorticity component $\langle|\omega_y|\rangle_{xz}$ integrated over horizontal planes for the event C indicated in Figures 5.10, with comparison of the three cases $N = 0,\ 0.5,\ 1.0$. b) Analogous data for the transverse component $\langle|\omega_x|\rangle_{xz}$. c) The same data for the component $\langle|\omega_z|\rangle_{xz}$.

In summary, the applied vertical magnetic field particularly reduces the transverse velocities u and w and therefore indirectly the vertical component of the vorticity ω_y. This streamwise vorticity is the direct cause of the zig-zag trajectory which is consequently reduced when ω_y is smaller.

5.5.3 Energy spectra

Using the simulation data, energy spectra were obtained for the velocity components v and u, i.e. corresponding to the direction of ascent and the predominant direction of the zig-zag, respectively. These spectra are spatial spectra and were determined along vertical lines through the bubble center (x_p, z_p) and along additional vertical lines shifted by $\pm r_{eq}$ in x and z. The spectra resulting from these five lines were ensemble-averaged, as well as time-averaged in an interval of $\Delta t_{\langle\rangle} = 1.25$ around event C. A byproduct of the immersed boundary method applied here is an artificial, weak flow field inside the bubble [136]. The velocity field therefore is continuously differentiable in the entire domain and sampling data through the bubble will not affect the convergence of the spectra. The spectra E_{vv} and E_{uu} are shown in Figure 5.15 over the spatial wave number ξ_y for the case without magnetic field, $N = 0$, and the strongest vertical magnetic field applied, $N = 1$. The results for $N = 0.5$ lie in between and have been removed for readability.

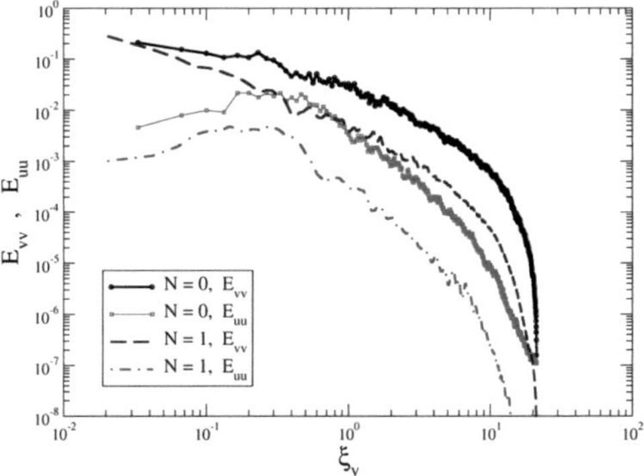

Figure 5.15 Spatial energy spectra E_{vv} of vertical velocity (v) and E_{uu} of the horizontal velocity (u) along vertical lines with ξ_y the spatial wave number in y for $N = 0$ and $N = 1$.

For $N = 0$, the spectrum of u exhibits an increase for low wave numbers, a maximum and a regular decay over more than two decades beyond which a fine-scale range with steeper decay is observed. The spectrum of the vertical velocity component behaves similarly. The overall amplitude is larger, particularly for the lower wave numbers, as this component is in the direction of the bubble rise velocity. Again, the high-frequency end of the spectrum decays fast and does not exhibit any sign of unphysical behavior as it would occur from under-resolution due to aliasing etc. On the other hand, it cannot be excluded that the finest scales are influenced by the grid resolution. In addition to the grid study presented above, the regular decay in both spectra over a large range of wave numbers demonstrates that the flow indeed is well resolved in its energy-containing range and well beyond.

The second type of information which can be extracted from Figure 5.15 relates to the application of the vertical magnetic field. It is apparent that the u-component, which is perpendicular to the field, is damped by an almost constant factor in the entire mid-to-high-frequency range. The amplitudes of the large wave numbers are also uniformly damped, but by a somewhat smaller factor. For the vertical component, a similar observation is made, except for the low wave numbers where the slope of the spectrum is changed. As a result, the largest scales are less influenced by the magnetic field. Overall, the damping by the magnetic field is seen to be stronger in the spectrum of u compared to the spectrum of v. This is coherent with the understanding of the action of the Lorentz force affecting predominantly the velocity component perpendicular to the field as discussed above. The spectra are instructive in this respect as they reveal damping, albeit less, also for the v-component of the velocity.

5.6 Examination of the employed numerical modeling and the experimental data

The scope of this section is to scrutinize the assumptions employed in the numerical modeling, but also to provide a comparative view on the experimental data for the ascent of a single argon bubble in the liquid metal GaInSn. During the course of this thesis, the numerical methodology was gradually improved allowing, e.g., for a more complex bubble shape. With the numerical model described above, already a very good agreement with the experimental data was achieved. All dominant effects of the magnetic field observed in the experiments were reproduced, and could be analyzed in more depth based on the available computational data of high detail. However, there is still room for improvement in the quantitative agreement. Furthermore, awareness of possible sources of error is important and an estimate about the magnitude of the error is a valuable information. Therefore, the following aspects are examined:

- The influence of the bubble shape representation is studied.
- A cross-comparison of experimental data obtained with different techniques is conducted.
- The impact of the phase-dependent electric conductivity is inspected.

5.6.1 Influence of bubble shape representation

In the previous study, the bubble shape was approximated as an oblate ellipsoid where the ellipsoid aspect ratio was correlated to the instantaneous bubble Weber number, $X(t) = f(We(t))$. For instance, the wake instability of a fixed axisymmetric bubble of realistic shape [30] shows a perceptible difference to a bubble of oblate ellipsoidal shape [169]. Consequently, an additional simulation is conducted where the bubble shape is represented by axisymmetric spherical harmonics (SH) with $N_{SH} = 12$ and the shape is computed directly from the fluid loads by the SH shape algorithm described in Section 4.4. All other parameters of the simulation remain unchanged. The non-dimensional numbers describing the simulation are $G = 2825$, $Eo = 2.5$, $N = 0$, i.e. no magnetic field is applied.

Figure 5.16 shows the bubble aspect ratio over time as well as the time-averaged bubble shape for both runs. The average shapes are nearly identical whereas the SH bubble has a moderate front-aft asymmetry being a bit front-flattened. In the case of the SH bubble, the aspect ratio is computed from $X = 2\max(x_S)/(\max(z_S) - \min(z_S))$ in the local reference frame resulting in a slightly lower value for the time-averaged aspect ratio. The projected area of the bubble during the ascent is however basically the same. Also the amplitude and frequency in the shape oscillation agree very well.

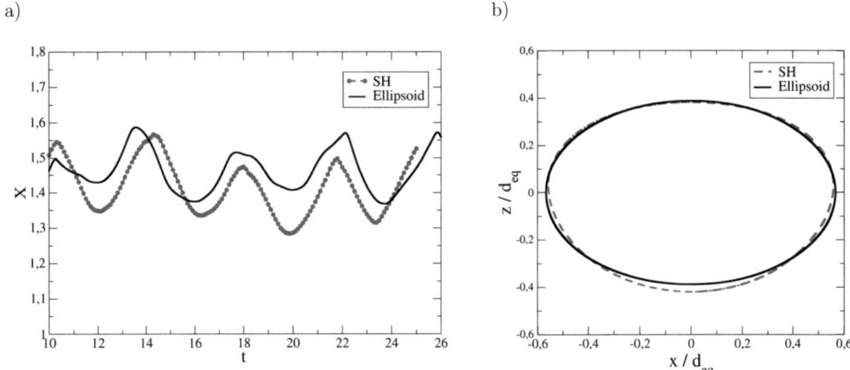

Figure 5.16 Comparison of results obtained with SH algorithm and with ellipsoidal shape from $X(t) = f(We(t))$, for parameters $G = 2825$, $Eo = 2.5$, $N = 0$: a) Aspect ratio over time. b) Time-averaged bubble shape.

Table 5.17a) lists the main figures describing the bubble dynamics and compares the results obtained with an ellipsoidal bubble shape, the SH representation, and the experimental data. Figure 5.17b) again shows the history of the bubble rise Reynolds number for the three cases. With the SH shape representation, a slightly higher average rise Reynolds number, Re_t, is observed compared to the ellipsoidal shape. The standard deviation, σ_{Re}, increases noticeable towards the experimental value whereas the frequency, f_{Re}, in the oscillation remains almost unchanged. The maximum inclination, $|\phi_z|_{max}$, decreases somewhat to 32°, while the measure of the zig-zag, Δ_{xz}/d_{eq}, is basically unaltered.

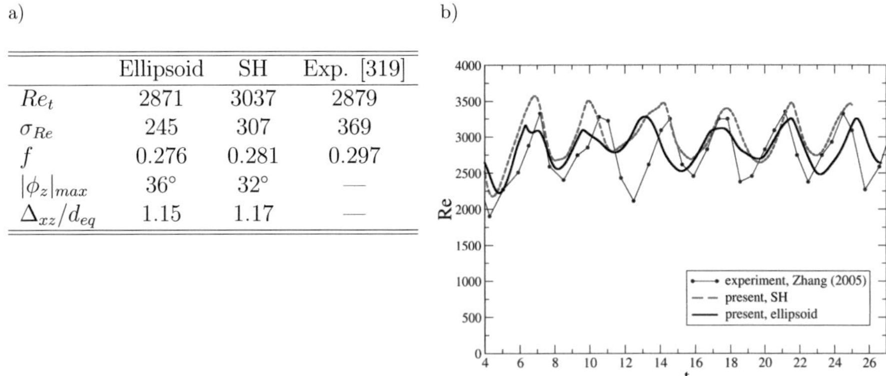

a)

	Ellipsoid	SH	Exp. [319]
Re_t	2871	3037	2879
σ_{Re}	245	307	369
f	0.276	0.281	0.297
$\|\phi_z\|_{max}$	36°	32°	—
Δ_{xz}/d_{eq}	1.15	1.17	—

Figure 5.17 Comparison of SH algorithm and ellipsoidal shape for parameters $G = 2825$, $Eo = 2.5$, $N = 0$. a) Bubble dynamics and b) history of bubble rise Reynolds number.

In summary, the approximation of the bubble shape as an oblate ellipsoid was well justified in the present case. The more sophisticated approach with the bubble shape represented by spherical harmonics and coupled to the hydrodynamic forces yields a very similar bubble shape and well comparable bubble dynamics. A non-axisymmetric shape might further improve the results. The correlation of the bubble aspect ratio to the instantaneous Weber number did also yield good agreement for the bubble shape oscillation. The additional simulation provides an a posteriori justification of the previous assumptions towards the bubble shape.

5.6.2 Cross-comparison of experimental data

In order to access the uncertainty in the experimental data that was used as a reference in the present study, further measurements and theoretical considerations are gathered. A cross-comparison of this data is conducted concerning the average rise Reynolds number of a single argon bubble of various bubble diameter in GaInSn without a magnetic field. The data comprises:

- Measurements by ultrasound Doppler velocimetry (UDV) [319]
- Measurements by ultrasound-transit time technique (UTTT) [6, 74]
- Measurements by X-ray radioscopy [14, 74]
- Theory in form of the Mendelson equation [176]
- Present simulation data

The UDV measurements [319] in the non-transparent liquid were already discussed before. With respect to the UTTT experiments [6, 74], only the two lowest argon flow rates (for two nozzle diameters) were considered from that reference to ensure sufficient distance between the bubbles and adequate single bubble motion. X-ray experiments [14, 74] were conducted, where the dimension of the container perpendicular to the main bubble motion ws rather narrow to allow the X-rays to penetrate the medium. The configuration is described in detail in Chapter 6. An additional simulation of a single bubble with $d_{eq} = 6$ mm was performed prior to the simulations of a bubble chain described in that chapter. The narrow dimension between the walls seems to have little influence on the bubble rise velocity as long as now direct bubble-wall contact is observed. This was also examined in the simulation using both no-slip and free-slip walls with very similar results for Re_t.
The Mendelson equation [176], which is derived from an analogy to the propagation speed of inviscid waves traveling over deep water, yields the terminal velocity of the bubble

$$u_t = \sqrt{\frac{2\sigma}{\rho_f\, d_{eq}} + \frac{g\, d_{eq}}{2}}\,. \tag{5.3}$$

Figure 5.18 shows the average rise Reynolds number versus the bubble diameter. Note that d_{eq} enters linearly into Re_t. The plot layout was chosen to again emphasize the high Reynolds number in liquid metals and also to illustrate that the computational costs increase with larger bubble sizes. Unfortunately, it seems quite difficult to create smaller bubbles using an injection nozzle in the experiments in this gas-liquid system.

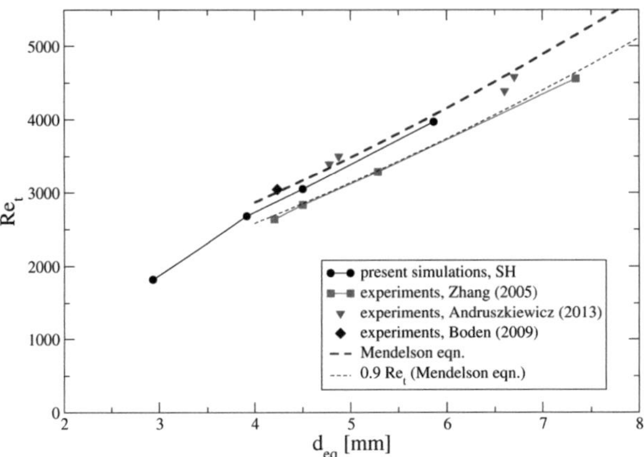

Figure 5.18 Average rise Reynolds number versus bubble diameter for argon bubbles in GaInSn without magnetic field. Comparison of present results achieved with the SH algorithm and experimental data obtained with UDV by Zhang et al. [319], with UTTT by Andruszkiewicz et al. [6, 74] and with X-ray radioscopy by Boden et al. [14, 74], as well as theoretical data calculated via the Mendelson equation (5.3) [176].

In general, good agreement is found for Re_t over d_{eq} for all experimental techniques, the simulations and the theory with deviations in the order of 10 - 15%. The data spans a corridor with the X-ray and UTTT measurements basically collapsing with the prediction from the Mendelson equation at the upper border. The simulation results have a slight offset towards lower Re_t. At the lower border of that corridor, we find the UDV measurements at approximately $0.9\,Re_t$ obtained from the Mendelson equation. The cross-comparison provides an interesting, but rough estimate of the error bounds related to the average rise Reynolds number of single bubbles in GaInSn. It should be emphasized again at this point, that the liquid metal GaInSn is very difficult to deal with and reproducibility is not easily achieved, e.g. due to the oxidation process: To quote one of the experimenters: "One day you measure this, the next day something else."

5.6.3 Influence of phase-dependent electric conductivity

One of the rather precarious assumptions of the previous study on the ascent of a single bubble under the influence of a magnetic field was the constant electric conductivity. In this section, the electric conductivity is made phase-dependent, $\sigma_e(\alpha)$, and the bubble is treated as an local insulator. No one-to-one comparison again using the ellipsoidal shape is conducted because the simulations are quite expensive. Instead the bubble shape is represented by spherical harmonics to achieve as much improvement as possible. All other parameters remained as outlined above. The study is further extended towards larger magnetic interaction parameters. With the modified numerical modeling, the electric current cannot penetrate the phase boundary and there is zero electric current within the insulating bubble where $\sigma_e(\alpha = 1) = 0$. Figure 5.19 shows an instantaneous, qualitative contour of the current density component, j_z, which is proportional to the Lorentz force component,

$-f_{L,x}$, in the present case of a vertical, homogeneous magnetic field. The non-dimensional numbers describing the simulation are $G = 2825$, $Eo = 2.5$, $N = 1$. The current paths have to close due to charge conservation which leads to the formation of circular patterns in front of the bubble and distorted loops in the bubble wake. The corresponding Lorentz force distribution leads to a damping of the transverse velocity components and indirectly affects the rise velocity, e.g. by altering the pressure field around the bubble.

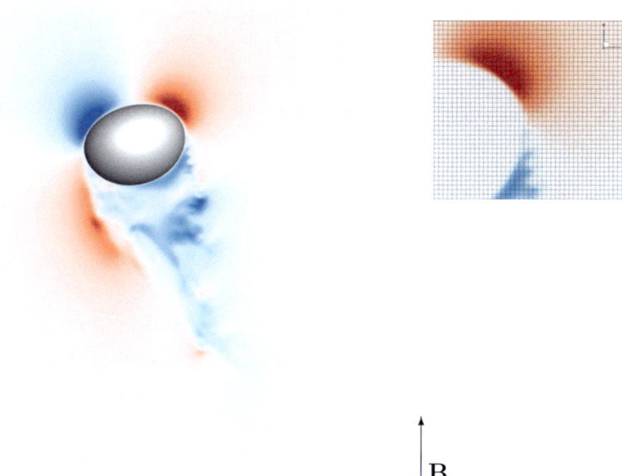

Figure 5.19 Instantaneous, qualitative contour of current density component, j_z, and Lorentz force component, $-f_{L,x}$, where $f_{L,x} \sim -j_z$ for the present case of a vertical, homogeneous magnetic field. Detailed view of phase boundary and mesh. Parameters: $G = 2825$, $Eo = 2.5$, $N = 1$ with $\sigma_e(\alpha)$, and SH bubble shape. The contour is an xy-plane through the particle center.

Table 5.4 summarizes the main results obtained with the SH bubble shape and $\sigma_e(\alpha)$ for various magnetic interaction parameters. The numbers in brackets indicate the previous simulation results with ellipsoidal bubble shape and constant electric conductivity. For $N = 1$, the average rise Reynolds number, Re_t, with phase-dependent electric conductivity $\sigma_e(\alpha)$ and SH bubble shape is almost identical to the value obtained with an ellipsoidal bubble and constant electric conductivity. The relative decrease in the standard deviation, σ_{Re}, compared to the case without magnetic field is similar for both runs with larger absolute values in the simulation with $\sigma_e(\alpha)$. The most significant change is that there is a substantially less pronounced damping in the dominant frequency, f_{Re}, in the case with $\sigma_e(\alpha)$. Now excellent agreement with the experimental data is found for the damping in the bubble Strouhal number. At an interaction parameter of $N = 1$, we obtain $Sr/Sr(N = 0) = 0.775$ and the value calculated from the experimental data is 0.779 (Figure 5.20). With constant electric conductivity and ellipsoidal shape, the quantitative agreement was less convincing giving a relative change in Strouhal number of $Sr/Sr(N = 0) = 0.656$. The reference values at $N = 0$ were obtained with the corresponding shape representation.
The extension of the study towards larger values of N supports the physical explanation of the effects of an longitudinal magnetic field on the bubble dynamics given in the previous sections. The data is plotted in Figure 5.20 in linear scaling. Indeed, there is a local max-

Table 5.4 Summary of simulation results with SH bubble shape and $\sigma_e(\alpha)$. Numbers in brackets indicate previous simulation results with ellipsoidal bubble shape and constant electric conductivity. $Eo = 2.5$, $G = 2825$

N	Re_t		σ_{Re}		f_{Re}	
0	3037	(2871)	307	(245)	0.281	(0.276)
0.5	—	(2957)	—	(166)	—	(0.233)
1.0	3122	(3132)	125	(90.3)	0.218	(0.181)
2.0	3009		144		0.138	
4.0	2639		148		0.072	

imum in Re_t over N for larger bubbles. With further increasing the magnetic interaction, the drag increases and the bubble rises slower. This is in agreement with the observation of an monotonously increasing drag with N for the flow around a fixed sphere or ellipsoid [173, 314]. The local increase in Re_t for low and moderate N stems from the damping of the lateral dynamics and the more rectilinear trajectory compared to the case without magnetic field. A roughly linear decrease in the bubble Strouhal number with N is observed for small to moderate interaction parameters which then seems to saturate at $N = 4$ where the damping is less pronounced. Note that the statistics for the simulation with $N = 4$ were obtained for two crossings of the periodic domain, but still only three quasi-periods of the oscillation in $Re(t)$ could be used. The uncertainty in the frequency and standard deviation of the oscillation therefore is rather high. For the magnetic interaction parameters studied here, the lateral dynamics of the high Reynolds bubble were not fully suppressed by the magnetic field.

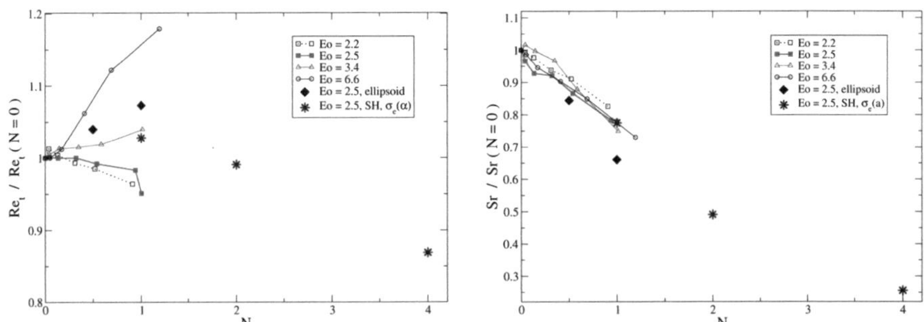

Figure 5.20 Relative change in average rise Reynolds number and Strouhal number: Present simulations with SH bubble shape and $\sigma_e(\alpha)$ compared to simulations with ellipsoidal bubble shape and $\sigma_e = const.$ for $Eo = 2.5$, $G = 2825$, as well as compared to experimental data of [319] for various Eo.

In conclusion, using a phase-dependent electric conductivity does improve the quantitative agreement with the experiments. The qualitative effect of the longitudinal magnetic field on the bubble dynamics, however, remains the same as in the studies reported earlier. With the improved modeling, further simulations at higher magnetic interaction were conducted and the physical explanations on the impact of the field on the rise Reynolds number are ascertained.

5.7 Conclusions for the rise of a single bubble affected by a longitudinal magnetic field

Phase-resolving simulations of single bubbles rising in liquid metal have been conducted. In summary, the following effects are observed, when increasing the magnetic interaction parameter N compared to the case without magnetic field:

The time-averaged bubble rise velocity increases for large bubbles (high Eo) reaches a local maximum and then decreases. For small bubbles (low Eo) it decreases for all N studied. The amplitude of oscillations in $v_p(t)$ decreases. The dimensionless characteristic frequency f_{Re} of oscillations in $Re(t)$ and the resulting Strouhal number Sr decrease for all bubbles. The amplitude of oscillation in lateral bubble positions $x_p(t)$, $z_p(t)$ decreases, i.e. the trajectory is more rectilinear. Also the amplitude of oscillation in tilting angles $\phi_i(t)$ decreases. The integral of the absolute value of the vertical vorticity component over cross sectional planes in the bubble wake decreases. Similar observations are made for the transverse components, but the vertical component of vorticity is affected most by the damping due to the vertical magnetic field.

The results are in good agreement with the corresponding experiments presented in [319]. The present results for the instantaneous vertical bubble and fluid velocity support the findings from these experiments where the velocity component along a line was measured by ultrasound Doppler velocimetry. Furthermore, additional data are now available from the simulations elucidating the full three-dimensional bubble trajectory, flow structures in the bubble wake and wake vorticity as well as energy spectra. These data provide valuable insight into the considered three-dimensional multiphase flow and into the dynamics of a single bubble in liquid metal under the impact of a longitudinal magnetic field which can so far not be obtained by experiments.

Simulations of bubble chains and swarms in liquid metal will be conducted to provide insight into the influence of a magnetic field on collective effects in bubble driven flows. Another interesting direction of research is the influence of a magnetic field on the flow through a relatively tight cluster of bubbles as presented in [102].

6 Bubble Chain Influenced by a Magnetic Field

6.1 Introduction to the influence of a magnetic field on a chain of bubbles in liquid metal

The injection of bubbles into liquid metal, forming bubble chains, swarms, or plumes, is of particular relevance for metallurgical applications, such as the continuous casting process and other industrial applications like argon refining. The bubbles enhance mixing of the melt, prevent blocking in the entry nozzle, and influence chemical reactions. Magnetic fields are used in metallurgical applications to alter and control the flow. Numerical simulations and experiments on both a laboratory and an industrial scale [272, 271] are conducted to gain understanding of the nature of these flows. Analytical considerations of the impact of a transverse magnetic field on a liquid metal jet or an axisymmetric vortex were developed in [41, 43]. The dynamics of single bubbles and spheroids in a longitudinal magnetic field were studied by simulations [244, 245] and by experiments [319] revealing the substantial and complex impact of the field. Even without a magnetic field, the complexity of the flow increases as the number of bubbles, and thus the gas fraction, increases. The dynamic behavior of bubbles rising in line and the mechanisms leading to instability are discussed in [227, 228]. A general review of gravity-driven bubbly flows is given in [188] with emphasis on bubble-induced turbulence in bubble columns reaching from large-scale circulation to small-scale vortical structures. The bubble-induced agitation and liquid velocity fluctuations behind a bubble swarm at intermediate to high Reynolds numbers are studied in more detail in [223]. A chain of nitrogen bubbles in water was experimentally investigated in [235] with a high-speed camera in a large container. Reynolds numbers covered $Re = 300 - 650$, yielding distances between the bubbles of $6.5 - 300$ diameters. Nearly identical bubble paths were found at low frequencies, while at higher frequencies a scatter in paths is observed after a certain height is achieved due to hydrodynamic interaction of the bubbles.

The behavior of gas bubbles in a turbulent liquid metal magnetohydrodynamic (MHD) flow for longitudinal and transverse magnetic fields was studied in [55, 56]. Local resistance probes showed a significant change in the bubble distribution over the channel cross-section under the impact of a magnetic field. Ultrasound Doppler measurements were performed in [320] to investigate a bubble-driven liquid metal jet in a confined cylindrical vessel with a transverse magnetic field. The spatio-temporal mapping of the liquid velocity distribution showed a restructuring of the convective motion when a magnetic field is applied. The previously axisymmetric stable flow becomes anisotropic with an oscillating flow pattern for moderate magnetic fields, in contrast to the general expectation of magnetic braking. A simulation of this configuration was conducted by Miao et al. [178] using an Euler-Euler

approach and a RANS-SST turbulence model. The multiphase ansatz uses correlations for the interfacial force, including analytic expressions for drag, lift, turbulent dispersion, etc., which were obtained without magnetic interaction and therefore do not consider the impact of the magnetic field on the bubble dynamics. The turbulence model was altered to account for bubble-induced turbulence and turbulence anisotropy due to the magnetic field.
In the present work, a three-dimensional phase-resolving simulation of a bubble chain in liquid metal is performed to study the impact of a vertical magnetic field. The current configuration comprises a channel of high aspect ratio, resembling X-ray experiments reported in [14]. The standard X-ray measurement technique experiences difficulties if a magnetic field is applied simultaneously. The simulations are capable of providing highly resolved, three-dimensional data of the fluid and the bubbles and give valuable insight into this liquid metal multiphase flow under the influence of a vertical magnetic field. Part of this manuscript was submitted to the European Physical Journal and appeared in [247] and [74].

6.2 Configuration and estimate of discretization error

The present study deals with a chain of bubbles ascending in a container of high aspect ratio filled with initially quiescent liquid metal. More specifically, separate individual argon bubbles are injected into the metal alloy GaInSn, which is liquid at room temperature. The material properties are provided in [319]. The configuration was originally studied experimentally by Boden et al. [14] using a container with rectangular cross section and dimensions $\mathbf{L} = (L_x, L_y, L_z) = (200, 300, 12)$ mm with a free surface. With $L_z = 12$ mm, the depth of the container was only approximately two bubble diameters to allow X-ray measurements. The experiments were performed without magnetic fields for technical reasons. The experimental apparatus allowed for various injection positions, including injection from one side of the container, its center, or simultaneous injection from both sides with different gas flow rates. Here, the injection near the left wall is considered. The configuration is sketched in Figure 6.1a. A preliminary simulation with relatively coarse resolution was performed with the original experimental setup for a volumetric gas flow rate of 100 sccm (cm^3 per minute at standard conditions) and without a magnetic field. An instantaneous contour plot of the vertical velocity obtained from this simulation is displayed in Figure 6.1c revealing the formation of a large standing vortex driven by the bubble chain. The time-averaged vertical velocity of the bubbles and its standard deviation are compared to the data from the X-ray measurements in Figure 6.1b showing overall good agreement. Although the discrepancies between the data sets are generally small, the simulation results yield a higher value for the rise velocity. Averaged quantities are denoted by $\langle\ \rangle$ throughout.
A grid refinement study was conducted considering the laps of time during which the first nine bubbles are injected. All numerical parameters correspond to the simulations presented below, but a mono-disperse bubble size distribution was used. From the coarse-grid simulation, the resolution was successively refined by a factor of two and again by a factor of two in all three directions, also requiring a corresponding reduction of the time-step size. Grids with $\mathbf{N} = (480, 480, 40), (960, 960, 80), (1920, 1920, 160)$ cells were used, corresponding to 20, 40, 80 points per volume-equivalent diameter d_{eq}, respectively. Based on the finest grid, the medium and coarse grid yield 3% and 8% larger average rise velocities $\langle v_p \rangle$, 9% and 34% smaller absolute values of the transverse velocity $\langle |u_p| \rangle$, as well as 6% and 22% increase in average aspect ratio, respectively. The medium grid was then used for the simulations presented in this study. The numerical resolution provides a good compromise between

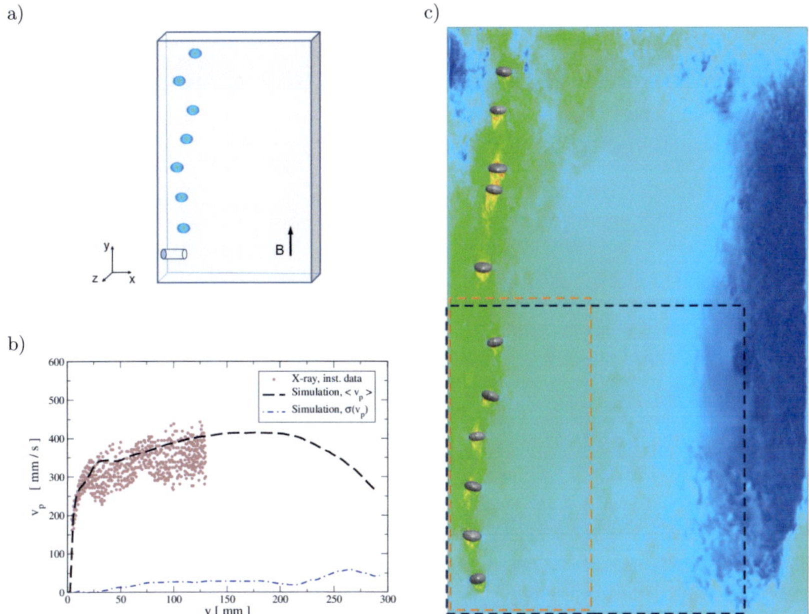

Figure 6.1 Original experimental configuration of [14] and initial coarse-grid simulation of this case. a) Schematic view of the configuration. The origin of the coordinate system is in the lower left, back corner of the container. b) Time-averaged vertical velocity $\langle v_p \rangle$ and its standard deviation $\sigma(v_p)$ from a preliminary coarse simulation and instantaneous experimental data [14]. c) Vertical velocity without a magnetic field in the plane $z = 0.5L_z$. Window of X-ray measurements (orange dashed line) and reduced computational domain used in the present study (dark dashed line).

accuracy and computational effort. Further refinement would exceed the computational resources available. The chosen resolution therefore does not correspond to a full DNS, but will be adequate to provide valuable and detailed insight into the physics of this magnetohydrodynamic, multiphase flow. Due to the limited computational resources, also a reduction of the container size was necessary to obtain converged statistical data on the bubble driven flow. Each computation, although computed on 64 - 128 processors, took months in wall-clock time, even for the reduced computational domain. The smaller computational box is comparable to the X-ray window also shown in Figure 6.1c. Corresponding experiments in this smaller container and with the application of a magnetic field are planned in the near future.

6.3 Results for the influence of a magnetic field on a bubble chain

6.3.1 Physical and numerical parameters

In the present work, the gravitational velocity $u_{ref} = \sqrt{|\pi_\rho - 1|\, g\, d_{eq}}$ is used as the reference velocity, and d_{eq} is employed as a reference length. Here, d_{eq} is the diameter of a volume-equivalent spherical bubble. This yields the time scale $t_{ref} = \sqrt{d_{eq} \,/\, (|\pi_\rho - 1|\, g)}$.
The parameters which govern the ascent of a *single bubble* in quiescent, contaminated, unbounded fluid are the Galilei number G, the Eötvös number Eo, and the density ratio π_ρ. The respective definitions were given in Section 1.1.1. In the simulations, the physical parameters characterizing each single bubble are $G = 4200$, $Eo = 4.2$, $\pi_\rho = 6 \cdot 10^{-4}$ and these correspond to a $d_{eq} = 6$ mm argon bubble in GaInSn. A no-slip condition is enforced at the phase boundary, and the bubble is electrically insulating.
In order to characterize a *bounded domain*, the dimensions of the container $\mathbf{L}$ are set in relation to the size of the bubbles d_{eq}. The computational domain extends over $\mathbf{L} = (24, 24, 2)\, d_{eq}$ and is visualized in Figure 6.1c (dark dashed line). No-slip boundary conditions are applied at all walls, and a free slip condition approximates a free surface at the top of the container. Note that the role of boundary conditions at a free surface in liquid metal flows is unsettled [58] due to the complex formation of an oxide layer. All boundaries in the present setup are electrically insulating. The computational domain is discretized with an equidistant grid $\mathbf{N} = (960, 960, 80)$, i.e. 73.7 million cells. The time-step size is determined adaptively to yield $CFL = 0.8$.
A disperse *bubble chain* can be described by the following parameters. The position of the injection nozzle inside the container yields the initial position of the bubbles $\mathbf{x}_0$. The nozzle is situated at $\mathbf{x}_0 = (3.3, 0.8, 1.0)\, d_{eq}$. Other conditions at the nozzle are the initial bubble velocity, the direction of injection, and the initial bubble shape. The latter, however, could not be measured from the experiments [14]. At the injection nozzle, an initial aspect ratio of $X_0 = 1.0$ and an initial bubble velocity of $\mathbf{u}_{p,0} = 0$ are prescribed.
The surface of each individual bubble is represented by $N_L = 9039$ Lagrangian forcing points. Axisymmetric spherical harmonic functions with a polynomial degree up to twelve are used to describe the bubble surface [246, 22]. The bubble shape is determined from the loads applied by the surrounding flow [246]. Therefore, collective effects on the bubble shape in swarms, such as wake entrapment or drafting, as well as wall effects, are resolved. This improves the bubble shape representation in comparison to bubble shapes imposed *a priori*

or via shape correlations valid for single bubble ascent only, discussed in [244, 2].
The bubble detachment frequency f_b is here assumed to be constant and can be used to calculate the gas flow rate. A bubble detachment frequency of $f_b\,t_{ref} = 0.3644$, and a variable bubble size according to a Gaussian distribution are prescribed at the nozzle. The latter is characterized by a mean diameter and a standard deviation corresponding to $\langle d_{eq} \rangle = 6$ mm and $\sigma_d = 0.48$ mm. The parameters were obtained by fitting experimental data which was provided by the authors of [14] (private communication). In the experiments, the mean bubble diameter and its standard deviation depended substantially on the wetting conditions at the nozzle. In the simulation, the same bubble size distribution at the injection nozzle is realized by means of simple random sampling [205]. The bubbles are 'removed' from the simulation before reaching the upper boundary by switching off the forcing in the IBM, i.e. the phase coupling, and therefore the bubbles do not interact with the free surface.
A collision model based on a repelling potential is employed to represent inter-bubble collisions and bubble-wall collisions as outlined in [102, 104] and Section 7.4 (CM-2). A coalescence model was applied towards the end of the simulation which allows phase-resolved merging of two bubbles. Such coalescence events, however, are very rare for the setup and can also be neglected. Detailed information on bubble coalescence can be found in Chapter 7.
Finally, the *magnetic interaction* parameter is introduced representing the ratio of magnetic forces to inertial forces [319, 139] In the present study, a homogeneous magnetic field $\mathbf{B}$ is applied in the y-direction, i.e. anti-parallel to gravity. Two simulations were conducted studying the influence of a vertical magnetic field on the bubble chain, one without a magnetic field, $N = 0$, and one with a magnetic interaction parameter of $N = 1$. The total wall-clock time of the simulation was about 4 months for each case on $O(100)$ cores of an SGI Altix and IBM iDataPlex dx360M2, corresponding to about $2{\cdot}10^5$ CPU hours.

6.3.2 Results for the continuous phase

A view into the nature of this bubble-driven flow is given in Figure 6.2, which shows the instantaneous vertical velocity in the plane $z = 0.5L_z$ and a visualization of the bubbles for the two cases, with and without a magnetic field. The lowest, spherical bubble indicates the position of the nozzle and is represented in its initial state. Its surface is penetrable until it is released at $t = a\,f_b^{-1}$, $a \in \mathbb{N}$.

The pictures show a disperse bubble chain remaining close to the wall in both cases. The bubbles rise on zig-zag paths, and each bubble follows its predecessor to a large extent. The bubble chain drives a large-scale vortex which covers the entire domain and yields a large region of negative vertical velocity near the right boundary for $N = 0$. Significant modifications of the flow field are visible in the presence of a vertical magnetic field. Without a magnetic field, intense turbulent fluctuations can be observed in all parts of the flow: In the upward bubble-driven region, including the bubble wakes, in the transverse flow at the free surface, as well as in the downward and backward transverse flow at the right and lower walls. In contrast, substantial damping of fluctuations is observed when a vertical magnetic field is applied, as also observed in homogeneous turbulence [139] or turbulent MHD channel flow [15, 145]. The rather small-scale isotropic vortices are reduced, and instead, large-scale structures aligned with the field can be discerned, especially further away from the bubble chain.
To estimate the end of the initial transient of the flow, the circulation $\Gamma = \oint \mathbf{u}{\cdot}d\mathbf{s}$ is calculated

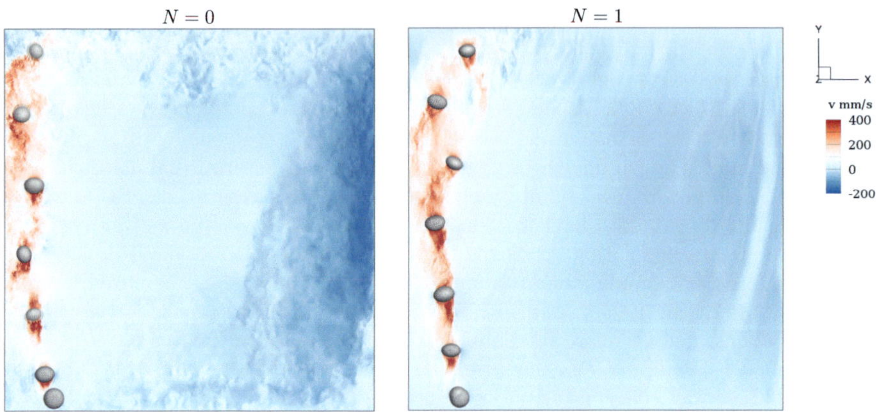

Figure 6.2 Instantaneous vertical velocity in the plane $z = 0.5L_z$ and visualization of bubbles by a subset of forcing points on their surface without magnetic field ($N = 0$) and with magnetic field ($N = 1$).

along a closed rectangular curve at $z = 0.5L_z$ parallel to the boundaries and $0.5\, d_{eq}$ apart from these. It quantifies the overall amount of vorticity ω_z contained in the area bounded by the described curve. It could also be measured using four ultrasound Doppler sensors which allow the measurement of the velocity component along a single line [319, 320, 58]. The circulation can also be utilized as an input parameter for flow control by magnetic fields. The temporal evolution of the circulation in the container is plotted in Figure 6.3. The case without a magnetic field was started with the fluid entirely at rest. As we are only interested in the steady state, the simulation with a magnetic field was initialized with results obtained on a coarse grid.

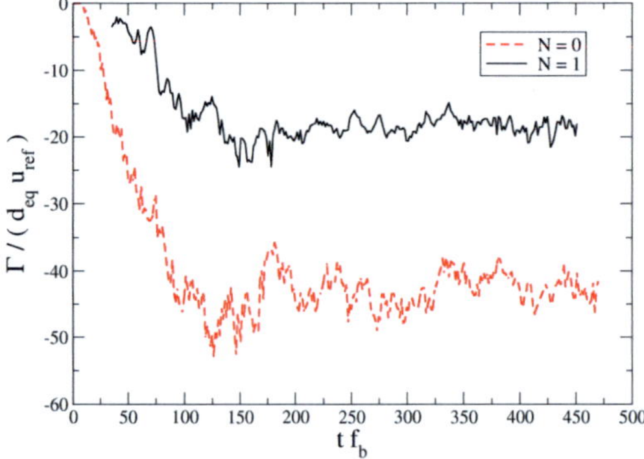

Figure 6.3 Temporal evolution of circulation Γ without magnetic field ($N = 0$) and with magnetic field ($N = 1$).

Statistics were then obtained starting at $t \cdot f_b = 200$ for more than 250 rising bubbles. The history of Γ indicates a vortex which rotates clock-wise with a circulation fluctuating around a constant mean for the developed flow in both cases. The mean circulation, i.e. the overall strength of the central vortex, is reduced substantially, by 57%, due to the vertical magnetic field with $N = 1$. The time-averaged velocity field visualized by the absolute value of velocity and average path lines are shown in Figure 6.4 with and without a magnetic field.

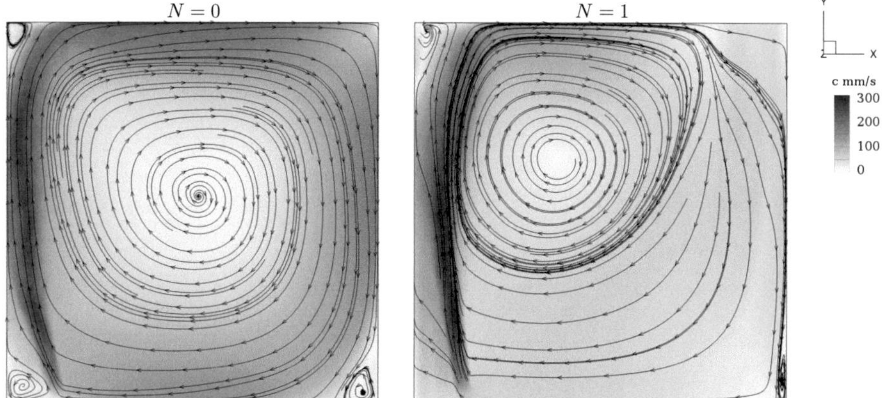

Figure 6.4 Time-averaged absolute value of velocity in the plane $z = 0.5L_z$ and path lines without a magnetic field ($N = 0$) and with a magnetic field ($N = 1$). Spiraling path lines are due to secondary flows in the z-direction.

An almost symmetric flow field is found in the case without the magnetic field. The center of the dominating vortex is located approximately in the center of the domain. With the magnetic field, the velocity magnitude at the free surface is reduced, which would be favorable in industrial applications to avoid slag entrainment. Consequently, mass conservation is realized on a shorter circuit leading to a shrinkage of the vortex and a displacement of its center towards the upper left of the domain. The fluid in the lower right part of the container becomes almost stagnant. A straightening of the bubble-induced jet is observed, and thus the jet moves away from the wall. The secondary vortex in the upper left corner grows. A more central position of the injection nozzle or even just more straightening of the jet due to stronger magnetic interaction can trigger downward flow also close to the left wall. This can lead to long-term oscillations of the entire jet as encountered in simulations with different parameters not shown here and observed in [320, 178] for a bubble-driven jet in a transverse magnetic field.

6.3.3 Results for the disperse phase

Statistics for the disperse phase were determined in the time period indicated above by dividing the domain in xy-bins of size $0.125\, d_{eq} \times 0.125\, d_{eq}$ and by calculating the probability of the bubble center being observed in each bin. Variations in the z-direction were not considered. Using the calculated probabilities, average bubble quantities can be determined as a function of column height. The time-averaged bubble trajectory and its standard deviation are plotted in Figure 6.5 for the two cases, $N = 0$ and $N = 1$, respectively. In addition, the right plot shows several sample paths to give an impression of the trajectories of individual bubbles during the simulation.

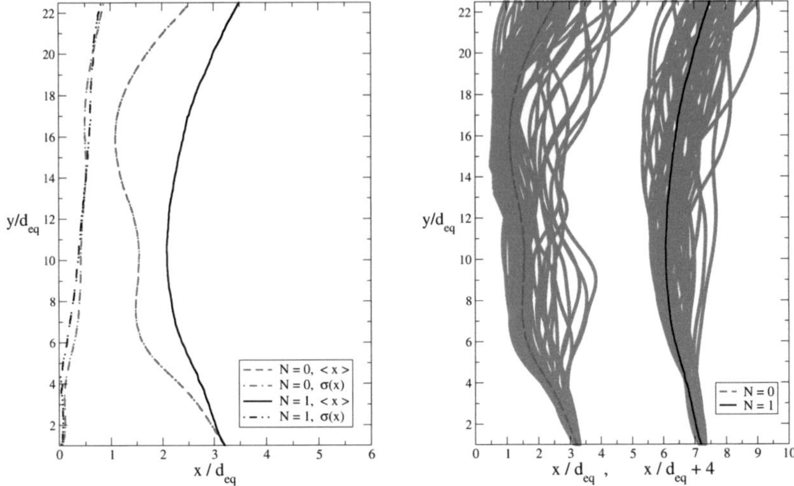

Figure 6.5 Time-averaged bubble trajectory, its standard deviation $\sigma(x)$ and sample bubble paths without magnetic field ($N = 0$) and with magnetic field ($N = 1$).

The distinct zig-zag trajectory of an individual bubble, observed in the case without a magnetic field, is stretched out in the vertical direction by the magnetic field, leading to a more rectilinear path [244, 319]. This also yields a straighter average trajectory and reduced transverse dispersion measured by the standard deviation $\sigma(x)$ in the trajectory, especially in the lower part of the container. In the upper region, the standard deviation in the transverse bubble position is similar for both cases. The bubbles are pushed closer to the wall by the more intense vortex in the case without a magnetic field. Note the truncated sample paths for $y \gtrsim 15d_{eq}$ where the bubbles collide with the left wall for $N = 0$. Figure 6.6 reports the time-averaged vertical bubble velocity with and without magnetic field. The average rise velocity decreases by approximately 10% under the impact of the vertical magnetic field.

The regular oscillations in $\langle v_p \rangle$ and $\langle x_p \rangle$ for $N = 0$ do not entirely vanish by averaging over the simulated period of time. The oscillations are related to the zig-zag pattern in the bubble trajectory, and the bubbles follow the path of their predecessors to a large extent. The oscillations are also present in the experimental data in the large container (Fig. 6.1b). In the presence of the magnetic field, these oscillations are removed as expected from the studies of single bubble ascent [244, 319].

Although the average rise velocity is lower, a flatter bubble shape is observed at the same Eo when a magnetic field is present. The time-averaged bubble shape is pictured for both cases at half the container height in Figure 6.7. The actual influence of magnetic fields on the bubble shape is still in dispute. In [256], the rise of a single bubble in a narrow enclosure under vertical magnetic fields was simulated by means of a volume of fluid approach with a

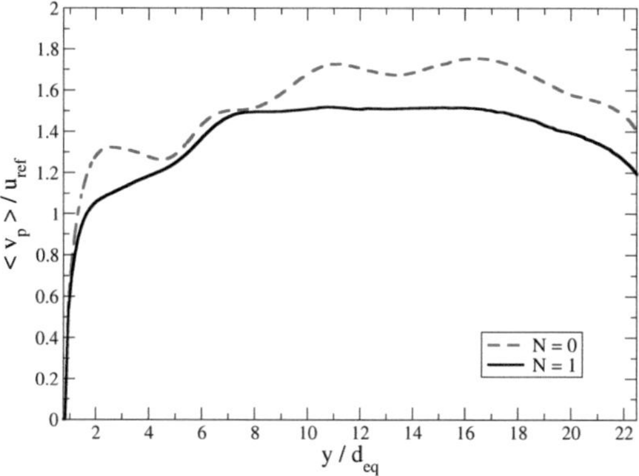

Figure 6.6 Time-averaged vertical bubble velocity without magnetic field ($N = 0$) and with magnetic field ($N = 1$). The primes indicate a bubble reference frame.

significantly reduced viscosity ratio, density ratio and Galilei number compared to a realistic liquid metal system. For very strong magnetic fields in the direction of the ascent, resulting in a complete suppression of recirculation in the bubble wake, an elongation of the bubble along the magnetic field lines is reported. A longitudinal magnetic field modifies the pressure field around the bubble and consequently the bubble shape. At moderate interaction parameters, $N = O(1)$, and high Reynolds numbers, $Re > 100$, especially the strength and size of the stagnation pressure region at the bubble front increase [173, 252, 245] leading to a flatter bubble shape. Note that the depicted axisymmetric bubble shape is represented in the local reference frame of the bubble. The orientation of the bubble is computed in each time step and fluctuates substantially. For the present regime, X-ray visualizations indeed show spheroidal bubble shapes in the xy-plane [14].

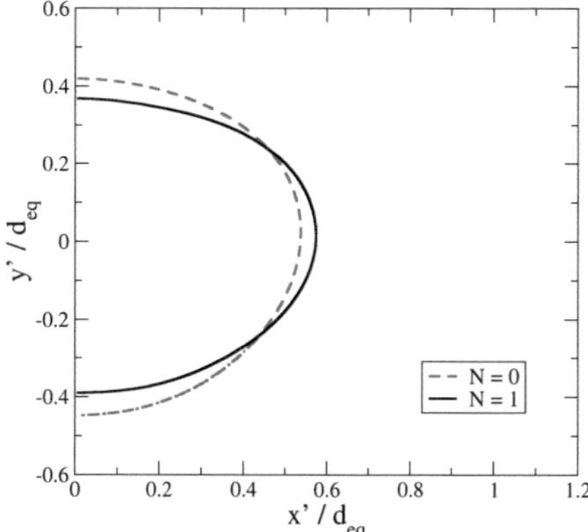

Figure 6.7 Time-averaged bubble shape at half the container height $y = 0.5\,L_y$ without magnetic field ($N = 0$) and with magnetic field ($N = 1$).

6.4 Conclusions on the influence of a magnetic field on a single bubble chain

The flow driven by a bubble chain in liquid metal has been simulated using an immersed boundary method. The influence of a homogeneous magnetic field acting anti-parallel to gravity has been quantified statistically for the continuous phase as well as for the dispersed bubbles. Three-dimensional, time-averaged and instantaneous data of high temporal resolution provide valuable insight into the physics of a bubble chain under the influence of a magnetic field. With a magnetic field, a restructuring of the flow field is found, comprising reduced overall circulation in the container, a more rectilinear average bubble trajectory and a reduction of the average rise velocity of the bubbles. Damping of turbulent fluctuations is observed, especially away from the jet. The issue of numerical resolution for the direct numerical simulation of liquid metal multiphase flows was addressed in Section 6.2. With the given computational resources, elucidating details about the flow were gathered that confirmed expectations from single bubble simulations and experiments, as well as experiments on bubble jets with transverse magnetic field.

7 Bubble Collision and Coalescence Modeling

7.1 Introduction to bubble interaction

The ascent of single bubbles is an interesting phenomenon and has been investigated for many years [168]. Additionally, the issue of interaction between bubbles becomes important when considering bubble chains, swarms, or bubble columns. If two bubbles approach they can either bounce or undergo coalescence depending on the nature of the interaction. Coalescence, as well as its counterpart bubble breakup, alter the bubble size distribution, and hence the characteristics of the entire multiphase system. The impact of bubble size at identical void fraction was studied, for example, by Santarelli et al. [236] considering monodisperse, spherical bubbles by phase-resolving simulations in turbulent channel flow. For the same void fraction, a substantial influence of the bubble diameter on the turbulent statistics and coherent structures was found. This highlights the need for adequate representation of coalescence and breakup in this type of simulation and constitutes the motivation of the present work. Here, we will only consider coalescence as this occurs more frequently in the flows investigated.

In simulations of bubbly flows which do not resolve individual bubbles, i.e. the bubble size is smaller than the grid size, the prediction of the bubble size distribution is crucial for correct closures and the choice of correlations. Models for coalescence and breakup have been found to be one of the weakest points [71, 162]. A review of coalescence models for such simulations is provided by Liao and Lucas [161] and a comparison of their predictions to experimental data is given in [222, 162]. Besides the influence of coalescence on the bubble size distribution, bubble shape oscillations related to the merging of two bubbles induce turbulent fluctuations with velocities larger than the rise velocity of the bubble [264, 265, 269, 270, 268].

To this date, no consensus is found in the literature on how to predict realistic coalescence [118]. The latter review concentrates on the processes in the very thin film between two bubbles and the mechanisms that eventually lead to the rupture of this film and thus to coalescence. There is general agreement that van der Waals forces, which are inter-molecular attracting forces, need to become important to yield coalescence [118, 54]. These forces have a range of about $10 - 60$ nm. Compared to the size of a bubble, typically with diameter 100 μm -10 mm, a gap of several orders of magnitude would have to be bridged if the rise of the bubble and the film thinning were to be resolved in the same simulation.

The simulation of the processes leading to a single coalescence can be achieved using, e.g., Boundary Integral Methods [213]. The phase-resolved simulation of many bubbles can be conducted for instance with front-capturing methods [307] like the Volume of Fluid Method (VOF) or the Level Set Method. In many of these methods, a diffuse interface exists between

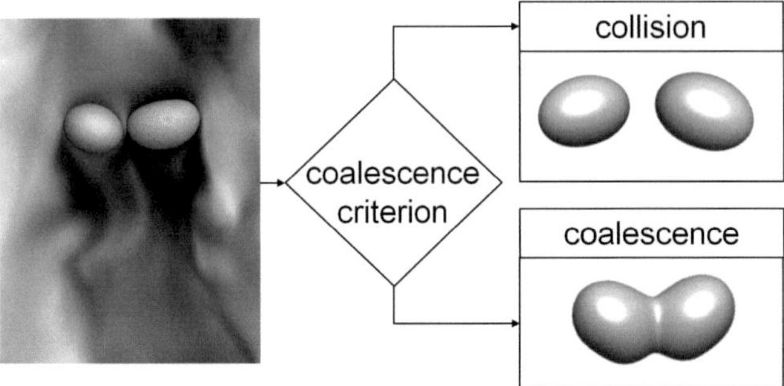

Figure 7.1 Coalescence modeling in phase-resolved simulations. During the bubble approach a coalescence criterion is used to decide whether a collision or a coalescence is modeled.

the gas and the liquid and numerical coalescence always occurs when the distance between the bubbles becomes smaller than the grid size. Due to the limitation in grid size resulting from limited computational resources such methods cannot resolve the processes in the liquid film between the bubbles and hence cannot accurately predict coalescence. Consequently, modeling is required. To create a model for bubble coalescence without resolving the microscopic processes inside the fluid film between the bubbles, but resolving the geometry of the bubbles, requires two features. One is a criterion for coalescence that accounts for the evolution of the film, the other is a representation of the bubble surface during coalescence. Figure 7.1 shows a schematic view of the modeling approach. In the following, we use the term 'interaction' when two bubbles come sufficiently close for the surfaces to touch (within the precision of the numerical grid). Interaction can then be subdivided into collisions, where the bubbles bounce back, and coalescence events. The coalescence process is understood as the merging of the bubbles, as well as the subsequent shape oscillations resulting from it. The present chapter is structured as follows:

- The inter-bubble distance is delt with in Section 7.2 with the focus on the numerical evaluation of the distance for different bubble shapes.
- For bubbles being sufficiently close, the coalescence criterion, introduced in Section 7.3, determines whether the bubbles undergo coalescence or bounce back.
- Section 7.4 provides an overview of the employed collision models.
- Section 7.5 introduces a coalescence model applicable to phase-resolving simulation of many bubbles in turbulent flows.
- Finally, the application of the developed coalescence model for the rise of two adjacent bubble chains is presented in Section 7.6.

7.2 Inter-particle and particle-wall distance

The modeling of bubble interaction, either collision or coalescence, requires the determination of the distance between the bubble surfaces. The bubble shapes are represented analytically. The shapes comprise spherical bubbles, ellipsoidal bubbles and bubbles represented by spherical harmonic functions (SH) as presented in Chapter 4. Consequently, the distance between two bubbles or the distance of a bubble surface and a wall or a free surface can also be determined with high accuracy.

7.2.1 Spherical bubbles

For *spherical* bubbles, the calculation of the inter-particle distance is trivial. Figure 7.2 provides a sketch of the configuration. The distance between the interfaces of particle A and particle B is given by $\zeta_{A,B} = \|\mathbf{x}_{p,A} - \mathbf{x}_{p,B}\| - (r_A + r_B)$. The straight line connecting the particle centers gives the location of the sub-contact points (base points) $\mathbf{x}_{sub,A}$ and $\mathbf{x}_{sub,B}$ on the respective particle surfaces. For two touching bubbles, $\zeta_{A,B} = 0$, these sub-contact points coincide and yield the contact point. The particle-wall distances are determined from Table 7.1.

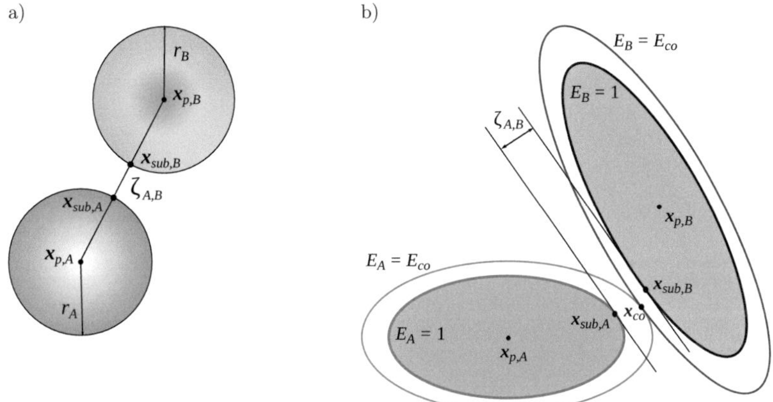

Figure 7.2 Schematic representation of distance determination between particle surfaces. a) Spheres. b) Ellipsoids.

7.2.2 Ellipsoidal bubbles

For *ellipsoidal* particles of arbitrary orientation, the calculation of the inter-particle distance and the determination of the sub-contact points is less trivial. Figure 7.3 shows three examples for two tri-axial ellipsoids illustrating the problem to solve. The inter-particle distance varies substantially based on the orientation of the ellipsoids. The ellipsoids would touch if the longest semi-axes a were aligned since the distance of particle centers is chosen as $\|\mathbf{x}_{p,A} - \mathbf{x}_{p,B}\| = 2a$.

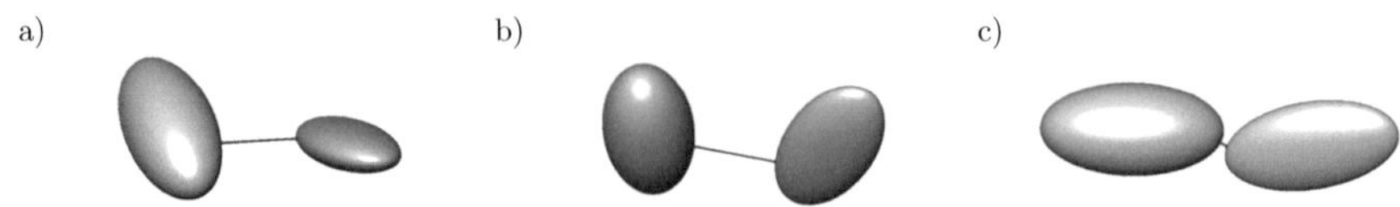

Figure 7.3 Inter-ellipsoid distance. a) - c) Three examples of closest distance for different orientations of two tri-axial ellipsoids. Distance of particle centers is $\|\mathbf{x}_{p,A} - \mathbf{x}_{p,B}\| = 2a$.

An approximation for the computation of the inter-particle distance was proposed by Paramonov and Yaliraki [206]. It is derived from the ellipsoid contact function. The approach was implemented and validated throughout the student thesis of Quering [217]. It is only very briefly recalled here. The main idea is to expand or shrink the two ellipsoids by a common factor until they touch. The virtual contact point of the touching, scaled ellipsoids is then used to construct two planes tangential to the unscaled ellipsoid surfaces from which then the distance is determined.
Isotropic scaling of a tri-axial, ellipsoid is realized by multiplying the semi-axes with the factor $\sqrt{E_{co}}$

$$\frac{x^2}{E_{co}a^2} + \frac{y^2}{E_{co}b^2} + \frac{z^2}{E_{co}c^2} = 1\,, \tag{7.1}$$

where $E_{co} > 1$ yields an expansion and $E_{co} < 1$ yields a shrinkage. The case of a non-rotated ellipsoid centered in the origin was used here.
To derive the contact function of two ellipsoids and determine E_{co} for the general case, it is useful to describe the ellipsoid by the quadratic form

$$E\left(\boldsymbol{x} - \boldsymbol{x}_p,\, \boldsymbol{\phi}_p,\, a,\, b,\, c\right) = \left(\boldsymbol{x} - \boldsymbol{x}_p\right)^T \boldsymbol{E}\left(\boldsymbol{x} - \boldsymbol{x}_p\right) \tag{7.2}$$

where the matrix $\boldsymbol{E}$ carries information on the orientation of the ellipsoid and its extents,

$$\boldsymbol{E}\left(\boldsymbol{\phi}_p,\, a,\, b,\, c\right) = \boldsymbol{A}(\boldsymbol{\phi}_p) \begin{pmatrix} a^{-2} & 0 & 0 \\ 0 & b^{-2} & 0 \\ 0 & 0 & c^{-2} \end{pmatrix} \boldsymbol{A}^T(\boldsymbol{\phi}_p)\,, \tag{7.3}$$

and $\boldsymbol{A}\left(\boldsymbol{\phi}_p\right)$ is the rotation matrix introduced above. The description is not limited to the surface of the ellipsoid, but also determines whether a point lies inside or outside the ellipsoid.

$$E\left(\boldsymbol{x} - \boldsymbol{x}_p,\, \boldsymbol{\phi}_p,\, a,\, b,\, c\right) \begin{cases} < 1 & \text{for } \boldsymbol{x} \text{ inside the ellipsoid} \\ = 1 & \text{for } \boldsymbol{x} \text{ on the surface of the ellipsoid} \\ > 1 & \text{for } \boldsymbol{x} \text{ outside the ellipsoid} \end{cases} \tag{7.4}$$

Two ellipsoids, A and B, are described by E_A and E_B, respectively. An affine combination of both is defined according to [206, 207] using the parameter $\lambda \in [0,\, 1]$. It reads as

$$S(\mathbf{x},\, \lambda) = \lambda E_A + (1-\lambda) E_B\,. \tag{7.5}$$

The minimum of $S(\mathbf{x},\, \lambda)$ for each value of λ describes a curve

$$\mathbf{x}(\lambda) = \left[\lambda \boldsymbol{E}_A + (1-\lambda)\boldsymbol{E}_B\right]^{-1} \cdot \left[\lambda \boldsymbol{E}_A \cdot \boldsymbol{x}_{p,A} + (1-\lambda)\boldsymbol{E}_B \cdot \boldsymbol{x}_{p,B}\right]^{-1} \tag{7.6}$$

which can be geometrically interpreted as a connection of the centers of the ellipsoids with the gradient vectors ∇E_A and ∇E_B being parallel along the curve. The virtual contact point

$\mathbf{x}_{co}$ is found along this curve where $S(\mathbf{x}(\lambda), \lambda)$ reaches its unique maximum. It is shown in [206] that the condition of a vanishing derivative of $S(\mathbf{x}(\lambda), \lambda)$ with respect to λ (denoted by S') in the virtual contact point yields

$$S'(\mathbf{x}_{co}, \lambda_{co}) = E_A(\mathbf{x}_{co}) - E_B(\mathbf{x}_{co}) = 0\,, \tag{7.7}$$

which can be rewritten as

$$E_A(\mathbf{x}_{co}) = E_B(\mathbf{x}_{co}) = E_{co}\,. \tag{7.8}$$

We therefore find the virtual contact point $\mathbf{x}_{co}$ and thus E_{co} by solving (7.7), which leads to a numerical determination of the root. It can be geometrically interpreted as approaching the surface $E_A(\mathbf{x}) = E_B(\mathbf{x})$ along the curve $\mathbf{x}(\lambda)$ given by (7.6). The root search is performed numerically, taking only a few iterations, using e.g. the bisectional algorithm, the secant method or the Newton's method, as $S'(\lambda)$ decreases monotonously.
The sub-contact points on the respective ellipsoids are then determined by

$$\mathbf{x}_{sub,A} = \mathbf{x}_{p,A} + E_{co}^{-1/2}\left(\mathbf{x}_{co} - \mathbf{x}_{p,A}\right), \quad \mathbf{x}_{sub,B} = \mathbf{x}_{p,B} + E_{co}^{-1/2}\left(\mathbf{x}_{co} - \mathbf{x}_{p,B}\right) \tag{7.9}$$

A good approximation of the closest distance between the ellipsoids is then achieved using the distance between two parallel planes being tangent to the respective ellipsoids in the sub-contact points. The normal vector describing these planes is calculated from

$$\mathbf{n}_{sub,AB} = \mathbf{n}_{A,B} = \frac{\nabla E_A\left(\mathbf{x}_{sub,A}\right)}{\|\nabla E_A\left(\mathbf{x}_{sub,A}\right)\|} = -\frac{\nabla E_B\left(\mathbf{x}_{sub,B}\right)}{\|\nabla E_B\left(\mathbf{x}_{sub,B}\right)\|}\,, \tag{7.10}$$

and the distance between the planes is finally computed by

$$\zeta_{A,B} \approx \left|\mathbf{n}_{sub,AB} \cdot \left(\mathbf{x}_{sub,A} - \mathbf{x}_{sub,B}\right)\right| \tag{7.11}$$

The computed distance slightly underpredicts the actual minimum distance between the two ellipsoids. A small overprediction would be achieved using the distance between the sub-contact points. The incorporated way of calculating the distance between the surfaces of two ellipsoids is very efficient and therefore also feasible when dealing with very large numbers of ellipsoids. The computed normal vector can then also be utilized in the computation of the normal collision force and the respective collision moment.
The particle-wall distances are determined by first solving for the sub-contact point from

$$\mathbf{n}_w = -\mathbf{n}_{sub,A} = \frac{\nabla E_A\left(\mathbf{x}_{sub,A}\right)}{\|\nabla E_A\left(\mathbf{x}_{sub,A}\right)\|}\,, \tag{7.12}$$

and the distances of the surface of particle A to the respective walls are then computed from Table 7.1.

Table 7.1 Particle-wall distance $\zeta_{A,w}$ employing compass notation.

West	$\zeta_{A,wW} = x_{sub,A}$	East	$\zeta_{A,wE} = L_x - x_{sub,A}$
South	$\zeta_{A,wS} = y_{sub,A}$	North	$\zeta_{A,wN} = L_y - y_{sub,A}$
Front	$\zeta_{A,wF} = z_{sub,A}$	Back	$\zeta_{A,wB} = L_z - z_{sub,A}$

7.2.3 Bubbles described by spherical harmonics

The inter-particle distance is defined by

$$\zeta_{A,B} = \min\left(\|\mathbf{x}_{S,A} - \mathbf{x}_{S,B}\|\right) , \tag{7.13}$$

with $\mathbf{x}_{S,A}$ and $\mathbf{x}_{S,B}$ being points on the surface of bubble A and B, respectively.
For *general and complex* bubble shapes (e.g. represented by SH), several approximations for the inter-particle distance were considered differing in accuracy and numerical costs.

A reasonably accurate approximation for SH bubbles that deviate only marginally from an ellipsoidal shape is using an *approximating ellipsoid* and the algorithm described above. This approach is in general the one with the least numerical costs. Since a large region in the bubble shape diagram [37] is covered by ellipsoidal bubbles, the approximation is quite useful. The only necessary step is the determination of the semi-axes of the approximating ellipsoid which then enter in (7.2). For axisymmetric spherical harmonics, an approximating, oblate ellipsoid is used with $a = c = \max(r(\theta))$ and $b = 0.5(\max(r(\theta)\cos(\theta)) - \min(r(\theta)\cos(\theta)))$.

Iterative procedures can be used to solve the minimization problem (7.13) and to determine the inter-particle distance. These procedures may employ the analytical description of the particle surfaces or its discrete representation. For SH bubbles, two surface grids are available - the unstructured triangular, basically equidistant Lagrangian forcing point grid and the structured grid of the spherical coordinates used for the spherical harmonic representation which is clustered towards the poles. The simplest approach is just *iterating over all points* of both surface grids until the minimum distance is found. Using the Lagrangian forcing point grid the inter-particle distance is computed by

$$\zeta_{A,B} = \min\left(\|\mathbf{x}_{fp,A} - \mathbf{x}_{fp,B}\|\right) . \tag{7.14}$$

The particle-wall distance, here written down only for the x-direction, follows from

$$\zeta_{A,wW} = \min\left(\mathbf{x}_{fp,A}\right) \qquad \zeta_{A,wE} = L_x - \max\left(\mathbf{x}_{fp,A}\right) . \tag{7.15}$$

This approach, however, can become quite costly for many particles, N_p, and large numbers of forcing points, N_L (or points of the SH grid). The computational costs scale with $0.5\,N_L^2\,N_p!/(N_p-2)!$. Consequently, it is recommended to narrow down the number of possible collision partners beforehand and then calculate the surface distance. This can be achieved using bounding spheres, with e.g. $r_A = \max(r(\theta,\,\phi))$, and the set of equations introduced for spherical bubbles above.

A more efficient approach uses an iterative procedure with a *multi-grid algorithm* on the structured SH surface grids of the bubbles, i.e. we search for $\zeta_{A,B} = \min\left(\|\mathbf{x}_{SH,A} - \mathbf{x}_{SH,B}\|\right)$. For clarity, we only discuss a two-stage multigrid algorithm here. An initial solution ζ_{coarse} is found using only, e.g., every tenth point in both directions of the spherical $(\theta,\,\phi)$ grid. The solution is then improved in the vicinity of the initial solution using every point of the SH grid. The principle of the algorithm is sketched in Figure 7.4.

A *directional iteration* using the tangential vectors of the SH shapes represents another way of calculating the inter-particle distance efficiently and accurately. For convex shapes, the

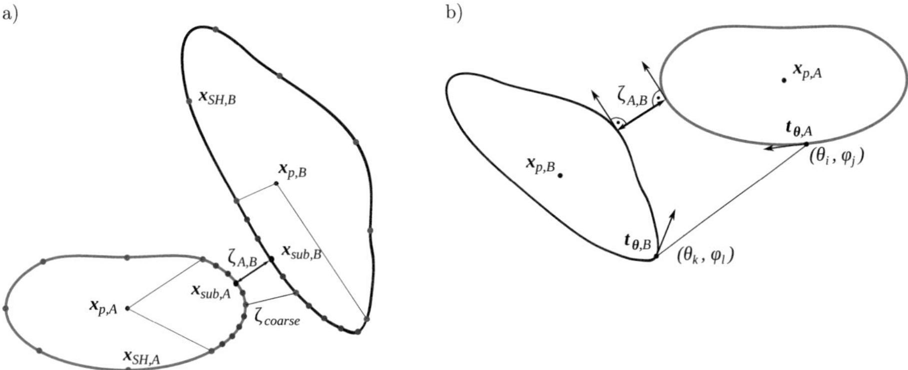

Figure 7.4 Schematic representation of distance determination between particle surfaces for complex bubble shapes. a) Multigrid approach. b) Directional iteration using tangential vectors.

distance vector connecting the two sub-contact points, as well as the two normal vectors in these points are collinear. Consequently, the scalar product of the distance vector $\boldsymbol{\zeta}_{A,B} = \mathbf{x}_{sub,A} - \mathbf{x}_{sub,B}$ with either one of the four tangential vectors $\boldsymbol{t}_\theta$ in the sub-contact point is zero. An iterative algorithm was developed in the diploma thesis by Tschisgale [280] which relates the value of this scalar product to a change in the angular coordinates. If the scalar product becomes zero also the change in the angular coordinates vanishes and the sub-contact points are found. The application of the algorithm to the continuous (in contrast to discrete) surface representation of the SH bubbles A and B leads to a very accurate determination of the distance between the bubble surfaces. However, it requires the calculation of the spherical harmonic basis functions, derivatives etc. at given values for (θ, ϕ), which becomes quite costly. It was therefore recommended using the discrete SH surface grids of the two bubbles instead, where the required quantities are already calculated. We group the indices of the SH grids of both bubbles as $\boldsymbol{I}_{SH} = [i, j, k, l]$ with the corresponding angular coordinates $\boldsymbol{\theta} = [\theta_i, \phi_j, \theta_k, \phi_l]^T$. Hence, the iterative algorithm towards the sub-contact points reads in the index space

$$\boldsymbol{I}_{SH}^{m+1} = \boldsymbol{I}_{SH}^{m} + \operatorname{sgn}\left(\boldsymbol{\zeta}^m \cdot \boldsymbol{t}_{\boldsymbol{\theta}}^m\right) , \tag{7.16}$$

$$\text{with} \qquad \operatorname{sgn}(x) = \begin{cases} +1 & x > 0 \\ 0 & x = 0 \\ -1 & x < 0 , \end{cases}$$

where m denotes the iteration step. The sketch illustrating the procedure is provided in Figure 7.4. The initial values at $m = 0$ are chosen from the intersection of the straight line connecting the bubble centers $\mathbf{x}_{p,A}$ and $\mathbf{x}_{p,B}$ and the respective surfaces. The algorithm (7.16) can be interpreted geometrically as successive diagonal jumps on the structured grid of each of the bubbles marching towards to respective sub-contact points. During this path the distance between the bubbles, as well as the scalar product between the distance vector and the tangential vectors are minimized. The algorithm is stopped when a fixpoint-iteration is detected and the final indices $\boldsymbol{I}_{SH}^{final}$ yield the sub-contact points $\mathbf{x}_{sub,A}$ and $\mathbf{x}_{sub,B}$.

Potential overlap, i.e. negative inter-particle or particle-wall distances, is directly accounted for when using spheres, ellipsoid contact function and algorithm based on tangential vectors for SH bubbles. Additional calculations are necessary when using the iterative procedures searching for the minimum of the absolute value of the distance, e.g. first one has to distinguish between a left and a right particle A and B (bottom and top, etc.), based on the particle centers, and then a search for the minimum of the signed distance between the surfaces is conducted.

7.3 Coalescence criteria

When two bubbles approach they either bounce or merge forming one larger bubble. Whether coalescence occurs is determined by the characteristics of the bubble interaction. Based on film thinning theory, the film thickness between two bubbles needs to become sufficiently small for van der Waals forces to become important and lead to coalescence [118]. A criterion can also be expressed relating the film drainage time and the interaction time between two bubbles [161]. If the interaction time is long enough, the film drains out and becomes thin enough to rupture. In other words, gentle bubble interactions yield a higher probability for coalescence to occur. Hence, the approach velocity must be below a critical value for coalescence to occur. If it is above that threshold, the bubbles bounce back. Figure 7.5 shows experimental data of Ribeiro and Mewes [221] on the occurrence of collision and coalescence events as a function of bubble size and the relative approach velocity. Air bubbles in water and ethanol were studied. Table 7.2 lists the physical properties of these systems. The proposed critical values for the approach velocity are also given in the plots, being 9.6 cm/s for air bubbles in water at 20°C and 4.8 cm/s for air bubbles in ethanol at 30°C. The relative approach velocity, u_{rel}, of the two bubbles is determined with respect to the bubble centers in the experiments and thus also in the simulation. Note that the actual relative velocity of the interfaces at the contact point might differ.

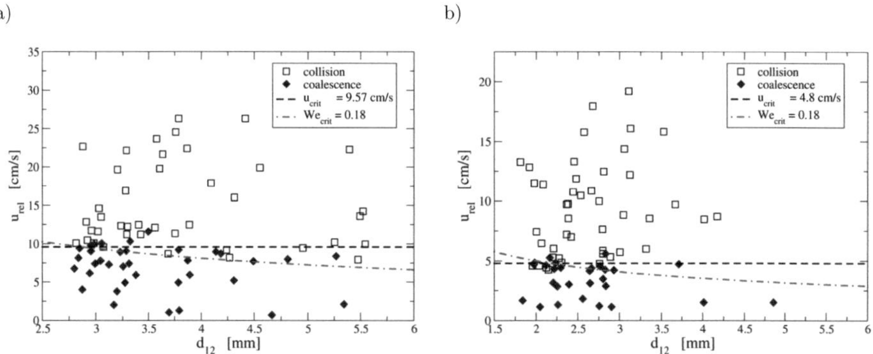

Figure 7.5 Occurrence of collision and coalescence for bubble pairs, data digitized from [221]. Critical relative velocities [221] compared to critical relative Weber number $We_{crit} = 0.18$ [54]. a) Air bubbles in water at 20°C. b) Air bubbles in ethanol at 30°C.

Duineveld [54] proposed another criterion for coalescence which is based on the relative approach Weber number, $We_{rel} = \rho_f u_{rel}^2 r_{12}/\sigma$ where r_{12} denotes the characteristic radius of the two interacting bubbles with $r_{12}^{-1} = \left(r_{1,eq}^{-1} + r_{2,eq}^{-1}\right)/2$ and d_{12} the respective diameter. Duineveld suggested a critical value of $We_{crit} = 0.18$ for air bubbles in water. This criterion is also traced in both graphs. The suggested value of We_{crit} is a good threshold for both gas-liquid systems. It also reproduces the slightly negative slope in the critical velocity observed in the data. Note that a slightly larger value of We_{crit} would even better fit the experiments of Ribeiro and Mewes [221]. There is, however, a substantial scatter in experimental data across the literature which can be related, e.g., to the purity of the considered gas liquid systems which limits the accuracy of a threshold in any case. For instance, Lehr et al. [156] report a critical relative velocity of 8 cm/s for air bubbles in water at the same temperature which is 17% lower than the value of [221].

Table 7.2 Physical properties of gas-liquid systems. Fluid density ρ_f, fluid kinematic viscosity ν and surface tension σ [40, 265, 319].

Gas-liquid system	ρ_f [kg/m^3]	ν [m^2/s]	σ [N/m]
Air - ethanol, 30°C	781.4	$1.255 \cdot 10^{-6}$	0.0217
Air - water, 20°C	995.5	$1.038 \cdot 10^{-6}$	0.0738
H_2 - 1-M H_2SO_4, 20°C	1066.1	$1.100 \cdot 10^{-6}$	0.0730
Ar - GaInSn, 20°C	6361.0	$3.459 \cdot 10^{-7}$	0.533

In summary, coalescence criteria can be formulated based on:

- Critical distance or film thickness between bubbles
- Drainage and interaction time
- Critical approach velocity
- Critical approach Weber number

Due to the extremely small length scales of the liquid film between the bubbles, the film cannot be resolved in the simulations presented below. A criterion based on a critical film thickness would therefore be rather inappropriate in the present simulations. Zenit and Legendre [318] show that the film drainage of deformable bubbles is dominated by inertial effects for millimeter sized air bubbles in water. Viscous effects are of secondary importance. The Weber number, as the ratio of inertial forces to surface tension forces, hence is a reasonable and general coalescence criterion.

The following coalescence criterion was implemented into the PRIME code. Based on the discussion of the criteria above, a two-step criterion is proposed here. As a primary criterion, the distance of the interfaces of the two bubbles needs to drop below a critical value which is selected to be twice the grid spacing here. This critical distance also corresponds to the range of the collision model employed and discussed below. The critical distance can also be specified in terms of the bubble diameter, e.g. as $0.1\,d_{12}$, to meet experimental considerations. The first step requires the calculation of the distance between bubbles of complex, three-dimensional shapes. The determination of the inter-particle distance is described in Section 7.2. For bubbles below the prescribed distance, the secondary criterion, the relative approach Weber number, is activated. If $We_{rel} < We_{crit} = 0.18$ coalescence is initiated, otherwise the bubbles are set to bounce, employing a collision model.

7.3.1 Relative Weber number criterion

The employed coalescence criterion is based on experimental observations. Here, derivations of the relative Weber number criterion based on lubrication theory are briefly re-called from the literature. A thorough prediction, whether the bubbles will bounce or get sufficiently close to undergo coalescence, can be obtained by solving the Navier-Stokes equations for the flow in the thin film between the bubbles and the evolution of this film. Separate treatment of substantially diverse length and time scales leads to the challenge of multi scale modeling, such as thin film models discussed in [276, 277, 267]. A review on film drainage and coalescence is found in [31] with specific discussion of the boundary conditions and assumptions for the thin film. In [54] a bouncing criterion is developed for bubbles of equal size. If the kinetic energy is transferred to surface energy before the film is drained, the bubbles will bounce back. Significant assumptions are posed including spherical bubbles, small deformations, a small rise Weber number, inviscid flow and a parabolic profile of the initial film thickness. For two equi-sized, trailing bubbles, i.e. the bubbles are aligned along the direction of gravity, a critical relative Weber number of $We_{crit} = 0.5$ was derived, as also found in [34]. If the bubbles rise side-by-side, the critical relative Weber number is $We_{crit} = 0.21$.
The interaction of a bubble with a free surface, corresponding to an infinite radius of one 'bubble' for the vertical interaction, was addressed in [34] and $We_{crit} = 0.125$ was derived under the same assumptions. In this study, the influence of viscosity is discussed, and very similar results are obtained if $Re_{rel} = u_{rel}r_{12}/\nu > 100$. Otherwise, the thinning rate of the film decreases leading to a somewhat smaller critical Weber number. It is stated in [127] that viscous forces are negligible until the very last stages of the film drainage. The bouncing of a bubble at a free surface was studied as above, but with a finite rise Weber number correction, and $We_{crit} = 0.24$ was obtained.
The critical Weber number range obtained from theoretical considerations agrees quite well with the chosen experiment-based value of $We_{crit} = 0.18$.

7.4 Collision model

This section is dedicated to outlining the collision models used throughout this work. Accurate representation of collisions in context of phase-resolving simulations of particle laden flows is very delicate even for spherical particles in viscous fluids as illustrated in [134, 135]. It becomes even more challenging when considering other particle shapes. It shall be stated clearly here, that the issue of collision modeling and its improvement is not the scope of this work. The intention here is to provide a very brief overview and a summary of existing models, which were applied during the course of the thesis with only minor modifications.
It is tempting to use similar modeling strategies for bubbles as for rigid particles, although the underlying physical mechanisms might differ. Indeed, there is experimental evidence that the description of the rebound process of a particle from a solid wall can be brought into a common form [154], for bubbles, drops and rigid spheres. This can be realized by relating the ratio of rebound to impact velocity to an appropriately defined Stokes number. The Stokes number is defined as the ratio of the hydrodynamic response time of the particle to a characteristic time scale of the flow as indicated below (7.20).
For bubbly flows, the inertia of the liquid phase clearly dominates over that of the dispersed phase, in contrast to immersed, heavy particles. Thus, the dynamics of two approaching

bubbles and their evasive motion are to large extent captured by the direct resolution of the hydrodynamic forces on the bubbles. Furthermore, bubbles undergo deformation and surface oscillations during collision which is directly accounted for by the SH shape coupling algorithm [246]. However, a short range collision model becomes necessary if the distance falls below critical threshold, e.g. twice the grid spacing. Then additional repulsive forces and the corresponding moments are introduced. The modeling accounts for inter-particle, as well as particle-wall collisions. Apart from being unphysical, overlapping of particles yields divergence issues and thus stability problems. Hence, one of the major goals of the collision modeling is to guarantee numerical robustness.

The main idea, discussed in [134, 135], is to decompose a general oblique collision into a normal and a tangential collision and to then calculate the respective collision forces and collision moments,

$$\mathbf{F}_{col} = \mathbf{F}_{col,n} + \mathbf{F}_{col,t}, \qquad \mathbf{M}_{col} = (\mathbf{x}_{sub} - \mathbf{x}_p) \times \mathbf{F}_{col}, \tag{7.17}$$

with $\mathbf{x}_{sub}$ being the subcontact point on the particle surface determined as described in Section 7.2. Tangential and normal collisions are then modeled according to experimental observations and findings from theoretical analysis. Furthermore, the forces in each direction can be decomposed into contributions originating from different physical mechanisms, as the elastic or plastic 'dry' collision of rigid particles, mechanisms related to lubrication and the liquid film between the particles, rolling or sliding of the particles, etc. The resulting collision force and moment acting on a particle, possibly stemming from collisions with multiple partners, are then introduced as respective source terms into the momentum equations (1.7) and (1.8) which describe the translational and rotational motion of the particle.

7.4.1 Normal collision

First, the modeling of the normal collision is addressed. The repulsive forces are determined from a mass-spring-damper system.
A general formulation of surface normal collision force $\mathbf{F}_{col,n}$ for a collision between particles A and B is given by

$$\mathbf{F}_{col,n}^{(A,B)} = \left(k_{col,n}\, \zeta_n^E - d_{col,n}\, u_{n,rel}\right)_{A,B} \mathbf{n}_{A,B}\,, \tag{7.18}$$

with $k_{col,n}$ being the normal stiffness, ζ_n the normal distance measure of the two colliding surfaces, and E a constant exponent. The normal damping coefficient is denoted by $d_{col,n}$ with damping being proportional to the normal relative velocity $u_{n,rel}$ of the two surfaces. The direction of the normal collision force is given by $\mathbf{n}_{A,B}$ being the normal vector in the subcontact points on the two particle surfaces. Additional collision forces are only introduced within the range of the collision model ζ_{col}, e.g. $\zeta_{col} = 2\Delta x$, which yields for the normal distance measure $\zeta_n = \zeta_{A,B} - \zeta_{col}$.
Depending on the choice of the model constants of (7.18), different models can be identified which are well documented in the references provided in Table 7.3.

Table 7.3 Overview of collision models employed.

Model	E	$d_{col,n}$	$k_{col,n}$	Applications
CM-1	2.0	0	*a priori*	Quadratic repelling potential of Glowinski [89, 90] with application to heavy, rigid ellipsoids in [203, 204, 202], problem specific, a priori choice of $k_{col,n}$.
CM-2	1.0	0	$2\pi\sigma$	Spring stiffness $k_{col,n} = 2\pi\sigma$ derived from surface tension force of spherical bubbles. Application to bubble clusters and foams [101, 102, 104].
CM-3	1.5	adaptive	adaptive	Adaptive Collision Model (ACM) for heavy, rigid spheres [135, 134], with $d_{col,n}$ and $k_{col,n}$ being determined adaptively for each collision to yield physically sound rebound.

In CM-1 and CM-2, usually a collision range of $\zeta_{col} = 2\Delta x$ is employed, whereas in CM-3 $\zeta_{col} = 0$ is used, i.e. the model is active in case of surface contact and inter-penetration.
In turbulent, particle laden flows confined by walls, collisions may comprise rapid impacts, gentle sliding or entrapment of one particle between multiple collision partners. It is therefore difficult to cover all these scenarios with *a priori* determined model constants. Given the spring stiffness is too low, the particles will overlap creating divergence issues due to the incompressibility constraint. On the other hand, a too high spring stiffness leads to a very short collision with large accelerations and possibly unphysically high rebound velocities.
For that reason, the main concept of the ACM (CM-3) is to determine the stiffness, $k_{col,n}$, and the damping, $d_{col,n}$, at the beginning of each collision by solving the differential equation

$$(\rho_p + C_{AM}\,\rho_f)\,V_p\frac{\mathrm{d}^2\zeta_n}{\mathrm{d}t^2} + d_{col,n}\frac{\mathrm{d}\zeta_n}{\mathrm{d}t} + k_{col,n}\zeta_n^E = 0 \tag{7.19}$$

to yield a given relative rebound velocity after a given collision time t_{col} when ζ_n is again zero. In the original ACM [135] for spherical, solid, heavy particles, the exponent $E = 1.5$ is used based on the Hertz contact theory [108] and no added mass, $C_{AM} = 0$, is considered as the model describes the 'dry' part of the collision. With the ACM, an additional lubrication model accounts for fluid forces on the particle during approach and rebound.
It appears appealing to also apply an adaptive collision model for bubbles. The remainder of this section is used to adapt the ACM for bubbles. To obtain $d_{col,n}$ and $k_{col,n}$ from the mass-spring-damper system (7.19), the approach and rebound velocity, the collision time, as well as the associated mass need to be determined. In the case of bubbles, an exponent of $E = 1.0$ is applied as motivated in the literature [102, 318, 154], yielding a linear mass-spring-damper system.
To globally describe the outcome of particle collisions, the coefficient of restitution e_n is introduced,

$$e_n = -\frac{u_{n,rel}^{out}}{u_{n,rel}^{in}}, \tag{7.20}$$

as the negative ratio of the normal, relative velocity of the rebound, $u_{n,rel}^{out}$, to the value of approach, $u_{n,rel}^{in}$. The exact definition of these latter two velocities is somewhat difficult. In model experiments, e.g. the normal collision of a bubble with a horizontal wall [318], $u_{n,rel}^{in}$ is chosen to be the terminal rise velocity of the bubble, i.e. the undisturbed velocity far away from the wall. The rebound velocity is defined as the minimum relative velocity of the bubble center after the impact, which in this case coincides with the velocity when the bubble loses 'wall contact'. Defined in this way, in general, the restitution coefficient quantifies the loss of kinetic energy during the collision due to viscous dissipation, material damping, shape oscillations, etc. The coefficient of restitution is well documented and tabulated in the literature for bubbles [318, 279, 278], drops [154, 224] and rigid spheres [92, 126, 311], and is therefore a suitable input parameter for the ACM. A fit of experimental data for e_n from bubble-wall collisions is provided in [318] with

$$e_n = \exp\left(-30\sqrt{\frac{Ca}{St^*}}\right) \tag{7.21}$$

where the dimensionless group $\sqrt{Ca/St^*} = Oh^*$ is a modified Ohnesorg number determined from the capillary number, $Ca = u_{n,in}\mu_f/\sigma$, and the modified Stokes number, $St^* = (\rho_p + C_{AM}\,\rho_f)\, d_{eq} u_{n,in}/(9\mu)$. The modified Stokes number accounts for the added mass of the bubble, where the added mass coefficient for an ellipsoid of rotation as a function of the aspect ratio X can be determined from [150, 318]

$$C_{AM} = \frac{\left(X^2-1\right)^{1/2} - \arccos\left(X^{-1}\right)}{\arccos\left(X^{-1}\right) - \left(X^2-1\right)^{1/2} X^{-2}}\,. \tag{7.22}$$

The well known value of $C_{AM} = 0.5$ is obtained in the limit $X = 1$ for a sphere. Thus, this equation for C_{AM} is derived for unbounded fluid, i.e. not taking into account the vicinity of a wall, which is in agreement with the definition of e_n. A rebound of bubbles is observed from equation (7.21) if inertial forces dominate over viscous forces. In contrast, the bubble sticks and possibly oscillates without notable bouncing for large dissipation.

The collision time, t_{col}, might be approximated for bubbles colliding with a wall by $t_{col} = \pi\sqrt{5/12}\sqrt{(\rho_p + C_{AM}\,\rho_f)\, r_{eq}^3/\sigma}$, which is derived in [154] from one half-period of the undamped, linear oscillator implied in (7.19). For inter-particle collisions, a suitable approximation is $t_{col} = \pi/\sqrt{12}\sqrt{\rho_f\, r_{eq}^3/\sigma}$, which corresponds to one half-period of the dominant mode of a bubble shape oscillation around the spherical shape [150] with assumptions outlined below equation (7.25).

Bubble-wall interactions in general have longer collision times than the corresponding particle-wall collisions of rigid spheres. Bubble deformation serves as an intermediate energy storage prolonging the approach, impact and rebound phase of the collision compared to rigid particles. Therefore, a stretching of the collision process to at least $t_{col} = 10\Delta t$, as applied in the original ACM [135], is usually not necessary. However, a respective lower limiter for t_{col} is used in production runs.

There are some remaining issues within the application of the ACM for bubbles. In production runs with many collisions in turbulent flow, the detection of the beginning of individual collisions, as it enters the respective definition of the restitution coefficient (7.20) and of the collision time, is challenging. As a compromise, the start of the collision is indicated when $\zeta_n = 0$, and a somewhat larger restitution coefficient is used, e.g. $e_n = 0.8$. Even for a single bubble-wall collision in quiescent fluid, the bubble shape and thus the added mass differ substantially during the course of the collision which limits the applicability of the mass-spring-damper system (7.19). In contrast to the original ACM, a force or safety range, $\zeta_{col} > 0$, has to be used to enable the pressure interpolation for the SH shape algorithm. Also employing the inter-particle or particle-wall distance (including the force range) as a measure of deformation, ζ_n, has to be scrutinized for large shape oscillations. Nevertheless, the modified ACM for bubbles provides good approximations for $d_{col,n}$ and $k_{col,n}$ and yields robust and physically sound bubble collisions as will be shown later.

7.4.2 Tangential collision

The tangential collision modeling is not discussed in detail here, as it has basically been neglected in the simulations presented in this thesis. A thorough review is found for spherical, rigid particles in [135, 134]. The nature of tangential contact between the collision partners determines whether rolling or sliding motion occurs. For pure rolling, the relative tangential velocity in the contact point is zero. In contrast, a sliding motion is characterized by tangential slip between the surfaces. A suitable modeling strategy, comprises setting the tangential collision force $\mathbf{F}_{col,t}$ proportional to the relative tangential velocity in the contact point [95]. For bubbles, rolling motion is not to be expected even in contaminated systems. Experimental data for bubble collisions with a tangential component are available in Podvin et al. [211] who examined the interaction of a bubble with an inclined wall and compared the results to model predictions derived from [180, 138]. The agreement between model and experiment is moderate. In the experiments, states of pure sliding or pure, repeated bouncing and intermediate states with transient bouncing and sliding were found depending on the inclination angle of the wall, as well as further parameters characterizing the bubble itself and the approach. Similar experiments can also be found in [279]. In both experiments, a critical angle of the wall measured from the horizontal of $55° - 60°$ is found for the transition from sliding to bouncing. A drop sliding along an inclined wall is examined employing a thin film model for phase-resolving simulations in [267, 276]. Since only collisions with vertical walls take place in the simulations of bubbly flows in wall-bounded domains presented below, the process of bouncing needs to be considered in more depth which is done in paragraph 7.4.3. The rebound process is dominated by the wall normal forces described above. In the present work, only a very weak tangential force is applied for bubbles, e.g. to damp numerical oscillations and sliding along a fully horizontal wall in the test case 7.4.3. Further studies should be conducted in the future focusing on the tangential interaction between multiple bubbles and bubbles with a solid wall. Tangential contact and collision modeling can be expected to be equally important to its normal counterpart in bubble foams and dense clusters.

7.4.3 Deformable bubble impacting against horizontal wall

Configuration

The dynamics of a bubble impinging on a horizontal wall are studied with focus on the performance of the bubble-wall collision modeling and the SH shape algorithm for strongly deformed bubbles. A principle sketch of the configuration is provided in the left part of Figure 7.6. The simulation captures the approach, impact and rebound of the bubble and the respective fluid motion. The simulation results are compared to experimental data by Zenit et al. [318] using the experimental run with the largest bubble deformation and the highest rebound velocity presented in that reference. The system considered is supposed to be fully wetting, i.e. at all times there remains a thin liquid film between the bubble and the wall.

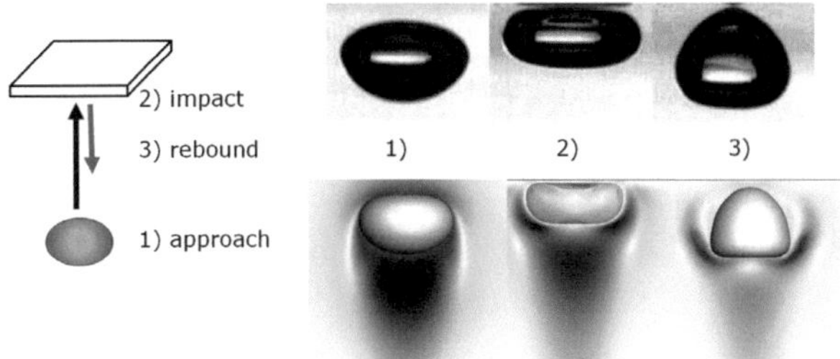

Figure 7.6 Collision of a bubble with a horizontal wall, sketch (left). Bubble shape during wall collision (right). Top row shows experimental data [318], bottom row shows present simulation data: 1) Terminal or equilibrium shape of bubble at $t\,v_{p,0}/d_{eq} \lesssim -1$ (beginning of approach, still far from wall). 2) Maximum deformation and zero velocity at $t\,v_{p,0}/d_{eq} \approx 0$ (impact). 3) Largest absolute rebound velocity and loss of 'contact' with wall at $t\,v_{p,0}/d_{eq} \approx 0.9$ (rebound). Contour of absolute value of fluid velocity illustrated in gray scale $(0,\ 1.5\,u_g)$ in a plane intersecting the bubble center.

Numerical parameters

The non-dimensional numbers characterizing the rise and shape of an individual bubble are chosen as $Eo = 4.2$, $G = 250$, $\pi_\rho = 1 \cdot 10^{-3}$ to match the shape and the velocity of approach $v_{p,0}$ from the experiment. An approach Reynolds number of $Re_{p,0} = v_{p,0} d_{eq}/\nu = 241$ is achieved and the collision process is characterized by the modified Stokes number of $St^* = 22.1$. Results from the simulation of a freely rising bubble in a large container (Section 4.4) are re-used here to initialize the simulation and enable a smaller size of the computational domain with $\mathbf{L} = (6.4, 5.0, 5.0)\,d_{eq}$. A local spatial resolution of $d_{eq}/\Delta x = 40$ is used and the time step is chosen to yield $CFL = 0.1$ based on $v_{p,0}$. A no-slip condition is applied at the upper wall and on the bubble surface. Periodic boundary conditions are prescribed in the lateral directions and the boundary condition at the lower wall is a free-slip one. The experimental data by [318] was obtained for a partially contaminated system as apparent from the drag coefficients provided in that reference, i.e. the boundary condition at the bubble surface is neither a pure no-slip, nor a pure free-slip condition. The bubble shape is represented by axisymmetric spherical harmonics with $N_{SH} = 12$ and constraints of constant bubble volume and no wall penetration of the bubble surface. A straight bubble trajectory is imposed during the approach. The ACM for bubbles is used as described above with $E = 1.0$ and $k_{col,n}$, $d_{col,n}$ are determined adaptively. The range of the collision model is chosen as $\zeta_{col} = 4\Delta x$. We stress again that the terminal rise velocity $v_{p,0}$, undisturbed by the wall, is used as an input of the model and in the definition of the restitution coefficient. Further, also the added mass coefficient is determined from the bubble shape still not influenced by the presence of the wall. The ACM for bubbles provides good estimates for $k_{col,n}$, $d_{col,n}$ if applied as described above and it improves the agreement with the experimental data for the chosen case substantially compared to the application of CM-2 (not shown). However, to match the experiment with better accuracy, as displayed in Figure 7.6 and 7.7, a parameter study was conducted yielding $k_{col,n} = 480$, $d_{col,n} = 0.9$. The results comprise an approximately four times stiffer spring constant compared to the ACM and 13 times stiffer compared to CM-2, respectively. The damping constant is about half of the value obtained from the ACM.

Bubble shape during wall collision

Figure 7.6 displays the bubble shape during the wall collision with the snapshots chosen to represent the approach, impact and rebound. The top row shows the experimental data of [318] and the lower row the present simulation data. A contour of the absolute value of the fluid velocity illustrates the flow field in a plane intersecting the bubble center. At the beginning of the approach phase (1), the bubble is still unaffected by the wall and its equilibrium shape and rise velocity correspond to those in unbounded fluid. In the present case, the bubble shape is basically steady ellipsoidal with a slightly flat front. As the bubble further approaches the wall, the pressure at the bubble front increases leading to a deceleration and a flatter bubble shape. The bubble front even becomes dimpled [34]. Fluid is squeezed out to the sides as the thickness of the film between bubble and wall further decreases [318]. The maximum deformation of the bubble is reached at wall impact (2) when the velocity of the bubble center comes to zero. Here, we observe also the formation of a dimple at the rear of the bubble due to the impinging wake fluid. The rebound (3) is characterized by a cap-shaped bubble of small aspect ratio. Here, a minimum in the bubble center velocity is reached, approximately as it loses the 'contact' with the wall. More wake fluid moves in counter-flow to the bubble and needs to be driven aside. The agreement between numerical and experimental bubble shapes is excellent with the simulation giving slightly larger bubble deformations.

Bubble dynamics during wall collision

The history of the vertical velocity of the bubble center normalized with the terminal rise velocity is plotted in Figure 7.7. The origin of the temporal axis is set at $v_p = 0$. Roughly one full bubble-wall interaction is shown consisting of the deceleration phase during approach, impact at $v_p = 0$ and reverted motion with $v_p < 0$. It is shown in [318] that the normalized plots of the bubble center velocity, $v_p(t)/v_{p,0}$ collapse for different modified Stokes numbers until the impact at $t\, v_{p,0}/d_{eq} \approx 0$. The simulation results are in good agreement with the experimental data of [318], especially for the restitution coefficient $e_n = 0.47$. Two vertical dashed lines indicate the temporal beginning and end of the collision model, i.e. the time when the particle-wall distance is within the range of the collision model. The particle center velocity has already decreased to approximately 60% of the approach velocity $v_{p,0}$ due to resolved viscous forces as collision modeling sets in. The slope in the velocity-time plot then becomes steeper due to the additional collision force. The minimum in particle center velocity is reached approximately at the same time after impact in both experiment and simulation. Shortly after the minimum in $v_p(t)$ the location of the bubble surface leaves the wall and thus the range of the collision model. An additional oscillation in $v_p(t)$ can be observed as the center velocity again becomes zero in the simulation and in a weaker form in the experiment. This instant corresponds to the maximum rebound height and the oscillation is caused by a reversal in bubble shape from rear to again front flattened. The maximum absolute rebound height measured from the upper wall is $0.70\, d_{eq}$ in the simulation and therefore slightly smaller than the corresponding value from the experiment with $0.75\, d_{eq}$.

Further findings, that were not addressed or plotted above, are documented now. The collision model CM-2 (Table 7.3) yields lower rebound velocities and lower rebound heights. If a collision is underresolved by the spatial or by the temporal discretization, lower deformation, lower rebound velocity and shorter collisions were observed, given the simulation was

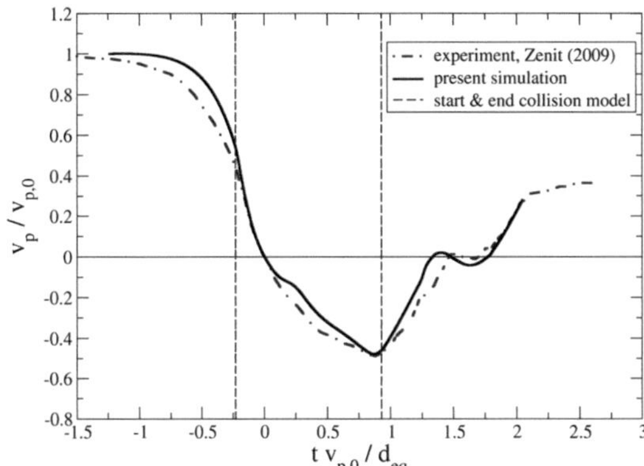

Figure 7.7 Bubble wall collision. History of the vertical velocity of the bubble center normalized with the terminal rise velocity. Origin of temporal axis set at $v_p = 0$.

still stable. In the student thesis of Hoffmann [109], simulations conducted without bubble deformation unveiled a general tendency towards an underprediction of collision time and rebound height. This tendency was more or less independent of the choice of the ansatz itself and the parameters in (7.18).
In conclusion of the present simulation, very good agreement with experimental data for bubble-wall collision can be achieved for properly chosen parameters of the collision model. The bubble shape and trajectory match the experimental observations. The very complex interplay of film drainage, bubble shape deformation, which serves as temporarily kinetic energy storage, and the mass-spring-damper collision model was captured successfully. Herein, the representation of the temporal evolution of bubble shape determines the collision dynamics to a large extent. There exists substantial room for improvements with the actual collision modeling of bubbles.

7.4.4 Towards improved bubble collision modeling

If high gas volume fractions and large collision rates characterize the configuration to be studied, the collision model can be estimated to have a significant influence on the results. Even though the idea of a unified collision model for rigid particles and deformable bubbles seems tempting, there are limitations to this approach. For solid particles, a very short contact of the two particle surfaces exists and the particle velocities prior and after this contact can be described by a 'dry' collision [135]. The shape of the rigid particle remains spherical at all times. The last stage of the particle approach and the early stage of the particle departure are captured by a lubrication model. For bubbles on the other hand, a very thin film remains in-between the two bubbles at all times if the bubbles are supposed to bounce back. The bubble shape deforms on the global scale and on the local scale near the point of closest inter-particle distance. A very important next step towards the improved collision modeling for bubbles is the inclusion of a suitable lubrication model for bubbles describing the physical processes in this thin film. The theoretical basis for this step and experimental data obtained by atomic force microscope measurements are gathered in [32].

The most challenging task is to distinguish between resolved and unresolved contributions during the collision. Other authors [318, 154] provide a model for the full collision process of a single bubble-wall collision in quiescent fluid on the global scale. The deformation, ζ_n^*, in their mass-spring-damper model can then be defined as the deviation from the spherical bubble shape. The definition of ζ_n in the present context is non-trivial and needs to be scrutinized. Within the IBM, the global deformation of the bubble is resolved to a large extend by the present method and the influence of the bubble deformation is thus inherently already included in the bubble's equation of motion. The collision model takes the role of a subgrid scale model and the collision forces and moments appear as additional source terms in the bubble momentum equations when the bubble surfaces get very close. A consistent formulation thus requires that the modeled collision force vanishes if the grid spacing of the simulation approaches zero, $\Delta x \rightarrow 0$, i.e. the collision is fully resolved. It is further required that $\mathbf{F}_{col,n} \rightarrow 0$ if $\zeta_n \rightarrow 0$, i.e. the closest inter-particle distance enters or leaves the specified range of the collision model.
The normal and tangential collision model have potential for improvement, also with regard to thin film modeling [267, 277], in order to get a better representation of inter-bubble and bubble-wall interaction at reasonable computational cost.

7.5 Coalescence model

If the coalescence criterion introduced in Section 7.3 is fulfilled, the bubbles do not bounce back, but undergo coalescence. A topological change from two individual bubbles to one larger bubble has to be realized in the phase-resolved simulation. This new object is described with a new local, spherical coordinate system $r(\theta, \phi)$ centered in the contact point of the two bubbles. The bubbles are then shifted marginally towards each other to initialize the coalescence process creating a minor overlap, thus removing any singularity at the contact point. The outer hull of the resulting large bubble is described by three-dimensional spherical harmonics [77] (Appendix J),

$$r(\theta, \phi, t) = \sum_{n=0}^{\infty} \sum_{m=-n}^{n} a_{nm}(t)\, Y_n^m(\theta, \phi) \approx \sum_{n=0}^{N_{SH}} \sum_{m=-n}^{n} a_{nm}(t)\, Y_n^m(\theta, \phi) \tag{7.23}$$

where $r(\theta, \phi, t)$ designates the distance of a point on the interface from the origin of the local coordinate system at a given time t, N_{SH} is the number of modes employed, a_{nm} are the shape coefficients and is the spherical harmonic function of grade n, m. When computing the resulting shape coefficients, a_{nm}, of the merged bubble, these are adjusted such that the volume of the large bubble equals the sum of the single bubble volumes.

7.5.1 Detailed simulation of bubble shape during coalescence

Configuration and numerical parameters

In the following, a detailed simulation of the temporal evolution of the bubble shape after film rupture is presented using the SH shape algorithm [246] which computes the bubble shape from the fluid loads. The setup comprises two 375 μm diameter hydrogen bubbles coalescing in the gap between two plate electrodes in quiescent, aqueous 1-M H_2So_4 and corresponds to the experimental configuration of Stover et al. [265, 264]. The gas-liquid system has similar physical properties as air in water, as demonstrated in Table 7.2. The Ohnesorg number of the present configuration is $Oh = \nu\sqrt{\rho_f \,/\, r_{12}\sigma} = 9.7 \cdot 10^{-3}$ indicating low influence of viscous forces compared to inertial and surface tension forces.

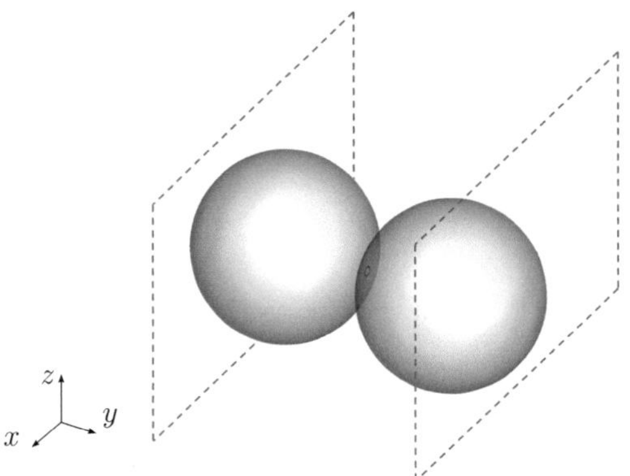

Figure 7.8 Coalescing bubbles between plate electrodes. Configuration and initial bubble shape. Only a fraction of the computational domain in x and z is shown.

In the simulation, the initial shape was defined by a so-called 'inverse ellipse' of rotation according to the definition in [112] reading

$$\begin{aligned} \gamma &= \arctan\left((1+m_{iE})/(1-m_{iE})\tan(\theta+\pi/2)\right) \\ x(\gamma) &= \frac{r_{iE}\,(1-m_{iE}^2)\,(1+m_{iE})\cos(\gamma)}{\sqrt{1+m_{iE}^2}\,\left(1+2m_{iE}\cos(2\gamma)+m_{iE}^2\right)} \\ y(\gamma) &= \frac{r_{iE}\,(1-m_{iE}^2)\,(1-m_{iE})\sin(\gamma)}{\sqrt{1+m_{iE}^2}\,\left(1+2m_{iE}\cos(2\gamma)+m_{iE}^2\right)}\,. \end{aligned} \tag{7.24}$$

The parameters were chosen with $m_{iE} = 0.99$ and πr_{iE}^2 determines the cross sectional area slicing through the axis of rotation (here $r_{iE}^2 \approx 2r_{12}^2$). This shape is a single closed object with differentiable surface and very close to two touching spheres. Each bubble is in contact with a plate electrode. The initial shape is displayed in Figure 7.8. The computational domain extends over $\mathbf{L} = (8,2,8)\,d_{12}$ and is discretized with a Cartesian grid $\mathbf{N} = (256,128,256)$, i.e. 8.4 million cells. The local resolution near the coalescing bubble is $d_{eq}/\Delta x = 80$ and the grid is stretched away from the bubble in x- and z-direction. No-slip walls bound the domain in y-direction; periodic boundary conditions are applied in x- and z-direction. A no-slip condition is enforced on the bubble surface. It was shown in [265] that the influence of hydrostatic pressure on the bubble shape is negligible which is why this component is not taken into account in the simulation. Constraints applied to the SH shape algorithm are constant bubble volume, a maximum extent in y-direction of $2d_{12}$ and an axisymmetric shape with the axis of rotation parallel to the y-axis. Translational and rotational degrees of freedom of the bubble are disabled, so that only the change in bubble shape generates a velocity in the surrounding liquid. The characteristic length scale of the problem is the radius r_{12} and the characteristic time scale is $t_{ref} = \sqrt{\rho_f r_{12}^3/\sigma}$. A constant time step size of $\Delta t/t_{ref} = 2.5 \cdot 10^{-4}$ is applied throughout the simulation. A large number of modes,

$N_{SH} = 30$, is used to describe the bubble shape and to accurately represent the large curvatures.

Results for the dynamics of the saddle point

The saddle point indicates the point where the bubbles touch initially. The saddle point radius r_s is measured as the distance of this point to the axis of rotation. The temporal evolution of r_s is shown in Figure 7.9 and the present results are compared to the experimental data of [265]. Initially, the saddle point moves outwards with high velocity. The initial motion of the saddle point radius scales approximately with $t^{1/2}$ [269]. This is very well reproduced by the simulation, as well as the experiment. The maximum slope of the saddle point motion near time zero, corresponding also to the maximum induced velocity, is 3.46 m/s in the present simulation, compared to 3.6 m/s in the experiment. This is substantially higher than the gravitational velocity scale of the bubble with $u_g = \sqrt{g\, d_{12}} = 6.1$ cm/s which is the characteristic velocity scale of the free rise of a single bubble. After the initial transient, the saddle point radius over-shoots the equilibrium position and a damped, harmonic oscillation around the equilibrium follows. The largest deviations to the experimental data are found around the first maximum in r_s. Note that in the experiments of Figure 7.9, the bubble detached from the electrodes and moved out of the focus which is apparent from the net shift in r_s. Apart from a slight difference in frequency, and bearing in mind the experimental uncertainty, the match of the simulation with theory and experiment is convincing. High frequency and low amplitude oscillations in $r_s(t)$ in Figure 7.9 indicate the occurrence of capillary waves in Figure 7.9.

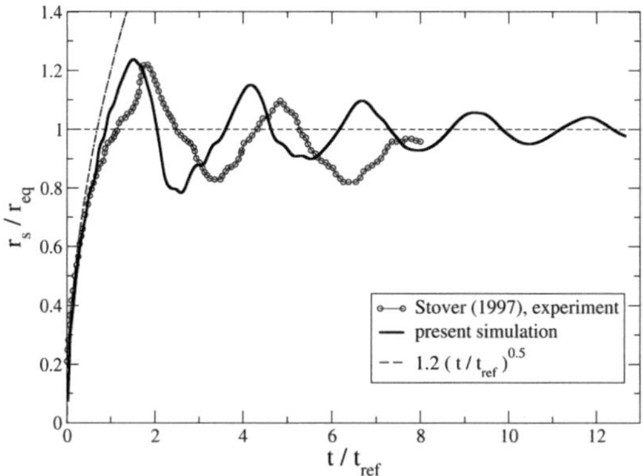

Figure 7.9 Temporal evolution of the saddle point radius r_s normalized with the radius of the coalesced bubble r_{eq} from the present simulation and comparison to experimental data [265].

Results for the bubble shape

Figure 7.10 displays a sequence of the bubble shapes extracted from the temporal evolution. At early times highly distorted bubbles and capillary waves are observed. These waves are created in the highly curved neck region and then travel outwards to the ends of the deformed bubble. The surface waves are gradually damped and at later stages a purely ellipsoidal oblate-prolate shape oscillation prevails. The computed bubble shapes are also consistent with the experimental observations by Menchaca-Rocha et al. [175].

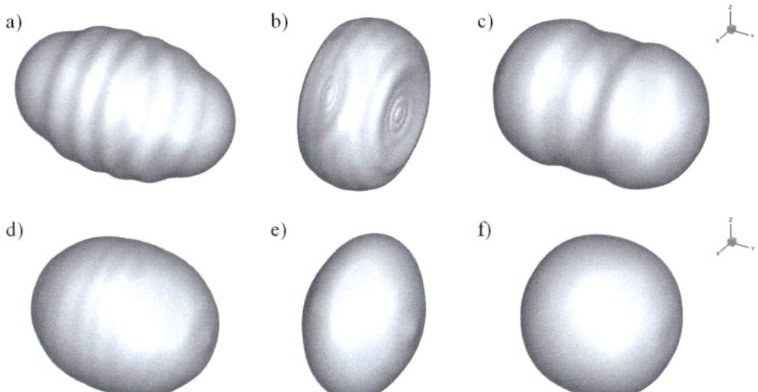

Figure 7.10 Evolution of bubble shape from detailed simulation of the configuration in Figure 7.8. a) - f) display the shape at $t/t_{ref} = 0.6$, 1.2, 2.4, 3.2, 4.2, 6.1.

As time proceeds, surface energy is converted to kinetic energy, back and forth, and is dissipated in the surrounding fluid. Initially high velocities are induced in the vicinity of the bubble. The rapid motion, however, is confined to a region close to the bubble and has only very limited influence on the far field. The bubble shape remains axisymmetric as imposed by the method. The flow field is predominantly axisymmetric, but also contains a superimposed three-dimensional contribution of the order of 0.1 m/s which is small compared to the initial velocity of the junction of 3.46 m/s mentioned above. Figure 7.11 shows one snapshot of the induced absolute value of the resulting fluid velocity, c, near the bubble. In this figure, the plotted range is limited to $c \in [0,\ 0.6]$ m/s for better visibility; $c \approx 0.6$ m/s is thus not the maximum velocity.

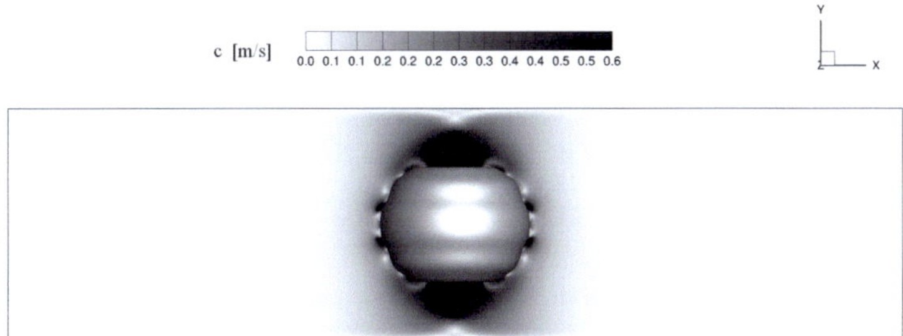

Figure 7.11 Bubble shape and absolute value of fluid velocity, c, in a plane $z = const.$ through the bubble center at $t/t_{ref} = 0.86$. The color scale is cut-off at 0.6 m/s for better visibility.

The frequencies of the bubble shape oscillation of mode n around the spherical shape with $r = const. = r_{eq}$ can be estimated according to Lamb [150]:

$$f_n = \frac{1}{2\pi}\sqrt{\frac{(n+1)(n-1)(n+2)\,\sigma}{\rho_f r_{eq}^3}}, \qquad f_2 = \frac{1}{2\pi}\sqrt{\frac{12\sigma}{\rho_f r_{eq}^3}} \tag{7.25}$$

These expressions were derived in the inviscid limit for small deviations from the spherical shape and in an unbounded domain. However, good agreement is also observed for shape oscillations of rising and coalescing bubbles in viscous fluids [54, 22, 233]. The characteristic frequency in the present simulation was determined from $r_s(t)$, as shown in Figure 7.9, and from the temporal evolution of the cross sectional area in the xy-plane, both providing almost identical results. The present simulation yield a leading frequency of 1.26 kHz which agrees very well with the theoretical value of 1.26 kHz according to equation 7.25. In the experiments [265], a frequency of 1.1 kHz was obtained. The damping rate was 0.52 1/ms in the present simulation, whereas 0.54 1/ms was reported in [265]. The agreement hence is very good so that this case constitutes a sound validation of the method to compute the instantaneous bubble shape.

7.5.2 Modeled evolution of bubble shape

Although the agreement of the simulated coalescence dynamics with both experimental data and theory is quite good, several difficulties are apparent from the direct simulation. A very small time step is necessary to compute the initial phase of the coalescence process. With insufficient temporal resolution, especially the occurrence of capillary waves poses stability problems for the present weak, explicit coupling scheme. Furthermore, a small time step size would impair the performance when dealing with turbulent, bubbly flows. We therefore motivate a further modeling of the bubble shape evolution. This motivates a higher level of modeling when representing the evolution of the bubble shape during coalescence. The idea is the following. Once coalescence is initiated by the fact that the criterion defined above is fulfilled, the evolution of the bubble shape is not computed from the balance of forces at the interface, but judiciously modeled for the initial phase of the evolution. This allows for a larger time step and the usage of less modes, N_{SH}, for the representation of the bubble shape. Principally, the same method is used as before, employing a local coordinate system positioned in the contact point and a tiny shift of the bubbles towards each other. But then, the evolution of the coefficients $a_{nm}(t)$ is imposed by the following procedure. After the merging at $t = t_0$, the initial values $a_{nm}(t_0) = a^0_{nm}$ are known. Then, an appropriate state for the end of phase being modeled is chosen, defined by the time $t_0 + \Delta T_e$ and the coefficients $a_{nm}(t_0 + \Delta T_e) = a^e_{nm}$. Instead of using the local balance of forces along the surface, the evolution $a_{nm}(t)$ is then computed via

$$\frac{a_{nm}(t) - a^0_{nm}}{a^e_{nm} - a^0_{nm}} = f(t), \ t \in (t_0, \ t_0 + \Delta T_e) \tag{7.26}$$

The size of the time step during the coalescence takes into account the CFL criterion. The time interval of the coalescence model ΔT_e is chosen based on the used shape evolution function (SEF) $f(t)$. The shape at the end of the modeling interval is approximated as an ellipsoid of revolution based on the covariance ellipsoid [20] computed from the initially touching bubbles. This ansatz also provides the orientation of the ellipsoid in the laboratory system at the end of the modeling interval. An example for the bubble shape at the beginning and the end of the coalescence model is provided in Figure 7.12 from [280].

Two shape evolution functions (SEF) $f(t)$ were implemented and tested:

$$\text{SEF-1} \qquad f(t) \ = \ \Delta T_e^{-1/2} \, (t - t_0)^{1/2} \tag{7.27}$$

$$\text{SEF-2} \qquad f(t) \ : \ \text{damped, harmonic oscillator} \tag{7.28}$$

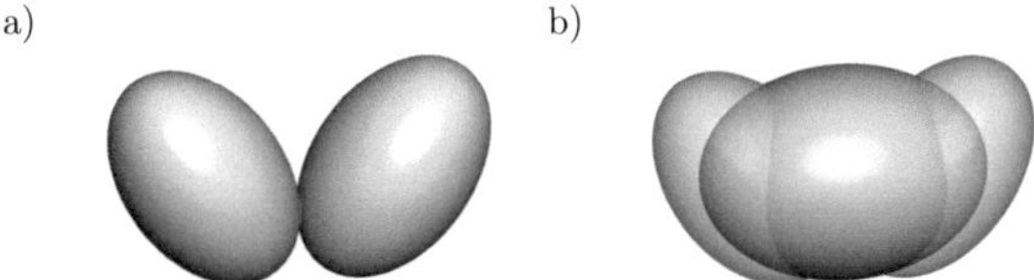

Figure 7.12 Bubble shape a) at beginning of the coalescence with $a_{nm}(t_0) = a^0_{nm}$ and b) at the end of the coalescence modeling with $a_{nm}(t_0 + \Delta T_e) = a^e_{nm}$ [280].

The model SEF-1 is inspired by the initial transient, resulting from theory and experiment. The constant was determined such that $f(t_0 + \Delta T_e) = 1$. It requires shorter modeling intervals, e.g. $\Delta T_e = \max(100\Delta t,\ t_{ref})$. The model SEF-2 is motivated by the damped shape oscillations observed in Figure 7.9 and 7.10. Experimental and theoretical knowledge, as well as simulation results, are used to determine the oscillation frequency and damping rate. Furthermore, $f(t)$ can be made dependent on the indices n and m which would, e.g., allow to damp higher modes stronger than lower modes, if desired. The reason for considering variant SEF-2 is to use a larger duration of the time interval during which modeling is applied, e.g. $\Delta T_e = \max(100\Delta t,\ 4t_{ref})$. The production runs below were conducted with variant SEF-1.

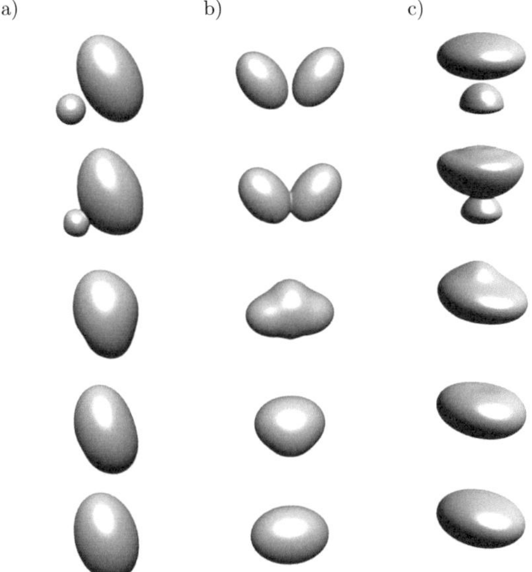

Figure 7.13 Modeled evolution of bubble shape determined by $a_{nm} \sim f(t)$. a) Small spherical and large ellipsoidal bubble. b) Side-to-side coalescence of two ellipsoidal bubbles. c) Front-to-back coalescence of ellipsoidal bubble and cap-shaped bubble. Time evolves from top to bottom with non-constant time intervals selected to highlight the different intermediate shapes.

The model allows for a general and robust description of the bubble shape evolution during coalescence. It covers various coalescence scenarios, like side-to-side coalescence, and wake capturing for bubbles of different size and shape. Figure 7.13 shows the modeled evolution of the bubble shape for three examples using SEF-2. During the coalescence the constraint of constant bubble volume is fulfilled avoiding mass losses as they would usually occur with VOF methods. With the proposed model, the velocities created in the surrounding liquid are well represented since the evolution of the bubble shape is well captured. Afterwards the bubble shape is again computed from the force balance.

7.6 Simulation of two adjacent bubble chains

7.6.1 Configuration and setup

A simulation of two adjacent bubble chains in a container of high aspect ratio was performed using the coalescence model with shape evolution function two (7.28). The container extends over $\mathbf{L} = (24,\, 24,\, 2)\, d_{eq}$ (Figure 7.14). The configuration is chosen in analogy to the simulation of a single bubble chain in liquid metal without representing coalescence in Chapter 6 and [247]. The narrow extension in z-direction also resembles the experimental configuration of Ribeiro and Mewes [221] and it promotes coalescence events. Recall that this experiment was conducted for single bubble pairs. The gravitational velocity scale $u_{ref} = u_g$ and the sphere-volume equivalent diameter d_{eq} serve as reference velocity and length scale, respectively. The average bubble diameter at the nozzle is denoted as d_{eq} here. The non-dimensional numbers governing the ascent of an individual bubble are the Galilei number, $G = 420$, the Eötvös number, $Eo = 4.2$, and the particle to liquid density ratio, $\pi_\rho = 10^{-3}$. The chosen values correspond to 3.5 mm air bubbles in ethanol. For comparison, the Ohnesorg number is determined to be $Oh = 5.710^{-3}$ and thus similar to the detailed simulation of coalescence presented above. During the course of the simulation, the bubbles are injected in pairs with a detachment frequency of $f_b = 0.364\, u_g \,/\, d_{eq}$. At the nozzles, a Gaussian size distribution is prescribed for the bubbles with a mean diameter of 3.5 mm and a standard deviation of 0.28 mm, and the bubble size selected independently for the two bubbles simultaneously released. The injection nozzles are located at $\mathbf{x}_{p,0}^{(1)} = (11,\, 0.8,\, 1)\, d_{eq}$ and $\mathbf{x}_{p,0}^{(2)} = (13,\, 0.8,\, 1)\, d_{eq}$. The spacing between the nozzles is thus two bubble diameters. The bubbles have spherical shape when they are inserted and have zero velocity. The bubble shape is represented with $N_{SH} = 16$ spherical harmonics and coupled to the loads applied by the surrounding fluid. Bubble collisions are accounted for by the adaptive model. The computational domain is discretized with $\mathbf{N} = (480,\, 480,\, 40)$ cells of a Cartesian equidistant grid and the time-step is adapted to yield $CFL = 0.5$. No-slip boundary conditions are applied at all walls, apart from the top boundary where a free-slip condition was used to represent a free surface. In an experiment, the bubbles can leave the fluid through the upper free surface. In the simulation, this was mimicked by removing the bubbles just before they reached the upper boundary. When an IBM is used, where the interior of each bubble is filled with 'artificial fluid' of the same density as the surrounding fluid [282], this can be implemented easily by just setting the forcing terms in the momentum equation resulting from the IBM to zero. Still, conservation of mass and momentum of the fluid are verified without further need for adjustment. The initial condition of the simulation for the continuous phase was quiescent fluid.

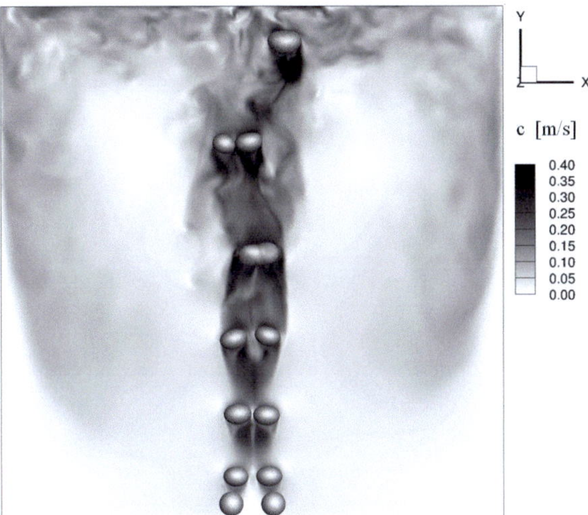

Figure 7.14 Adjacent chains of air bubbles in ethanol ($Eo = 4.2$, $G = 420$). Instantaneous contour of the absolute value of velocity, c, in the center plane.

7.6.2 Results for the continuous phase

Two dominant vortices develop in the container created by the bubbles ascending in the center. Figure 7.15 depicts the average streamlines in the center plane, $z = L_z/2$, together with the magnitude of the velocity vector. The upward vertical jet generated by the ascending bubbles is very well visible. The flow descends along the left and the right wall of the container (without bubbles, as these are removed at the top). Close to the bottom, the fluid is entrained towards the center of the container by the vertical jet, so that a closed recirculation region with low velocity is formed in each of the lower corners. Note that the spiraling of the streamlines is not an imprecision of the graphic software, nor is it a violation of continuity. It is rather a reflection of the secondary flow of the first kind [200] generated by streamline curvature. Centrifugal forces act in the center plane driving the fluid outwards while new fluid enters the vortex centers perpendicularly to this plane.

Time-averaged profiles of the vertical liquid velocity v are plotted over x in Figure 7.16 for different vertical locations (0.25, 0.5, 0.75 times the container height, and $z = L_z/2$). Additional spatial averaging was applied using the symmetry in x. The v-profiles show the spreading of the bubble jet with height, as well as the downward flow to the left and to the right of the center jet. Slightly positive velocities near the walls at $y/d_{eq} = 6$are caused by the corner vortices.

7.6.3 Results for the disperse phase

The large-scale vortices drive the two bubble chains towards each other and lead to subsequent bubble interactions. Figure 7.14 displays an instantaneous contour plot of the absolute value of velocity, c, in the center plane. Coalescing bubbles can be observed at approximately half of the container height and one large, already coalesced bubble is apparent near the free surface.

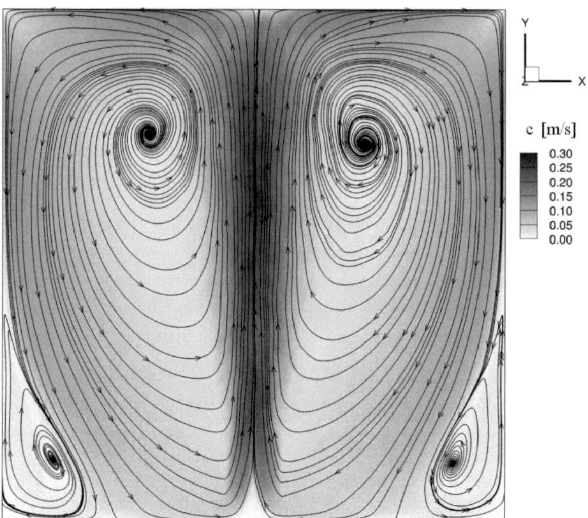

Figure 7.15 Flow pattern in container visualized by time average of absolute value of velocity c ($z = L_z/2$).

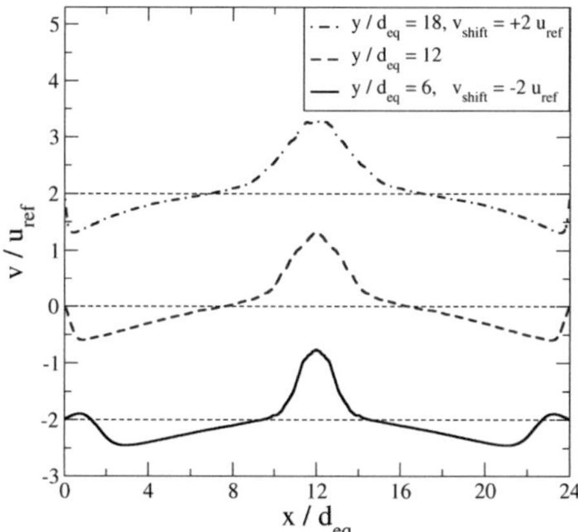

Figure 7.16 Averaged profiles of the vertical liquid velocity v induced by the bubble jet at different vertical locations 0.25, 0.5, 0.75 times the container height ($z = L_z/2$), shift of $\pm u_{ref}$ for the upper and lower position, respectively, to enhance visibility.

A sample of individual bubble trajectories is plotted in Figure 7.17. The bubble paths reveal a zig-zag motion of the bubbles. The center of individual bubbles is tracked. Interrupted paths are a consequence of bubble coalescence. A larger sample of coalescence events is indicated by triangular symbols. Most coalescence events take place very close to the nozzles at first contact as also reported by Duineveld [54] for individual bubble pairs. A second region where coalescence occurs in the present setting is found at approximately half of the container height. Coalescence events in the upper third of the vessel are rare.

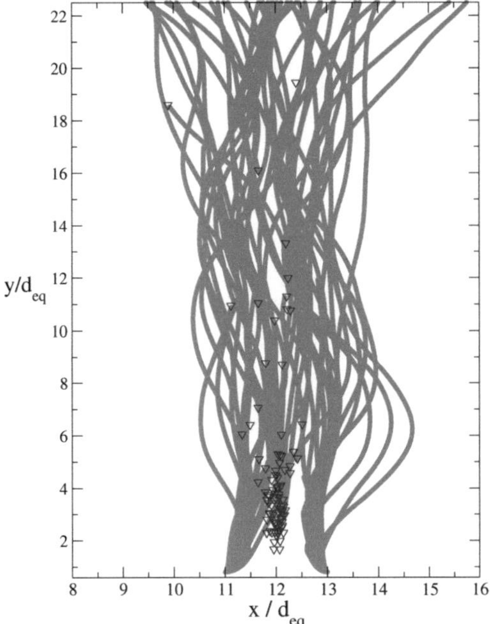

Figure 7.17 Sample of individual bubble trajectories showing zig-zag motion of individual bubbles. Also shown is a larger sample of coalescence events indicated by triangular symbols.

Quantitative results on bubble coalescence were determined from the simulation over a duration of $t = 268 f_b^{-1}$, i.e. covering the injection of 268 bubble pairs. In this time interval 361 interactions were counted which provides the average frequency of interaction. Along the trajectory, the first interaction with a second bubble may already lead to coalescence, otherwise a collision takes place often leading to a subsequent second and possibly third interaction. Note that the start and the end of individual collisions are difficult to detect. In the present study, the criterion applied for the detection of the end of a collision was that, after contact, the distance of the bubble surface is again larger than $0.1 d_{12}$, i.e. the bubbles have bounced back to an amount which is also measurable in experiments. Almost all interactions occurred between the bubbles of one pair simultaneously released (see Figure 7.14 for an example). Less than 3% of the interactions occurred with trailing or leading bubbles, often after one of these had previously undergone coalescence thus having a larger volume than the other. In total, 124 coalescence events were gathered for the observed 361 interactions yielding a coalescence efficiency of $\lambda_{12} = 0.343$. Only a single coalesced, large bubble underwent a second coalescence.

7.6.4 Comparison to semi-empirical model predictions

Comparing the present results to predictions from 'point bubble' coalescence models is difficult. These models were reviewed thoroughly by Liao and Lucas [161] and they are conceived for simulations with the bubble size being smaller than the step size of the numerical grid. Most of these models are also based on the film thinning theory. Film drainage models then often compute the coalescence efficiency from $\lambda_{12} = \exp(-t_{drainage}/t_{contact})$. The time scales are then derived from dimensional analysis, physical reasoning or semi-empirically taking into account one or more mechanisms that may lead to bubble interaction. These comprise turbulent motion and eddy capture, bubble contact due to mean velocity gradients, and relative bubble velocity due to different rise velocities stemming from different bubble size or due to wake interaction. A model taking into account all the mechanisms is presented in Liao et al. [162], but the model constants are poorly documented. Note that even for identical input, these models may give substantially different results with deviations of 50% and more (see Fig. 6 in [161] or Figure 7.18 providing a comparison of two models for the same coalescence mechanism).

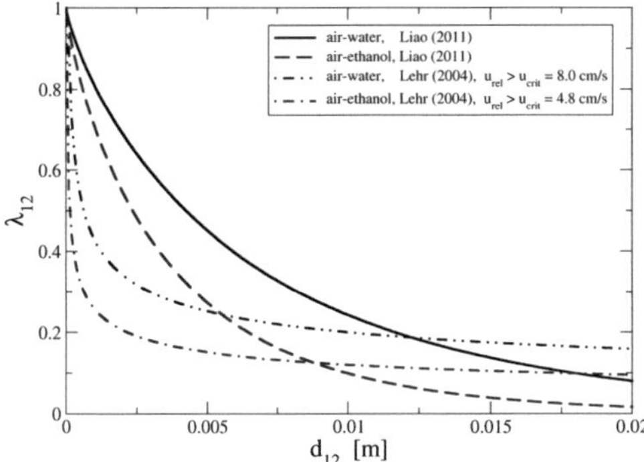

Figure 7.18 Prediction of coalescence efficiency λ_{12} as a function of diameter d_{12} by the models of Liao et al. [162] and Lehr et al. [156] for the two systems air-water and air-ethanol with $u_{rel} = u_\varepsilon$ and $\varepsilon = 0.8\ \text{m}^2/\text{s}^3$.

A different parametrization is used in the model of Lehr et al. [156] with $\lambda_{12} = \min(u_{crit}/u_{rel},\ 1)$. Here, u_{crit} is the critical approach velocity (e.g. from data as displayed in Figure 7.5) where the authors report 8 cm/s for air-water in contrast to 9.57 cm/s observed by Ribeiro and Mewes [221]. The relative velocity in the model of Lehr et al. is computed from

$$u_{rel} = \max\left(\sqrt{2}\,\varepsilon^{1/3}\sqrt{d_{eq,1}^{2/3} + d_{eq,2}^{2/3}}\,,\ |v_{p,1} - v_{p,2}| \right) . \tag{7.29}$$

The first term describes the relative motion due to turbulence; the second one is the relative velocity due to different rise velocities. In the present case, the bubbles rise side-by-side with very similar rise velocities. The dissipation rate of turbulent kinetic energy ε used in the correlation is not easily brought into accordance with the high resolution data of the simulation. The dissipation rate is first determined as a temporal average at each grid point of the domain from the average and fluctuating part of the velocity field. Additional spatial averaging over the entire domain yields 0.8 $\mathrm{m}^2/\mathrm{s}^3$, a local average in $10 \leq x/d_{eq} \leq< 14$, i.e. the main corridor of the bubble paths, yields 4.0 $\mathrm{m}^2/\mathrm{s}^3$. Using the global average of ε, the model of Liao et al. [162] predicts $\lambda_{12} = 0.381$, whereas the model of Lehr et al. [156] yields $\lambda_{12} = 0.177$. If the local average is used, both models underestimate the coalescence efficiency ($\lambda_{12} = 0.193$ and $\lambda_{12} = 0.099$, respectively). Prediction for the coalescence efficiency λ_{12} as a function of diameter d_{12} of both models are shown in Figure 7.18 for the two systems air-water and air-ethanol. Note that both models yield non-vanishing coalescence efficiencies even for large relative velocities or relative Weber numbers which is in contrast to Figure 7.5 and might lead to an over-prediction of bubble sizes as apparent in the simulation of bubbly, vertical pipe flow in [162]. In general, there is significant room for improvement in the prediction of coalescence efficiencies.

7.6.5 Comparison to results without coalescence model

It was stated in the introduction that the prediction of the correct size distribution as a consequence of coalescence and breakup is one of the most crucial points in the simulation of bubble-laden flows [71, 161]. To quantify the deviations related to the coalescence modeling for the present case, the previous simulation was repeated with all parameters unchanged, but without a coalescence model. This means that no merge of bubbles will occur forming larger bubbles and that every bubble interaction yields a collision.

Figure 7.19 presents histograms of normalized equivalent bubble diameter d_b/d_{eq} obtained at the nozzle and near the free surface. The bubble size distribution at the nozzle is a prescribed Gaussian one. The plot shows a realization obtained from the 268 bubble pairs of the run with coalescence and additional 256 bubble pairs from the simulation without coalescence model. In the case without coalescence, the bubble size distribution remains unchanged over the height of the container. In contrast, coalescence leads to the formation of larger bubbles which is clearly visible from the histogram obtained near the free surface in Figure 7.19b). The new mean bubble diameter is $1.096\, d_{eq}$ at $y_p/d_{eq} = 22$. Two bubbles with d_{eq} form a bubble with twice the volume, and the new diameter increases to $\sqrt[3]{2}\, d_{eq} \approx 1.26\, d_{eq}$. Larger bubbles have a higher rise velocity than smaller ones [37]. The bubble shape and dynamics, as well as the generated wake structures are determined by larger bubble Reynolds and Weber number. Table 7.4 lists the changes in the results obtained with and without coalescence model. The absolute value of the horizontal velocity component, which is distinctive for the zig-zag trajectory, is increased by 16.2% in average if a coalescence model is applied. With a coalescence model, the average bubble rise velocity is increases by 7.3%.

For comparison, the effect of coalescence is set in relation to other impact parameters influencing the rise velocity studied throughout this thesis. The strongest magnetic field altered the rise velocity of a single bubble ascending in quiescent liquid metal by 13.1% as studied in Chapter 5. About 10% change in average rise velocity were obtained for a single bubble chain under the influence of a magnetic field with magnetic interaction of $N_y = 1$ in Chapter 6. Again for the rise of a single bubble, a discretization error in bubble rise velocity of about 10% compared to the reference solution was determined using only half the number of grid points in all three directions as in the production runs. This corresponds to 1/16 of the computational effort since also the time step needs to be adjusted.

The global flow patter remains unchanged also without coalescence model, as the double bubble chain creates two large scale vortices to the left and right of the bubble jet. A higher

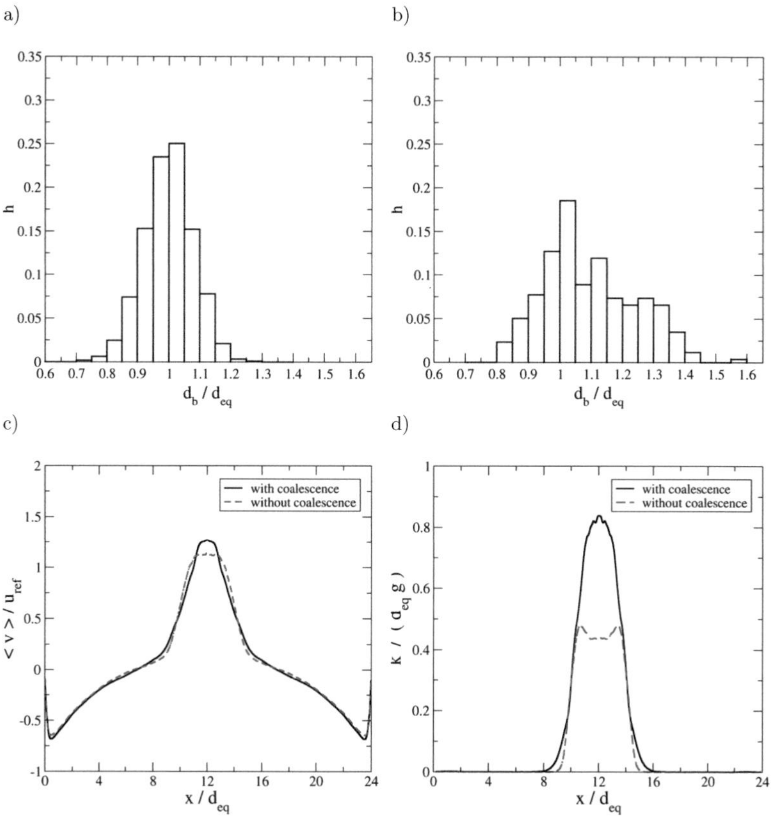

Figure 7.19 Bubble size distribution, fluid velocity profiles and bubble-induced turbulent kinetic energy with and without coalescence model. Histograms of normalized equivalent bubble diameter d_b/d_{eq} a) at the nozzles ($y_p/d_{eq} = 0.8$), b) near the free surface ($y_p/d_{eq} = 22$) with coalescence. c) Averaged profiles of the vertical liquid velocity $\langle v \rangle$, and d) bubble-induced turbulent kinetic energy $k/u_g^2 = k/(gd_{eq})$ at 0.75 times the container height ($z = L_z/2$).

peak fluid velocity is observed in Figure 7.19c) in the center of the jet for the case with coalescence model due to the higher average bubble velocity. Significant changes are apparent with respect to velocity fluctuations. With coalescence, the local average ($10 \leq x/d_{eq} \leq< 14$) and the global average of the dissipation rate of turbulent kinetic energy $\langle\varepsilon\rangle_{loc}$ and $\langle\varepsilon\rangle_{glob}$ increase by 21.4% and 22.8%, respectively. The pointwise statistics were not fully converged at the end of the simulation. However, the preliminary results indicate that the turbulent kinetic energy increases substantially compared to the simulation without coalescence as shown in 7.19d) for profiles obtained at 0.75 times the container height and $z = L_z/2$.
Thus, the deviations caused by bubble coalescence are indeed substantial even at the moderate coalescence efficiency of the present setup.

Table 7.4 Comparison of simulation results obtained with and without coalescence model for average bubble rise velocity $\langle v_p\rangle$, average, absolute, horizontal velocity $\langle|u_p|\rangle$, local and global average of dissipation rate of turbulent kinetic energy $\langle\varepsilon\rangle_{loc}$ and $\langle\varepsilon\rangle_{glob}$.

	$\langle v_p\rangle$ [m/s]	$\langle\lvert u_p\rvert\rangle$ [m/s]	$\langle\varepsilon\rangle_{loc}$ [m^2/s^3]	$\langle\varepsilon\rangle_{glob}$ [m^2/s^3]
without coalescence	0.262	0.0357	3.32	0.663
with coalescence	0.281	0.0415	4.03	0.814
deviation	7.3%	16.2%	21.4%	22.8%

7.7 Towards bubble break-up

Bubble break-up is mainly caused due to turbulent fluctuations and collisions, viscous shear forces and interfacial instability, where the mechanisms and the theoretical models covering break-up are thoroughly reviewed in [160].
The modeling of bubble break-up in the context of an IBM can be achieved in a similar way as for the bubble coalescence. A break-up criterion needs to be stated which activates the topology change. A mother-daughter-size distribution can be described or can be obtained from the resolved deformation. As a further modeling step, a defined surface evolution could be included. A preliminary combined breakup criterion is defined. Break-up occurs if a specified critical Weber number is exceeded, e.g. $We_{crit} = 12$ [210], or if substantial deformation, e.g. due to shear, is measured. The latter can be quantified for instance by the ratio of a higher spherical harmonic mode to the zeroth 'spherical' mode. The critical Weber number stated above roughly corresponds to the occurrence of the maximum bubble aspect ratio of four in uniform flow in Figure 4.6.
Successful early tests have been conducted with a very simple model, where the surface evolution during the break-up is neglected and two equi-sized, spherical daughter bubbles are initialized near the removed mother bubble's position once the criterion is fulfilled. As the bubble is substantially deformed and elongated prior to a realistic break-up, a redistribution or adding of forcing points is probably necessary in a future model.
With a representation of coalescence and break-up, a self-determined bubble size-distribution can be obtained in large scale simulations also capturing the main physics during coalescence and breakup.

7.8 Conclusions and outlook with respect to bubble interaction

After reviewing coalescence criteria, which determine whether bubbles merge or bounce back, an overview on collision modeling was given and the involved computation of inter-particle distances was elaborated. A coalescence model for the phase-resolving simulation of bubbly flows was developed. The change in shape of the coalescing bubble is prescribed by a temporal evolution of the shape coefficients in a spherical harmonic expansion judiciously chosen to match fairly well the physical behavior. The choice of a shape evolution function is backed by direct simulation of the shape oscillation, as well as by experimental data and theoretical considerations. The coalescence model was then applied in a simulation of two adjacent bubble chains. Statistics were obtained for the liquid phase, and also for the interaction frequency and the coalescence efficiency. The latter were compared to model predictions used for closure in 'industrial scale CFD'. The influence of the coalescence model was quantified by comparison to results obtained for a simulation where coalescence was disabled and the deviations are indeed substantial. The dissipation rate of turbulent kinetic energy appears as a significant parameter in all 'point bubble' coalescence models. Liquid metal systems should be examined to study the influence of turbulence on bubble coalescence. As an outlook we mention that at present simulations are planned which consider the same geometry as described above, but where air and ethanol are replace by argon and the liquid metal GaInSn. This pairing is relevant to model experiments devoted to liquid metal technologies. A substantial body of research in this area is reviewed in Fröhlich et al. [74].

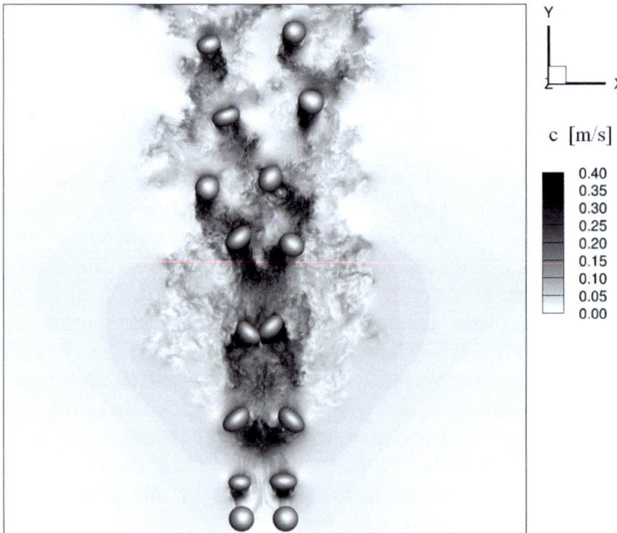

Figure 7.20 As Figure 7.14, but for 6 mm argon bubbles in the liquid metal GaInSn ($Eo = 4.2$ and $G = 4200$) - initial phase.

Figure 7.20 provides a snapshot of argon bubbles in the liquid metal GaInSn to illustrate the qualitative changes in the fluid dynamics compared to the air-ethanol systems studied above. The snapshot was taken in the initial phase and the dominant vortices are not developed yet. The turbulent character of the bubble wakes and the induced flow, however, is already quite distinct [244, 247]. Gas-liquid metal systems are characterized by large Reynolds numbers due to the low kinematic viscosity of the liquid metal. Similar Weber numbers as for air-water systems are achieved as the high density of the liquid metal is accompanied by a large surface tension. The generality of the critical relative Weber number as coalescence criterion needs to be double checked for very high surface tension systems. Further experiments and numerical simulations should be conducted to get more insight into the mechanisms that lead to coalescence. Additionally, a bubble breakup model needs to be developed as a counterpart to estimate bubble size distributions realistically. Similar strategies as for coalescence could be pursued employing a criterion for the occurrence of the breakup [160], followed by an interface-resolved topology change.

8 Summary and Outlook

An immersed boundary method for the phase-resolving simulation of rigid particles and deformable bubbles in magnetohydrodynamic flows has been developed. The Navier-Stokes equations of the incompressible Newtonian fluid were coupled to the electromagnetic equations in the quasi-static approximation by the Lorentz force. The single phase MHD was rigorously validated and extended by a representation of immersed insulating objects.
As a central aspect of this thesis, the IBM was extended towards more general immersed geometries including the development of a suitable data structure and a modification of the forcing procedure at the interface. The description of the orientation and motion of non-spherical particles was introduced and the IBM is now capable of dealing with very light particles. A structured study of the flow around spheroidal particles was conducted to examine the mechanisms leading to path oscillations and to illustrate the complex interaction of particle shape, wake and dynamics. The study was completed by a characterization and quantification of the impact of a magnetic field on the flow and the particle trajectory, revealing increased drag and damped lateral dynamics of the particle in the presence of an aligned magnetic field.
The representation of bubble shapes by the IBM was realized by a classification of observed bubble shapes and matching analytical numerical descriptions. A large share of all possible bubble shapes can be depicted by rather simple shape parametrizations which comprise spherical or ellipsoidal bubble shapes. More complex bubble shapes are represented by spherical harmonic functions which analytically describe the interface between the phases and enable the accurate computation of surface normals and curvature. Only very distorted bubble shapes, e.g. strongly dimpled with geometric undercuts, cannot be accounted for. A numerical algorithm was developed to capture bubble deformation by the flow and it was validated and applied towards quasi-steady and time-dependent bubble shapes.
The rise of a single bubble in liquid metal under the impact of an aligned magnetic field was studied in detail and the results were compared to experimental data showing good agreement. The experimental findings on the impact of the magnetic field on the bubble wake and bubble dynamics could be backed by highly resolved, three-dimensional numerical data giving elucidating insight into the otherwise opaque system.
A bubble chain exposed to an aligned magnetic field was simulated to quantify the influence of the field on the bubbles and the induced global flow field in the mold. The average bubble trajectory was found to be more rectilinear at a decreased average rise velocity leading to a reduced overall circulation in the container. The results allow for implications towards electromagnetic flow control in industrial applications.
A view on particle interaction was provided in the last chapter with the specific focus on bubble collisions and bubble coalescence. The computation of the inter-particle distance was realized for complex particle shapes and the current status towards bubble collision modeling was surveyed. A new coalescence model was developed to account for the merge of two bub-

bles given a specified coalescence criterion is fulfilled. The model was applied towards the quantification of the impact of bubble coalescence and a leading order influence was found for the test case considered.

Throughout this thesis, the boundary condition at the bubble surface was chosen to be a no-slip condition. This is well-satisfied for contaminated systems [2, 168], and specifically liquid metal systems are always substantially contaminated and furthermore a thin oxide-layer forms at the bubble surface. However, the forcing procedure of the IBM is not limited to the forcing of a no-slip condition. The forcing of a free-slip condition, the appropriate boundary condition in hyper-clean systems, can also be achieved. With the surface grid on the bubble, either the triangular mesh or the structured spherical harmonics grid, there is also a sound basis for more advanced boundary conditions in-between the two extreme cases. An additional transport equation of the contaminants could be solved in the fluid and on the bubble surface to compute the level of contamination and an appropriate, spatially varying partial slip.

With the present IBM, a sound basis is provided for the study of systems with many bubbles, as well as bubble-turbulence interaction by means of computationally efficient, phase-resolving simulation. Based on the present IBM, studies on the clustering of spherical bubbles and the formation of wet metal foam were conducted [102, 101, 104]. Many spherical and ellipsoidal light particles in turbulent channel flow were examined with the extended IBM for example in [236, 238, 237]. Figure 8.1 provides a snapshot from a present simulation of an upward turbulent channel flow, where the bubbles are represented by spherical harmonics and gravity points downwards.

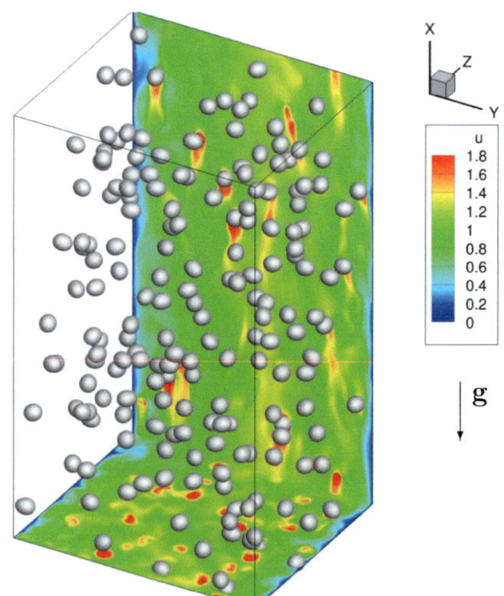

Figure 8.1 Bubbles in upward turbulent channel flow. Instantaneous streamwise velocity contours u, where the bulk velocity is $u_b = 1$. Snapshot of the distribution of bubbles, where each deformable bubble is represented by spherical harmonic functions.

Without going into detail on the simulation, the configuration shall be briefly described. No-slip walls bound the flow in y-direction and periodic boundary conditions are applied in the other directions. The system is rather dilute with a gas volume fraction of one percent.

The bulk Reynolds number is kept constant at $Re_b = 5263$ based on the channel height L_y. The Eötvös number of the equi-sized, deformable bubbles is $Eo = 4.2$, with $L_y/d_{eq} = 18.9$. Coalescence and break-up are inhibited to achieve a temporally constant bubble size distribution. A rather small computational domain was chosen for this illustration. It can be stated that the mean flow, the turbulence statistics and coherent structures are significantly altered by the presence of the deformable bubbles compared to the unladen turbulent channel flow [16, 236, 166]. The bubbles form clusters and take preferential orientations [238, 237, 39]. The turbulence modulation by the bubbles is of significant importance for many applications, such as the reduction of frictional drag bubbles for ships [190]. With the present IBM, a powerful tool has been provided that enables efficient fundamental research by phase-resolving simulation. Open issues were addressed straightforward throughout the thesis. Future research should follow the idea of multiscale modeling [276, 135] to the capture physics taking place at strongly diverse length and time scales efficiently, as it was discussed for bubble collision and coalescence above. Another key aspect for future research is the LES of turbulent particulate flows. The development of subgrid scale models needs to be advanced for both, the fluid momentum equation and the particle momentum equation, accounting for the particle-turbulence interaction. The present IBM can be employed to gather important phase-resolving reference data for the LES development.

Acknowledgment

First of all, I would like to thank my friends and family for their continuous support, their understanding, and for providing essential distractions when my mind had been trapped in an infinite loop. Furthermore, I send a big thank you to the entire team of the Professur für Strömungsmechanik at TU Dresden. It was a pleasure to work with you! Thanks for all the fruitful discussions, especially in the coffee breaks. Thank you to Prof. J. Fröhlich for the supervision of my thesis, and all the efforts that came along with it. The PRIME code, which this thesis is based on, was developed and is still improved by T. Kempe. Without his work, this thesis would not have been possible. PD Dr. T. Boeck is acknowledged for accepting the part as the second examiner of my thesis. The present work was partially funded by DFG through SFB 609, project A9, and some of the results were obtained within the Helmholtz-Alliance LIMTECH. I would also like to thank DAAD and Freunde und Förderer der TU Dresden for additional grants enabling me to visit international conferences. Computation time was provided by ZIH Dresden. S. Eckert, S. Boden, C. Zhang, M. Jenny and J. Degroote are acknowledged for providing their data in electronic form. Finally, I would like to thank all the students that I have been supervising, as well as everyone who was willing to proofread my thesis.

Bibliography

[1] H. Abe, H. Kawamura, and Y. Matsuo. Direct numerical simulation of fully developed turbulent channel flow with respect to the Reynolds number dependence. *J. Fluids Eng*, 123:382–393, 2001.

[2] S. Aland, S. Schwarz, J. Fröhlich, and A. Voigt. Modeling and numerical approximations for bubbles in liquid metal. *Eur. Phys. J. ST*, 220(1):185–194, 2013.

[3] H. Alfvén. Existence of electromagnetic-hydrodynamic waves. *Nature*, 150:405–406, 1942.

[4] G.P. Almeida, D.F.G. Durao, and M.V. Heitor. Wake flows behind two-dimensional model hills. *Exp. Therm Fluid Sci.*, 7(1):87–101, 1993.

[5] E. Anderson, Z. Bai, C. Bischof, S. Blackford, J. Demmel, J. Dongarra, J. Du Croz, A. Greenbaum, S. Hammarling, A. McKenney, and D. Sorensen. *LAPACK: a portable linear algebra library for high-performance computers, Users' Guide*. Society for Industrial and Applied Mathematics, www.netlib.org/lapack, third edition, 1999.

[6] A. Andruszkiewicz, K. Eckert, S. Eckert, and S. Odenbach. Gas bubble detection in liquid metals by means of the ultrasound transit-time-technique. *Eur. Phys. J. ST*, 220(1):53–62, 2013.

[7] S.V. Apte and J.R. Finn. A variable-density fictitious domain method for particulate flows with broad range of particle-fluid density ratios. *J. Comput. Phys.*, 243(0):109–129, 2013.

[8] S. Balachandar and J.K. Eaton. Turbulent dispersed multiphase flow. *Annu. Rev. Fluid Mech.*, 42(1):111–133, 2010.

[9] S. Balay, W.D. Gropp, L.C. McInnes, and B.F. Smith. *Efficient Management of Parallelism in Object Oriented Numerical Software Libraries*. Birkhäuser Press, 1997. http://www.mcs.anl.gov/petsc.

[10] D.W. Bechert, M. Bruse, and W. Hage. Experiments with three-dimensional riblets as an idealized model of shark skin. *Exp. Fluids*, 28:403–412, 2000. 10.1007/s003480050400.

[11] D.W. Bechert, M. Bruse, W. Hage, J.G.T. Van der Hoeven, and G. Hoppe. Experiments on drag-reducing surfaces and their optimization with an adjustable geometry. *J. Fluid Mech.*, 338:59–87, 1997.

[12] L. Beetz. *Nachlaufinduzierte Oszillation eines Ellipsoiden in Queranströmung*. Technische Universität Dresden, Fakultät für Maschinenwesen, Institut für Strömungsmechanik, Diploma thesis, 2012.

[13] D. Bhaga and M.E. Weber. Bubbles in viscous liquids: shapes, wakes and velocities. *J. Fluid Mech.*, 105:61–85, 1981.

[14] S. Boden, S. Eckert, G. Gerbeth, M. Simonnet, M. Anderhuber, and P. Gardin. X-ray visualisation of bubble formation and bubble motion in liquid metals. *Electromagn. Process. Mater.*, 2009.

[15] T. Boeck, D. Krasnov, and E. Zienicke. Numerical study of turbulent magnetohydrodynamic channel flow. *J. Fluid Mech.*, 572:179–188, 2007.

[16] I.A. Bolotnov, K.E. Jansen, D.A. Drew, A.A. Oberai, R.T. Lahey Jr., and M.Z. Podowski. Detached direct numerical simulations of turbulent two-phase bubbly channel flow. *Int. J. Multiphase Flow*, 37(6):647–659, 2011.

[17] T. Bonometti and J. Magnaudet. An interface-capturing method for incompressible two-phase flows. validation and application to bubble dynamics. *Int. J. Multiphase Flow*, 33:109–133, 2007.

[18] I. Borazjani, L. Ge, and F. Sotiropoulos. Curvilinear immersed boundary method for simulating fluid structure interaction with complex 3D rigid bodies. *J. Comput. Phys.*, 227(16):7587–7620, 2008.

[19] G. Bouchet, M. Mebarek, and J. Dušek. Hydrodynamic forces acting on a rigid fixed sphere in early transitional regimes. *Eur. J. Mech. B/Fluids*, 25:321–336, 2006.

[20] S. Brandt and G. Gowan. *Data analysis: Statistical and computational methods for scientists and engineers.* Springer, 1998.

[21] H. Branover. *Magnetofluiddynamic Flow in Ducts.* John Wiley, 1978.

[22] G. Brenn, V. Kolobaric, and F. Durst. Shape oscillations and path transition of bubbles rising in a model bubble column. *Chem. Eng. Sci.*, 61:3795–3805, 2006.

[23] M. Breuer, M. Alletto, and F. Langfeldt. Sandgrain roughness model for rough walls within Eulerian-Lagrangian predictions of turbulent flows. *Int. J. Multiphase Flow*, 43(0):157–175, 2012.

[24] M. Breuer, N. Peller, C. Rapp, and M. Manhart. Flow over periodic hills - numerical and experimental study in a wide range of Reynolds numbers. *Comput. Fluids*, 38:433–457, 2009.

[25] W.-P. Breugem. A second-order accurate immersed boundary method for fully resolved simulations of particle-laden flows. *J. Comput. Phys.*, 231(13):4469–4498, 2012.

[26] D. Broday, M. Fichman, M. Shapiro, and C. Gutfinger. Motion of spheroidal particles in vertical shear flows. *Phys. Fluids*, 10(1):86–100, 1998.

[27] C. Brücker. Structure and dynamics of the wake of bubbles and its relevance for bubble interaction. *Phys. Fluids*, 11(7):1781–1796, 1999.

[28] E. Buckingham. On physically similar systems; illustrations of the use of dimensional equations. *Phys. Rev.*, 4:345–376, 1914.

[29] P. Burattini, M. Kinet, D. Carati, and B. Knaepen. Anisotropy of velocity spectra in quasistatic magnetohydrodynamic turbulence. *Phys. Fluids*, 20(6):065110, 2008.

[30] J.C. Cano-Lozano, P. Bohorquez, and C. Martínez-Bazán. Wake instability of a fixed axisymmetric bubble of realistic shape. *Int. J. Multiphase Flow*, 51:11–21, 2013.

[31] D.Y.C. Chan, E. Klaseboer, and R. Manica. Film drainage and coalescence between deformable drops and bubbles. *Soft Matter*, 7:2235–2264, 2011.

[32] D.Y.C. Chan, E. Klaseboer, and R. Manica. Theory of non-equilibrium force measurements involving deformable drops and bubbles. *Adv. Colloid Interface Sci.*, 165:70–90, 2011.

[33] C. Chan-Braun, M. Garcia-Villalba, and M. Uhlmann. Force and torque acting on particles in a transitionally rough open-channel flow. *J. Fluid Mech.*, 684:441–474, 2011.

[34] A.K. Chesters and G. Hofman. Bubble coalescence in pure liquids. *Appl. Sci. Res.*, 38:353–361, 1982.

[35] A.J. Chorin. A numerical method for solving incompressible viscous flow problems. *J. Comput. Phys.*, 2:12–26, 1967.

[36] M. Chrust, G. Bouchet, and J. Dušek. Parametric study of the transition in the wake of oblate spheroids and flat cylinders. *J. Fluid Mech.*, 665:199–208, 2010.

[37] R. Clift, J.R. Grace, and M.E. Weber. *Bubbles, Drops, and Particles.* Dover Publications, 1978.

[38] C.T. Crowe. *Multiphase Flow Handbook.* CRC Press, 2005.

[39] S. Dabiri and G. Tryggvason. Simulation of a rising bubble near vertical walls. *Int. Conf. Multiphase Flow, Jeju, South Korea*, (570), 2013.

[40] T.E. Daubert and R.P Danner. Data compilation tables of properties of pure compounds. *American Institute of Chemical Engineers*, 1985.

[41] P.A. Davidson. Magnetic damping of jets and vortices. *J. Fluid Mech.*, 299:153–186, 1995.

[42] P.A. Davidson. Magnetohydrodynamics in materials processing. *Annu. Rev. Fluid Mech.*, 31(1):273–300, 1999.

[43] P.A. Davidson. *An Introduction to Magnetohydrodynamics.* Cambridge University Press, 2002.

[44] A.W.G. de Vries, A. Biesheuvel, and L. van Wijngaarden. Notes on the path and wake of a gas bubble rising in pure water. *Int. J. Multiphase Flow*, 28(11):1823–1835, 2002.

[45] J. Degroote, P. Bruggeman, R. Haelterman, and J. Vierendeels. Bubble simulations with an interface tracking technique based on a partitioned fluid-structure interaction algorithm. *J. Comput. Appl. Math.*, 234(7):2303–2310, 2010.

[46] J. Degroote, P. Bruggeman, and J. Vierendeels. A coupling algorithm for partitioned solvers applied to bubble and droplet dynamics. *Comput. Fluids*, 38(3):613–624, 2009.

[47] J.C. del Álamo and J. Jiménez. Spectra of the very large anisotropic scales in turbulent channels. *Phys. Fluids*, 15(6):L41, 2003.

[48] J.C. del Álamo, J. Jiménez, P. Zandonade, and R.D. Moser. Scaling of the energy spectra of turbulent channels. *J. Fluid Mech.*, 500:135–144, 2004.

[49] J.C. del Álamo, J. Jiménez, P. Zandonade, and R.D. Moser. Self-similar vortex clusters in the turbulent logarithmic region. *J. Fluid Mech.*, 561:329–358, 2006.

[50] P. Deuflhard and F. Bornemann. *Numerische Mathematik II.* de Gruyter Lehrbuch, 1994.

[51] J. Diebel. Representing attitude: Euler angles, unit quaternions, and rotation vectors. Stanford University. Techn. report. 2006.

[52] S. Dong, D. Krasnov, and T. Boeck. Secondary energy growth and turbulence suppression in conducting channel flow with streamwise magnetic field. *Phys. Fluids*, 24(7):074101, 2012.

[53] F. Ducros, F. Nicoud, and T. Poinsot. Wall-adapting local eddy-viscosity models for simulations in complex geometries. *6th ICFD Conf. Numer. Meth. Fluid Dyn.*, 1:293–299, 1998.

[54] P.C. Duineveld. Bouncing and coalescence of bubble pairs rising at high Reynolds number in pure water or aqueous surfactant solutions. *Appl. Sci. Res.*, 58:409–439, 1998.

[55] S. Eckert, G. Gerbeth, and O. Lielausis. The behaviour of gas bubbles in a turbulent liquid metal magnetohydrodynamic flow. Part I: Dispersion in quasi-two-dimensional magnetohydrodynamic turbulence. *Int. J. Multiphase Flow*, 26:45–66, 2000.

[56] S. Eckert, G. Gerbeth, and O. Lielausis. The behaviour of gas bubbles in a turbulent liquid metal magnetohydrodynamic flow. Part II: Magnetic field influence on the slip ratio. *Int. J. Multiphase Flow*, 26:67–82, 2000.

[57] S. Eckert, G. Gerbeth, W. Witke, and H. Langenbrunner. MHD turbulence measurements in a sodium channel flow exposed to a transverse magnetic field. *Int. J. Heat Fluid Flow*, 22:358–364, 2001.

[58] S. Eckert, V. Shatrov, C. Zhang, and G. Gerbeth. Oscillating melt flow driven by a rotating magnetic field in a cylindrical column with oxidized surface. *8th PAMIR Int. Conf. on Fundamen. Appl. MHD, Borgo, Corsica, France*, 2:611–616, 2011.

[59] K. Ellingsen and F. Risso. On the rise of an ellipsoidal bubble in water: Oscillating paths and liquid-induced velocity. *J. Fluid Mech.*, 440:235–268, 2001.

[60] M. Emans. *Numerische Simulation des unterkühlten Blasensiedens in turbulenter Strömung: Ein Euler-Lagrange-Verfahren auf orthogonalen Gittern.* PhD thesis, TU München, 2003.

[61] M. Emans and C. Zenger. An efficient method for the prediction of the motion of individual bubbles. *Int. J. Comp. Fluid D.*, 19:347–357, 2005.

[62] P. Ern, P.C. Fernandes, F. Risso, and J. Magnaudet. Evolution of wake structure and wake-induced loads along the path of freely rising axisymmetric bodies. *Phys. Fluids*, 19(11):113302, 2007.

[63] P. Ern, F. Risso, D. Fabre, and J. Magnaudet. Wake-induced oscillatory paths of bodies freely rising or falling in fluids. *Annu. Rev. Fluid Mech.*, 44(1):97–121, 2012.

[64] D. Fabre, F. Auguste, and J. Magnaudet. Bifurcations and symmetry breaking in the wake of axisymmetric bodies. *Phys. Fluids*, 20:051702, 2008.

[65] R.D. Falgout, J.E. Jones, and U.M. Yang. *The Design and Implementation of hypre, a Library of Parallel High Performance Preconditioners*, volume 51. Springer-Verlag, 2006. https://computation.llnl.gov/casc/hypre.

[66] R. Bel Fdhila and P.C. Duineveld. The effect of surfactant on the rise of a spherical bubble at high Reynolds and Peclet numbers. *Phys. Fluids*, 8(2):310–321, 1996.

[67] P.C. Fernandes, P. Ern, F. Risso, and J. Magnaudet. Dynamics of axisymmetric bodies rising along a zigzag path. *J. Fluid Mech.*, 606:209–223, 2008.

[68] J.H. Ferziger and M. Peric. *Computational Methods for Fluid Dynamics.* Springer, 3rd edition, 2001.

[69] S.B. Field, M. Klaus, M.G. Moore, and F. Nori. Chaotic dynamics of falling disks. *Nature*, 388:252–254, 1997.

[70] K.A. Flack and M.P. Schultz. Review of hydraulic roughness scales in the fully rough regime. *J. Fluids Eng.*, 132:041203, 2010.

[71] T. Frank, P.J. Zwart, E. Krepper, H.-M. Prasser, and D. Lucas. Validation of CFD models for mono- and polydisperse air-water two-phase flows in pipes. *Nucl. Eng. Des.*, 238:647–659, 2008.

[72] J. Fröhlich. *Large Eddy Simulation turbulenter Strömungen.* Teubner, 2006.

[73] J. Fröhlich, C.P. Mellen, W. Rodi, L. Temmerman, and M.A. Leschziner. Highly resolved large-eddy simulation of separated flow in a channel with streamwise periodic constrictions. *J. Fluid Mech.*, 526:19–66, 2005.

[74] J. Fröhlich, S. Schwarz, S. Heitkam, C. Santarelli, C. Zhang, T. Vogt, S. Boden, A. Andruszkiewicz, K. Eckert, S. Odenbach, and S. Eckert. Influence of magnetic fields on the behavior of bubbles in liquid metals. *Eur. Phys. J. ST*, 220(1):167–183, 2013.

[75] Deutsches Institut für Normung e.V. *DIN 9300-1+2: Luft- und Raumfahrt; Begriffe, Größen und Formelzeichen der Flugmechanik; Bewegung des Luftfahrzeuges gegenüber der Luft.* Beuth-Verlag, 1990.

[76] V. Galindo, K. Niemietz, O. Pätzold, and G. Gerbeth. Numerical and experimental modeling of vgf-type buoyant flow under the influence of traveling and rotating magnetic fields. *J. Cryst. Growth*, 360:30–34, 2012.

[77] E.J. Garboczi. Three-dimensional mathematical analysis of particle shape using X-ray tomography and spherical harmonics: Application to aggregates used in concrete. *Cem. Concr. Res.*, 32(10):1621–1638, 2002.

[78] R. Garcia-Mayoral. *The interaction of riblets with wall-bounded turbulence.* PhD thesis, Universidad Politécnica de Madrid, 2011.

[79] R. Garcia-Mayoral and J. Jiménez. Drag reduction by riblets. *Phil. Trans. R. Soc. A*, 369:1412–1427, 2011.

[80] M. Garcia-Villalba and J.C. del Álamo. Turbulence modification by stable stratification in channel flow. *Phys. Fluids*, 23(4):045104, 2011.

[81] D. Gaudlitz. *Numerische Untersuchung des Aufstiegsverhaltens von Gasblasen in Flüssigkeiten.* PhD thesis, Technische Universität München, Fakultät für Maschinenwesen, 2008.

[82] D. Gaudlitz and N.A. Adams. Two-phase flow computations by the hybrid particle-level-set method. *5th Int. Symp. Turbul. Shear Flow Phenom.*, Vol. 1, 2007.

[83] D. Gaudlitz and N.A. Adams. On improving mass-conservation properties of the hybrid particle-level-set method. *Comput. Fluids*, 37(10):1320–1331, 2008.

[84] D. Gaudlitz and N.A. Adams. The influence of magnetic fields on the rise of gas bubbles in electrically conductive liquids. *Direct and Large-Eddy Simulation VII, ERCOFTAC Series*, 13(6):465–471, 2010.

[85] E. Gavze and M. Shapiro. Particles in a shear flow near a solid wall: Effect of non-sphericity on forces and velocities. *Int. J. Multiphase Flow*, 23:155–182, 1997.

[86] E. Gavze and M. Shapiro. Motion of inertial spheroidal particles in a shear flow near a solid wall with special application to aerosol transport in microgravity. *J. Fluid Mech.*, 371:59–79, 1998.

[87] G. Gerbeth, K. Eckert, and S. Odenbach. Electromagnetic flow control in metallurgy, crystal growth and electrochemistry. *Eur. Phys. J. ST*, 220(1):1–8, 2013.

[88] A. Giesecke, F. Stefani, T. Wondrak, and M. Xu. Forward and inverse problems in fundamental and applied magnetohydrodynamics. *Eur. Phys. J. ST*, 220(1):9–23, 2013.

[89] R. Glowinski, T.-W. Pan, T.I. Hesla, and D.D. Joseph. A distributed Lagrange multiplier/fictitious domain method for particulate flows. *Int. J. Multiphase Flow*, 25(5):755–794, 1999.

[90] R. Glowinski, T.-W. Pan, T.I. Hesla, D.D. Joseph, and J. Périaux. A fictitious domain approach to the direct numerical simulation of incompressible viscous flow past moving rigid bodies: Application to particulate flow. *J. Comput. Phys.*, 169(2):363–426, 2001.

[91] H. Goldstein, Jr. C.P. Poole, and J.L. Safko. *Klassische Mechanik, 3. Auflage.* WILEY-VCH Verlag, Weinheim, 2006.

[92] P. Gondret, M. Lance, and L. Petit. Bouncing motion of spherical particles in fluids. *Phys. Fluids*, 14:643–652, 2002.

[93] B.E. Griffith, R.D. Hornung, D.M. McQueen, and C.S. Peskin. An adaptive, formally second order accurate version of the immersed boundary method. *J. Comput. Phys.*, 223:10–49, 2007.

[94] D.G.E. Grigoriadis, S.C. Kassinos, and E.V. Votyakov. Immersed boundary method for the MHD flows of liquid metals. *J. Comput. Phys.*, 228(3):903–920, 2009.

[95] P. Haff and B. Werner. Computer simulation of the mechanical sorting of grains. *Powder Technol.*, 48:239–245, 1986.

[96] M. Hanke. *Eine numerische Methode zur Bestimmung erweiterter flugdynamischer Derivata durch aerostrukturdynamische Simulation.* PhD thesis, RWTH Achen, Fakultät für Mathematik, Informatik und Naturwissenschaften, 2003.

[97] F.H. Harlow and J.E. Welch. Numerical calculation of time-dependent viscous incompressible flow of fluid with free surface. *Phys. Fluids*, 8(12):2182–2189, 1965.

[98] C. Hartmann and A. Delgado. Numerical simulation of stress distribution on deformable drops. *Eng. Life Sci.*, 2:11–15, 2002.

[99] J. Hartmann. Hg-dynamics I: Theory of the laminar flow of an electrically conductive liquid in a homogeneous magnetic field. *K. Dan. Vidensk. Selsk. Mat. Fys. Medd.*, 15(6):1–28, 1937.

[100] J. Hartmann and F. Lazarus. Hg-dynamics II: Experimental investigations on the flow of mercury in a homogeneous magnetic field. *K. Dan. Vidensk. Selsk. Mat. Fys. Medd.*, 15(7):1–45, 1937.

[101] S. Heitkam, W. Drenckhan, and J. Fröhlich. Packing spheres tightly: Influence of mechanical stability on close-packed sphere structures. *Phys. Rev. Lett.*, 108:148302, Apr 2012.

[102] S. Heitkam, S. Schwarz, and J. Fröhlich. Simulation of the influence of electromagnetic fields on the drainage in wet metal foam. *Magnetohydrodynamics*, 48(2):313–320, 2012.

[103] S. Heitkam, S. Schwarz, C. Santarelli, and J. Fröhlich. Artificial zero-gravity in wet metal foam by means of electromagnetic fields. *Int. Conf. Multiphase Flow, Jeju, South Korea*, (728), 2013.

[104] S. Heitkam, S. Schwarz, C. Santarelli, and J. Fröhlich. Influence of an electromagnetic field on the formation of wet metal foam. *Eur. Phys. J. ST*, 220(1):207–214, 2013.

[105] R. Hermann, M. Uhlemann, H. Wendrock, G. Gerbeth, and B.Büchner. Magnetic field controlled single crystal growth and surface modification of titanium alloys exposed for biocompatibility. *J. Cryst. Growth*, 318(1):1048–1052, 2011.

[106] C. Hertel, M. Schümichen, J. Lang, and J. Fröhlich. Using a moving mesh PDE for cell centres to adapt a finite volume grid. *Flow Turbul. Combust*, 90:785–812, 2013.

[107] M. Hertel, A. Spille-Kohoff, U. Füssel, and M. Schnick. Numerical simulation of droplet detachment in pulsed gas-metal arc welding including the influence of metal vapour. *J. Phys. D: Appl. Phys.*, 46:224003, 2013.

[108] H. Hertz. Über die Berührung fester elastischer Körper. *Reine Angew. Math.*, 92:156–171, 1882.

[109] A. Hoffmann. *Auswahl und Validierung eines geeigneten Kollisionsmodells für Blasen.* Technische Universität Dresden, Fakultät für Maschinenwesen, Institut für Strömungsmechanik, Großer Beleg, 2012.

[110] K. Höfler and S. Schwarzer. Navier-Stokes simulation with constraint forces: Finite-difference method for particle-laden flows and complex geometries. *Phys. Rev. E*, 61:7146–7160, 2000.

[111] A. Hölzer and M. Sommerfeld. New simple correlation formula for the drag coefficient of non-spherical particles. *Powder Technol.*, 184(3):361–365, 2008.

[112] R.W. Hopper. Coalescence of two equal cylinders: Exact results for creeping viscous plane flow driven by capillarity. *J. Am. Ceram. Soc.*, 67:262–264, 1984.

[113] M. Horowitz and C.H.K. Williamson. The effect of Reynolds number on the dynamics and wakes of freely rising and falling spheres. *J. Fluid Mech.*, 651:251–294, 2010.

[114] M.S. Howe. On the force and moment on a body in an incompressible fluid, with application to rigid bodies and bubbles at high and low Reynolds numbers. *Q. J. Mech. Appl. Math.*, 48(3):401–426, 1995.

[115] S. Hoyas and J. Jiménez. Scaling of the velocity fluctuations in turbulent channels up to Re = 2003. *Phys. Fluids*, 18:011702, 1006.

[116] H. Huang, X. Yang, M. Krafczyk, and X.-Y. Lu. Rotation of spheroidal particles in Couette flows. *J. Fluid Mech.*, 692:369–394, 2 2012.

[117] J.A.Koza, S. Mühlenhoff, P. Zabinski, P.A. Nikrityuk, K. Eckert, M. Uhlemann, A. Gebert, T. Weier, L. Schultz, and S. Odenbach. Hydrogen evolution under the influence of a magnetic field. *Electrochim. Acta*, 56(6):2665–2675, 2011.

[118] P.J.A. Janssen and P.D. Anderson. Modeling film drainage and coalescence of drops in a viscous fluid. *Macromol. Mater. Eng.*, 296:238–248, 2011.

[119] G.B. Jeffery. The motion of ellipsoidal particles immersed in a viscous fluid. *Proc. R. Soc. London, Ser. A*, 102(715):161–179, 1922.

[120] M. Jenny. *Etude de la transition au chaos d'une sphère en ascension ou en chute libre dans un fluide Newtonien.* PhD thesis, Université Louis Pasteur Strasbourg, 2003.

[121] M. Jenny, G. Bouchet, and J. Dušek. Nonvertical ascension or fall of a free sphere in a Newtonian fluid. *Phys. Fluids*, 15(1):L9–L12, 2003.

[122] M. Jenny and J. Dušek. Efficient numerical method for the direct numerical simulation of the flow past a single light moving spherical body in transitional regimes. *J. Comput. Phys.*, 194(1):215–232, 2004.

[123] J. Jiménez. Turbulent flows over rough walls. *Ann. Rev. Fluid Mech.*, 36:173–196, 2004.

[124] T. A. Johnson and V. C. Patel. Flow past a sphere up to a Reynolds number of 300. *J. Fluid Mech.*, 378:19–70, 1999.

[125] G. Joseph and M.L. Hunt. Oblique particle wall collisions in a liquid. *J. Fluid Mech.*, 510:71–93, 2004.

[126] G. Joseph, R. Zenit, M.L. Hunt, and A.M. Rosenwinkel. Particle wall collisions in a viscous fluid. *J. Fluid Mech.*, 433:329–346, 2001.

[127] A.M. Kamp, A.K. Chesters, C. Colin, and J. Fabre. Bubble coalescence in turbulent flows: A mechanistic model for turbulence-induced coalescence applied to microgravity bubbly pipe flows. *Int. J. Multiphase Flow*, 27:1363–1396, 2001.

[128] D.G. Karamanev. The study of free rise of buoyant spheres in gas reveals the universal behaviour of free rising rigid spheres in fluid in general. *Int. J. Multiphase Flow*, 27(8):1479–1486, 2001.

[129] D.G. Karamanev, C. Chavarie, and R.C. Mayer. Dynamics of the free rise of a light solid sphere in liquid. *AlChE J.*, 42:1789–1792, 1996.

[130] D.G. Karamanev and L.N. Nikolov. Free rising spheres do not obey Newton's law for free settling. *AlChE J.*, 38:1843–1846, 1992.

[131] A. Karnis, H.L. Goldsmith, and S.G. Mason. Axial migration of particles in Poiseuille flow. *Nature*, 200:159–160, 1963.

[132] H. Kawamura, H. Abe, and K. Shingai. DNS of turbulence and heat transport in a channel flow with different Reynolds and Prandtl numbers and boundary conditions. *3rd Int. Symp. Turbul., Heat Mass Transfer, Edited by Nagano, Y., Hanjalic, K., Tsuji, T*, 1:15–32, 2000.

[133] T. Kawamura and Y. Kodama. Numerical simulation method to resolve interactions between bubbles and turbulence. *Int. J. Heat Fluid Flow*, 23(5):627–638, 2002.

[134] T. Kempe. *A numerical method for interface-resolving simulations of particle-laden flows with collisions.* TUDpress Verlag der Wissenschaften GmbH, 2013.

[135] T. Kempe and J. Fröhlich. Collision modelling for the interface-resolved simulation of spherical particles in viscous fluids. *J. Fluid Mech.*, 709:445–489, 2012.

[136] T. Kempe and J. Fröhlich. An improved immersed boundary method with direct forcing for the simulation of particle laden flows. *J. Comput. Phys.*, 231:3663–3684, 2012.

[137] C.A. Kennedy, M.H. Carpenter, and R.M. Lewis. Low-storage, explicit Runge-Kutta schemes for the compressible Navier-Stokes equations. *Appl. Numer. Math*, 35(3):177–219, 2000.

[138] E. Klaseboer, J.P. Chevailier, A. Mate, O. Masbernat, and C. Gourdon. Model and experiments of a drop impinging on an immersed wall. *Phys. Fluids*, 13(1):45–57, 2001.

[139] B. Knaepen and R. Moreau. Magnetohydrodynamic turbulence at low magnetic Reynolds number. *Annu. Rev. Fluid Mech.*, 40:25–45, 2008.

[140] H. Kobayashi. Large eddy simulation of magnetohydrodynamic turbulent channel flows with local subgrid-scale model based on coherent structures. *Phys. Fluids*, 18(4):045107, 2006.

[141] C. Köhler. *Fluid-Struktur-Interaktion eines Ellipsoids in Queranströmung.* Technische Universität Dresden, Fakultät für Maschinenwesen, Institut für Strömungsmechanik, Großer Beleg, 2011.

[142] D. Koschichow. *Numerische Modellierung magnetohydrodynamischer Effekte.* Technische Universität Dresden, Fakultät für Maschinenwesen, Institut für Strömungsmechanik, Großer Beleg, 2009.

[143] D. Krasnov, M. Rossi, O. Zikanov, and T. Boeck. Optimal growth and transition to turbulence in channel flow with spanwise magnetic field. *J. Fluid Mech.*, 596:73–101, 2008.

[144] D. Krasnov, E. Zienicke, O. Zikanov, T. Boeck, and A. Thess. Numerical study of the instability of the Hartmann layer. *J. Fluid Mech.*, 504:183–211, 2004.

[145] D. Krasnov, O. Zikanov, and T. Boeck. Comparative study of finite difference approaches in simulation of magnetohydrodynamic turbulence at low magnetic Reynolds number. *Comput. Fluids*, 50(1):46–59, 2011.

[146] D. Krasnov, O. Zikanov, and T. Boeck. Numerical study of magnetohydrodynamic duct flow at high Reynolds and Hartmann numbers. *J. Fluid Mech.*, 704:421–446, 7 2012.

[147] D. Krasnov, O. Zikanov, J. Schumacher, and T. Boeck. Magnetohydrodynamic turbulence in a channel with spanwise magnetic field. *Phys. Fluids*, 20:095105, 2008.

[148] F. Kuipers. *Klassische Mechanik, 8. Auflage.* WILEY-VCH Verlag, Weinheim, 2008.

[149] M. Kwakkel, W.-P. Breugem, and B.J. Boersma. DNS of turbulent bubbly downward flow in a vertical plane channel with a coupled level-set / volume of fluid method. *Proc. Direct and Large-Eddy Simulation IX, Dresden, Germany*, 2013.

[150] H. Lamb. *Hydrodynamics.* Dover, New York, 1932.

[151] P. Lancaster and K. Salkauskas. Surfaces generated by moving least squares methods. *Math. Comput.*, 37(155):141–158, 1981.

[152] D. Lee and H. Choi. Magnetohydrodynamic turbulent flow in a channel at low magnetic Reynolds number. *J. Fluid Mech.*, 439:367–394, 2001.

[153] D. Leenov and A. Kolin. Theory of electromagnetophoresis 1. Magnetohydrodynamic forces experienced by spherical and symmetrically oriented cylindrical particles. *J. Chem. Phys.*, 22(4):683–688, 1954.

[154] D. Legendre, C. Daniel, and P. Guiraud. Experimental study of a drop bouncing on a wall in a liquid. *Phys. Fluids*, 17(9):097105, 2005.

[155] D. Legendre and J. Magnaudet. The lift force on a spherical bubble in a viscous linear shear flow. *J. Fluid Mech.*, 368:81–126, 1998.

[156] F. Lehr, M. Millies, and D. Mewes. Bubble-size distributions and flow fields in bubble columns. *AlChE J.*, 48(11):2426–2443, 2002.

[157] P. Leopardi. A partiton of the unit sphere into regions of equal area and small diameter. *Electron. Trans. Numer. Anal.*, 25:309–327, 2006.

[158] D. Levin. The approximation power of moving least-squares. *Math. Comput.*, 67(224):1517–1531, 1998.

[159] W. Z. Li, Y. Y. Yan, and J. M. Smith. A numerical study of the interfacial transport characteristics outside spheroidal bubbles and solids. *Int. J. Multiphase Flow*, 29(3):435–460, 2003.

[160] Y. Liao and D. Lucas. A literature review of theoretical models for drop and bubble breakup in turbulent dispersions. *Chem. Eng. Sci.*, 64(15):3389–3406, 2009.

[161] Y. Liao and D. Lucas. A literature review on mechanisms and models for the coalescence process of fluid particles. *Chem. Eng. Sci.*, 65(10):2851–2864, 2010.

[162] Y. Liao, D. Lucas, E. Krepper, and Schmidtke. Development of a generalized coalescence and breakup closure for the inhomogeneous MUSIG model. *Nucl. Eng. Des.*, 241:1024–1033, 2011.

[163] J.T. Lindt. On the periodic nature of the drag on a rising bubble. *Chem. Eng. Sci.*, 27(10):1775–1781, 1972.

[164] M. Longuet-Higgins, B. Kerman, and K. Lunde. The release of air bubbles from an underwater nozzle. *J. Fluid Mech.*, 230:365–390, 1991.

[165] E. Loth. Quasi-steady shape and drag of deformable bubbles and drops. *Int. J. Multiphase Flow*, 34:523–546, 2008.

[166] J. Lu and G. Tryggvason. Effect of bubble deformability in turbulent bubbly upflow in a vertical channel. *Phys. Fluids*, 20(4):040701, 2008.

[167] K. Lunde and R. Perkins. Shape oscillations of rising bubbles. *Appl. Sci. Res.*, 58:387–408, 1998.

[168] J. Magnaudet and I. Eames. The motion of high-Reynolds-number bubbles in inhomogeneous flows. *Annu. Rev. Fluid Mech.*, 32(1):659–708, 2000.

[169] J. Magnaudet and G. Mougin. Wake instability of a fixed spheroidal bubble. *J. Fluid Mech.*, 572:311–337, 2007.

[170] J. Magnaudet, M. Rivero, and J. Fabre. Accelerated flows past a rigid sphere or a spherical bubble. Part 1. Steady straining flow. *J. Fluid Mech.*, 284:97–135, 1995.

[171] A.J. Majda and A.L. Bertozzi. *Vorticity and incompressible flow.* Cambridge Texts in Applied Mathematics, 2002.

[172] M. Mandø and L. Rosendahl. On the motion of non-spherical particles at high Reynolds number. *Powder Technol.*, 202(1-3):1–13, 2010.

[173] T. Maxworthy. Experimental studies in magneto-fluid dynamics: Pressure distribution measurements around a sphere. *J. Fluid Mech.*, 31:801–814, 1968.

[174] D.M. McQueen and C.S. Peskin. A three-dimensional computational method for blood flow in the heart. II. Contractile fibers. *J. Comput. Phys.*, 82(2):289–297, 1989.

[175] A. Menchaca-Rocha, A. Martinez-Dávalos, R. Núñez, S. Popinet, and S. Zaleski. Coalescence of liquid drops by surface tension. *Phys. Rev. E*, 63:046309, 2001.

[176] H.D. Mendelson. The prediction of bubble terminal velocities from wave theory. *AlChE J.*, 13(2):250–253, 1967.

[177] V. Metan, K. Eigenfeld, D. Räbiger, M. Leonhardt, and S. Eckert. Grain size control in al-si alloys by grain refinement and electromagnetic stirring. *J. Alloys Compd.*, 487(1-2):163–172, 2009.

[178] X. Miao, D. Lucas, Z. Ren, S. Eckert, and G. Gerbeth. Numerical modeling of bubble-driven liquid metal flows with external static magnetic field. *Int. J. Multiphase Flow*, 48:32–45, 2013.

[179] R. Mittal and G. Iaccarino. Immersed boundary methods. *Annu. Rev. Fluid Mech.*, 37:239–261, 2005.

[180] F.J. Moraga, S. Cancelos, and R.T. Lahey Jr. Modeling wall-induced forces on bubbles for inclined walls. *Multiphase Sci. Technol.*, 17(4):483–505, 2005.

[181] N. Mordant and J.-F. Pinton. Velocity measurement of a settling sphere. *Eur. Phys. J. B*, 18:343–352, 2000.

[182] R. Moreau. *Magnetohydrodynamics*, volume 3. Springer, 1990.

[183] Y. Mori, K. Hijikata, and I. Kuriyama. Experimental study of bubble motion in mercury with and without a magnetic field. *J. Heat Transfer*, 99(3):404–410, 1977.

[184] Y. Morinishi, T.S. Lund, O.V. Vasilyev, and P. Moin. Fully conservative higher order finite difference schemes for incompressible flow. *J. Comput. Phys.*, 143:90–124, 1998.

[185] R.D. Moser, J. Kim, and N.N. Mansour. Direct numerical simulation of turbulent channel flow up to $\mathrm{Re}_\tau = 590$. *Phys. Fluids*, 11:943–945, 1999.

[186] G. Mougin and J. Magnaudet. The generalized Kirchhoff equations and their application to the interaction between a rigid body and an arbitrary time-dependent viscous flow. *Int. J. Multiphase Flow*, 28:1837–1851, 2002.

[187] G. Mougin and J. Magnaudet. Wake-induced forces and torques on a zigzagging/spiralling bubble. *J. Fluid Mech.*, 567:185–194, 2006.

[188] R.F. Mudde. Gravity-driven bubbly flows. *Annu. Rev. Fluid Mech.*, 37:393–423, 2005.

[189] U. Müller and L. Bühler. *Magnetofluiddynamics in Channels and Containers.* Springer Verlag, 2001.

[190] Y. Murai. Fundamentals in frictional drag reduction by bubble injection. *Int. Conf. Multiphase Flow, Jeju, South Korea*, (plenary), 2013.

[191] G. Mutschke, K. Tschulik, M. Uhlemann, A. Bund, and J. Fröhlich. Comment on magnetic structuring of electrodeposits. *Phys. Rev. Lett.*, 109:229401, 2012.

[192] G. Mutschke, K. Tschulik, T. Weier, M. Uhlemann, A. Bund, and J. Fröhlich. On the action of magnetic gradient forces in micro-structured copper deposition. *Electrochim. Acta*, 55(28):9060–9066, 2010.

[193] A. Nealen. An as-short-as-possible introduction to the least squares, weighted least squares and moving least squares methods for scattered data approximation and interpolation. *http://nealen.de/projects/mls/asapmls.pdf*, 2004.

[194] I. Nezu and H. Nakagawa. *Turbulence in Open-Channel Flows.* IAHR/AIRH Monograph, 1993.

[195] M.-J. Ni and J.-F. Li. A consistent and conservative scheme for incompressible MHD flows at a low magnetic Reynolds number. Part III: On a staggered mesh. *J. Comput. Phys.*, 231(2):281–298, 2012.

[196] M.-J. Ni, R. Munipalli, N.B. Morley, P. Huang, and M.A. Abdou. A current density conservative scheme for incompressible MHD flows at a low magnetic Reynolds number. Part I: On a rectangular collocated grid system. *J. Comput. Phys.*, 227(1):174–204, 2007.

[197] M.-J. Ni, R. Munipalli, N.B. Morley, P. Huang, and M.A. Abdou. A current density conservative scheme for incompressible MHD flows at a low magnetic Reynolds number. Part II: On an arbitrary collocated mesh. *J. Comput. Phys.*, 227(1):174–204, 2007.

[198] F. Nicoud and F. Ducros. Subgrid-scale stress modelling based on the square of the velocity gradient tensor. *Flow Turbul. Combust.*, 62:183–200, 1999.

[199] N. Noack. *Immersed Boundary Methode für MHD Mehrphasenströmungen.* Technische Universität Dresden, Fakultät für Maschinenwesen, Institut für Strömungsmechanik, Diploma thesis, 2010.

[200] H. Oertel. *Prandtl - Führer durch die Strömungslehre. Grundlagen und Phänomene.* Vieweg Verlag, 2002.

[201] T. Ohwada and P. Asinari. Artificial compressibility method revisited: Asymptotic numerical method for incompressible Navier-Stokes equations. *J. Comput. Phys.*, 229(5):1698–1723, 2010.

[202] T.-W. Pan, C.-C. Chang, and R. Glowinski. On the motion of a neutrally buoyant ellipsoid in a three-dimensional Poiseuille flow. *Comput. Methods Appl. Mech. Engrg.*, 197:2198–2209, 2008.

[203] T.-W. Pan, R. Glowinski, and G.P.Galdi. Direct simulation of the motion of a settling ellipsoid in Newtonian fluid. *J. Comput. Appl. Math.*, 149:71–82, 2002.

[204] T.-W. Pan, D.D. Joseph, and R. Glowinski. Simulating the dynamics of fluid-ellipsoid interactions. *Comput. Struct.*, 83:463–478, 2005.

[205] T. Pang. *Introduction to Computational Physics.* Cambridge University Press, 1997.

[206] S. N. Paramonov, L. und Yaliraki. The directional contact distance of two ellipsoids: Coarse-grained potentials for anisotropic interactions. *J. Chem. Phys.*, 123:194111, 2005.

[207] J. W. Perram, J. Rasmussen, E. Praestgaard, and J.L. Lebowitz. Ellipsoid contact potential: Theory and relation to overlap potentials. *Phys. Rev. E*, 54:6566–6572, 1996.

[208] C.S. Peskin. Numerical analysis of blood flow in the heart. *J. Comput. Phys.*, 25(3):220–252, 1977.

[209] C.S. Peskin and D.M. McQueen. A three-dimensional computational method for blood flow in the heart. I. Immersed elastic fibers in a viscous incompressible fluid. *J. Comput. Phys.*, 81(2):372–405, 1989.

[210] M. Pilch and C.A. Erdman. Use of breakup time data and velocity history data to predict the maximum size of stable fragments for acceleration-induced breakup of a liquid drop. *Int. J. Multiphase Flow*, 13(6):741–757, 1987.

[211] B. Podvin, F.J. Moraga, and D. Attinger. Model and experimental visualizations of the interaction of a bubble with an inclined wall. *Chem. Eng. Sci.*, 63:1914–1928, 2008.

[212] T. Pöschel and T. Schwager. *Computational granular dynamics.* Springer, 2006.

[213] C. Pozrikidis. *Boundary-Integral and Singularity Methods for Linearized Viscous Flow.* Cambridge U.P., 1992.

[214] A. Prosperetti. Bubbles. *Phys. Fluids*, 16(6):1852–1865, 2004.

[215] A. Prosperetti and G. Tryggvason. *Computational Methods for Multiphase Flow.* Cambridge University Press, 2007.

[216] D. Qi and L.-S. Luo. Rotational and orientational behaviour of three-dimensional spheroidal particles in Couette flows. *J. Fluid Mech.*, 477:201–213, 2003.

[217] K. Quering. *Kollisionsmodellierung ellipsoider Partikel in einer viskosen Strömung.* Technische Universität Dresden, Fakultät für Maschinenwesen, Institut für Strömungsmechanik, Großer Beleg, 2011.

[218] D. Räbiger, S. Eckert, G. Gerbeth, S. Franke, and J. Czarske. Flow structures arising from melt stirring by means of modulated rotating magnetic fields. *Magnetohydrodynamics*, 48(1):213–220, 2012.

[219] M.M. Rai and P. Moin. Direct simulations of turbulent flow using finite-difference schemes. *J. Comput. Phys.*, 96:15–53, 1991.

[220] M.R. Raupach, R.A. Antonia, and S. Rajagopalan. Rough-wall turbulent boundary layers. *Appl. Mech. Rev.*, 44(1):1–25, 1991.

[221] C.P. Ribeiro and D. Mewes. On the effect of liquid temperature upon bubble coalescence. *Chem. Eng. Sci.*, 61:5704–5716, 2006.

[222] M.M. Ribeiro, P.F. Regueiras, M.M.L. Guimarães, C.M.N. Madureira, and J.J.C. Cruz-Pinto. Optimization of breakage and coalescence model parameters in a steady-state batch agitated dispersion. *IEC Research*, 50:2182–2191, 2011.

[223] G. Riboux, F. Risso, and D.G. Legendre. Experimental characterization of the agitation generated by bubbles rising at high Reynolds number. *J. Fluid Mech.*, 643:509–539, 2010.

[224] D. Richard and D. Quéré. Bouncing water drops. *Europhys. Lett.*, 50:769–775, 2001.

[225] A.M. Roma, C.S. Peskin, and M.J. Berger. An adaptive version of the immersed boundary method. *J. Comput. Phys.*, 153(2):509–534, 1999.

[226] T. Roßbach. *Konzeption für experimentelle Untersuchungen an einem umströmten, beweglich gelagerten Ellipsoid im Wasserumlaufkanal.* Technische Universität Dresden, Fakultät für Maschinenwesen, Institut für Strömungsmechanik, Diploma thesis, 2012.

[227] M.C. Ruzicka. On bubbles rising in line. *Int. J. Multiphase Flow*, 26(7):1141–1181, 2000.

[228] M.C. Ruzicka. Vertical stability of bubble chain: Multiscale approach. *Int. J. Multiphase Flow*, 31(10-11):1063–1096, 2005.

[229] H.K. Moffatt S. Molokov, R. Moreau. *Magnetohydrodynamics: Historical Evolution and Trends*, volume 80. Springer, 2007.

[230] P.G. Saffman. On the rise of small air bubbles in water. *J. Fluid Mech.*, 1:249–275, 1956.

[231] H. Sakamoto and H. Haniu. A study on vortex shedding from spheres in a uniform flow. *J. Fluids Eng.*, 112:386–392, 1990.

[232] D.W. Sallet. The drag and oscillating transverse force on vibrating cylinders due to steady fluid flow. *Ing. Arch.*, 44:113–122, 1975.

[233] T. Sanada, A. Sato, M. Shirota, and M. Watanabe. Motion and coalescence of a pair of bubbles rising side by side. *Chem. Eng. Sci.*, 64:2659–2671, 2009.

[234] T. Sanada, M. Shirota, and M. Watanabe. Bubble wake visualization by using photochromic dye. *Chem. Eng. Sci.*, 62(24):7264–7273, 2007.

[235] T. Sanada, M. Watanabe, T. Fukano, and A. Kariyasaki. Behavior of a single coherent gas bubble chain and surrounding liquid jet flow structure. *Chem. Eng. Sci.*, 60(17):4886–4900, 2005.

[236] C. Santarelli and J. Fröhlich. Simulation of bubbly flow in a vertical turbulent channel. *Proc. Appl. Math. Mech.*, 12:503–504, 2012.

[237] C. Santarelli and J. Fröhlich. Characterisation of bubble clusters in simulations of upward turbulent channel flow. *Proc. Appl. Math. Mech.*, 13, 2013.

[238] C. Santarelli, S. Schwarz, and J. Fröhlich. Numerical simulation of light ellipsoidal particles in vertical turbulent channel flow. *Int. Conf. Multiphase Flow, Jeju, South Korea*, (746), 2013.

[239] I. E. Sarris, S.C. Kassinos, and D. Carati. Large-eddy simulations of the turbulent Hartmann flow close to the transitional regime. *Phys. Fluids*, 19(8):085109, 2007.

[240] R. Scardovelli and S. Zaleski. Direct numerical simulation of free surface flows and interfacial flows. *Annu. Rev. Fluid Mech.*, 31:567–603, 1999.

[241] H. Schlichting and K. Gersten. *Boundary layer theory.* Springer, 8th edition, 2000.

[242] L. Schouveiler, A. Brydon, T. Leweke, and M.C. Thompson. Interactions of the wakes of two spheres placed side by side. *Eur. J. Mech. - B/Fluids*, 23(1):137–145, 2004.

[243] A.L. Schwab and J.P. Meijaard. How to draw Euler angles and utilize Euler parameters. *Proc. of IDETC/CIE, ASME*, page 99307, 2006.

[244] S. Schwarz and J. Fröhlich. DNS of single bubble motion in liquid metal and the influence of a magnetic field. *7th Int. Symp. Turbul. Shear Flow Phenom., Ottawa, Canada*, 1, 2011.

[245] S. Schwarz and J. Fröhlich. Numerical simulation of MHD flow around fixed and freely ascending spheres and ellipsoids. *8th PAMIR Int. Conf. Fundamen. Appl. MHD, Borgo, Corsica, France*, I:223–227, 2011.

[246] S. Schwarz and J. Fröhlich. Representation of deformable bubbles by analytically defined shapes in an immersed boundary method. *Proc. ICNAAM, AIP Conf. Series, edited by T.E. Simos, G.Psihoyios, C.Tsitouras and Z. Anastassi*, 1479:104–108, 2012.

[247] S. Schwarz and J. Fröhlich. Simulation of a bubble chain in a container of high aspect ratio exposed to a magnetic field. *Eur. Phys. J. ST*, 220(1):195–205, 2013.

[248] S. Schwarz and J. Fröhlich. Numerical study of single bubble motion in liquid metal exposed to a longitudinal magnetic field. *Int. J. Multiphase Flow, IJMF-D-13-00172R2, accepted*, 2014.

[249] S. Schwarz and J. Fröhlich. A temporal discretization scheme to compute the motion of light particles in viscous flows by an immersed boundary method. *J. Comput. Phys., submitted*, 2014.

[250] S. Schwarz, S. Tschisgale, and J. Fröhlich. Modeling of bubble coalescence in phase-resolving simulations by an immersed boundary method. *Int. Conference on Multiphase Flows, Jeju, South Korea*, (750), 2013.

[251] T.V.S. Sekhar, R. Sivakumar, and T. V. R. R. Kumar. Flow around a circular cylinder in an external magnetic field at high Reynolds numbers. *Int. J. of Numer. Method H*, 16:740–759, 2006.

[252] T.V.S. Sekhar, R. Sivakumar, and T.V.R. Ravi Kumar. Magnetohydrodynamic flow around a sphere. *Fluid Dyn. Res.*, 37(5):357–373, 2005.

[253] J. A. Sethian. Evolution, implementation, and application of level set and fast marching methods for advancing fronts. *J. Comput. Phys.*, 169(2):503–555, 2001.

[254] J.A. Shercliff. *Textbook of Magnetohydrodynamics*. Pergamon Press, 1965.

[255] N. Shevchenko, S. Boden, S. Eckert, D. Borin, M. Heinze, and S. Odenbach. Application of X-ray radioscopic methods for characterization of two-phase phenomena and solidification processes in metallic melts. *Eur. Phys. J. ST*, 220(1):63–77, 2013.

[256] Y. Shibasaki, K. Ueno, and T. Tagawa. Computation of a rising bubble in an enclosure filled with liquid metal under vertical magnetic fields. *ISIJ Int.*, 50:363–370, 2010.

[257] M. Simcik, M.C. Ruzicka, and J. Drahos. Computing the added mass of dispersed particles. *Chem. Eng. Sci.*, 63:4580–4595, 2008.

[258] P.R. Spalart, R.D. Moser, and M.M. Rogers. Spectral methods for the Navier-Stokes equations with one infinite and two periodic directions. *J. Comput. Phys.*, 96:297–324, 1991.

[259] K.W. Spring. Euler parameters and the use of quaternion algebra in the manipulation of finite rotations: A review. *Mech. Mach. Theory*, 21(5):365–373, 1986.

[260] J.H. Spurk. *Strömungslehre: Einführung in die Theorie der Strömungen*. Springer, 2007.

[261] F. Stefani and G. Gerbeth. Asymmetric polarity reversals, bimodal field distribution, and coherence resonance in a spherically symmetric mean-field dynamo model. *Phys. Rev. Lett.*, 94:184506, May 2005.

[262] J. Stiller and K. Koal. A numerical study of the turbulent flow driven by rotating and travelling magnetic fields in a cylindrical cavity. *J. Turbul.*, 10:N44, 2009.

[263] J. Stiller, K. Koal, W.E. Nagel, J. Pal, and A. Cramer. Liquid metal flows driven by rotating and traveling magnetic fields. *Eur. Phys. J. ST*, 220(1):111–122, 2013.

[264] R.L. Stover. *Bubble coalescence dynamics and supersaturation in electrolytic gas evolution.* PhD thesis, Ernest Orlando Lawrence Berkeley National Laboratory, 1996.

[265] R.L. Stover, C.W. Tobias, and M.M. Denn. Bubble coalescence dynamics. *AlChE J.*, 43(10):2385–2393, 1997.

[266] L. Temmerman, M.A. Leschziner, C.P. Mellen, and J. Fröhlich. Investigation of wall-function approximations and subgrid-scale models in large-eddy simulation of seperated flow in a channel with periodic constrictions. *Int. J. Heat Fluid Flow*, 24:157–180, 2003.

[267] S. Thomas, A. Esmaeeli, and G. Tryggvason. Multiscale computations of thin films in multiphase flows. *Int. J. Multiphase Flow*, 36(1):71–77, 2010.

[268] S.T. Thoroddsen, T.G. Etoh, and K. Takehara. High-speed imaging of drops and bubbles. *Annu. Rev. Fluid Mech.*, 40(1):257–285, 2008.

[269] S.T. Thoroddsen, T.G. Etoh, K. Takehara, and N. Ootsuka. On the coalescence speed of bubbles. *Phys. Fluids*, 17(7):071703, 2005.

[270] S.T. Thoroddsen, K. Takehara, and T.G. Etoh. The coalescence speed of a pendent and a sessile drop. *J. Fluid Mech.*, 527:85–114, 2005.

[271] K. Timmel, S. Eckert, and G. Gerbeth. Experimental investigation of the flow in a continuous casting mould under the influence of a transverse, direct current magnetic field. *Metall. Mater. Trans. B*, 42:68–80, 2011.

[272] K. Timmel, S. Eckert, G. Gerbeth, and F. Stefani. Experimental modeling of the continuous casting process of steel using low melting point alloys - the LIMMCAST program. *ISIJ Int.*, 58(8):1134–1141, 2010.

[273] A.G. Tomboulides and S.A. Orszag. Numerical investigation of transitional and weak turbulent flow past a sphere. *J. Fluid Mech.*, 416:45–73, 2000.

[274] A. Tomiyama, G. P. Celatab, S. Hosokawaa, and S. Yoshida. Terminal velocity of single bubbles in surface tension force dominant regime. *Int. J. Multiphase Flow*, 28:1497–1519, 2002.

[275] G. Tryggvason, B. Bunner, A. Esmaeeli, D. Juric, N. Al-Rawahi, W. Tauber, J. Han, S. Nas, and Y.-J. Jan. A front-tracking method for the computations of multiphase flow. *J. Comput. Phys.*, 169:708–759, 2001.

[276] G. Tryggvason, S. Dabiri, B. Aboulhasanzadeh, and J. Lu. Multiscale considerations in direct numerical simulations of multiphase flows. *Phys. Fluids*, 25(3):031302, 2013.

[277] G. Tryggvason, S. Thomas, J. Lu, and B. Aboulhasanzadeh. Multiscale issues in DNS of multiphase flows. *Acta Mathematica Scientia*, 30(2):551–562, 2010.

[278] H.-K. Tsao and D.L. Koch. Collisions of slightly deformable, high Reynolds number bubbles with short-range repulsive forces. *Phys. Fluids*, 6(8):2591–2605, 1994.

[279] H.-K. Tsao and D.L. Koch. Observations of high Reynolds number bubbles interacting with a rigid wall. *Phys. Fluids*, 9:44–56, 1997.

[280] S. Tschisgale. *Modellierung von Blasenkoaleszenz mit sphärischen harmonischen Funktionen.* Technische Universität Dresden, Fakultät für Maschinenwesen, Institut für Strömungsmechanik, Diploma thesis, 2012.

[281] M. Uhlmann. Simulation of particulate flows on multi-processor machines with distributed memory. ISSN 1135-9420, 2003.

[282] M. Uhlmann. An immersed boundary method with direct forcing for the simulation of particulate flows. *J. Comput. Phys.*, 209:448–476, 2005.

[283] M. Uhlmann. Interface-resolved direct numerical simulation of vertical particulate channel flow in the turbulent regime. *Phys. Fluids*, 20(5):053305, 2008.

[284] Gambit user guide. Gambit 2.4.6, ANSYS FLUENT, www.fluent.com. 2013.

[285] M. Vanella and E. Balaras. A moving-least-squares reconstruction for embedded-boundary formulations. *J. Comput. Phys.*, 228(18):6617–6628, 2009.

[286] C.H.J. Veldhuis, A.Biesheuvel, L. van Wijngaarden, and D. Lohse. Motion and wake structure of spherical particles. *Nonlinearity*, 18:C1–C2, 2005.

[287] C.H.J. Veldhuis and A. Biesheuvel. An experimental study of the regimes of motion of spheres falling or ascending freely in a Newtonian fluid. *Phys. Fluids*, 33:1074–1087, 2007.

[288] C.H.J. Veldhuis, A. Biesheuvel, and D. Lohse. Freely rising light solid spheres. *Int. J. Multiphase Flow*, 35(4):312–322, 2009.

[289] C.H.J. Veldhuis, A. Biesheuvel, and L. van Wijngaarden. Shape oscillations on bubbles rising in clean and in tap water. *Phys. Fluids*, 20(4):040705, 2008.

[290] G.T. Vickers. The projected areas of ellipsoids and cylinders. *Powder Technol.*, 86(2):195–200, 1996.

[291] A. Viré and B. Knaepen. On discretization errors and subgrid scale model implementations in large eddy simulations. *J. Comput. Phys.*, 228(22):8203–8213, 2009.

[292] T. von Kármán and H. Rubach. Über den Mechanismus des Flüssigkeits- und Luftwiderstandes. *Phys. Z.*, 2:49–59, 1912.

[293] A. Vorobev, O. Zikanov, P.A. Davidson, and B. Knaepen. Anisotropy of magnetohydrodynamic turbulence at low magnetic Reynolds number. *Phys. Fluids*, 17(12):125105, 2005.

[294] B. Vowinckel and J. Fröhlich. Numerical simulation of bed load transport in open channel flow. *Proc. Appl. Math. Mech.*, 12:505–506, 2012.

[295] B. Vowinckel, T. Kempe, and J. Fröhlich. Impact of collision models on particle transport in open channel flow. *Proc. 7th Int. Symp. Turbul. Shear Flow Phenom., Ottawa, Canada*, 2011.

[296] B. Vowinckel, T. Kempe, J. Fröhlich, and V. Nikora. Direct numerical simulation of bed-load transport of finite-size spherical particles in a turbulent channel flow. *Proc. Direct and Large-Eddy Simulation IX, Dresden, Germany*, 2013.

[297] B. Vowinckel, T. Kempe, J. Fröhlich, and V.I.Nikora. Numerical simulation of sediment transport in open channel flow. *In Murillo, R., editor, River Flow*, ISBN 978-0-415-62129-8:507–514, 2012.

[298] D.R. Waigh and R.J. Kind. Improved aerodynamic characterization of regular three-dimensional roughness. *AIAA J.*, 36(6):1117–1119, 1998.

[299] L. Wakaba and S. Balachandar. On the added mass force at finite Reynolds and acceleration numbers. *Theor. Comput. Fluid Dyn.*, 21(2):147–153, 2007.

[300] Y. Wang and L. Zhang. Fluid flow-related transport phenomena in steel slab continuous casting strands under electromagnetic brake. *Metall. Mater. Trans. B*, 42:1319–1351, 2011.

[301] T. Weier and S. Landgraf. The two-phase flow at gas-evolving electrodes: Bubble-driven and Lorentz-force-driven convection. *Eur. Phys. J. ST*, 220(1):313–322, 2013.

[302] E.W. Weisstein. From MathWorld - a Wolfram web resource, http://mathworld.wolfram.com. 2013.

[303] R.M. Wellek, A.K. Agrawal, and A.H.P. Skelland. Shape of liquid drops moving in liquid media. *AlChE J.*, 12:854–862, 1966.

[304] W. Wilke. *Segmentierung und Approximation großer Punktwolken.* PhD thesis, TU Darmstadt, 2000.

[305] J. H. Williamson. Low-storage Runge-Kutta schemes. *J. Comput. Phys.*, 35:48–56, 1980.

[306] T. Wondrak, S. Eckert, G. Gerbeth, K. Klotsche, F. Stefani, K. Timmel, A. Peyton, N. Terzija, and W. Yin. Combined electromagnetic tomography for determining two-phase flow characteristics in the submerged entry nozzle and in the mold of a continuous casting model. *Metall. Mater. Trans. B*, 42:1201–1210, 2011.

[307] M. Wörner. Numerical modeling of multiphase flows in microfluidics and micro process engineering: A review of methods and applications. *Microfluid Nanofluid*, 12:841–886, 2012.

[308] A.A. Wray. Minimal storage time advancement schemes for spectral methods. NASA Ames Research Center, 1986.

[309] M. Wu and M. Gharib. Experimental studies on the shape and path of small air bubbles rising in clean water. *Phys. Fluids*, 14(7):L49–L52, 2002.

[310] E.J.W. Wynn. Simulations of rebound of an elastic ellipsoid colliding with a plane. *Powder Technol.*, 196(1):62–73, 2009.

[311] F. Yang and M. Hunt. Dynamics of particle-particle collisions in a viscous liquid. *Phys. Fluids*, 18:121506, 2006.

[312] B.A. Yergey, Beninati M.-L., and J. Marshall. Sensitivity of incipient particle motion to fluid flow penetration depth within a packed bed. *Sedimentology*, 57(2):418–428, 2010.

[313] C. Yin, L. Rosendahl, S. Knudsen Kaer, and H. Sørensen. Modelling the motion of cylindrical particles in a nonuniform flow. *Chem. Eng. Sci.*, 58(15):3489–3498, 2003.

[314] G. Yonas. Measurements of drag in a conducting fluid with an aligned field and large interaction parameter. *J. Fluid Mech.*, 30:813–821, 1967.

[315] Z. Yu, N. Phan-Thien, and R.I. Tanner. Rotation of a spheroid in a Couette flow at moderate Reynolds numbers. *Phys. Rev. E*, 76(2):026310, 2007.

[316] M. Zastawny, G. Mallouppas, F. Zhao, and B. van Wachem. Derivation of drag and lift force and torque coefficients for non-spherical particles in flows. *Int. J. Multiphase Flow*, 39:227–239, 2012.

[317] L. Zeng, S. Balachandar, and P. Fischer. Wall-induced forces on a rigid sphere at finite Reynolds number. *J. Fluid Mech.*, 536:1–25, 2005.

[318] R. Zenit and D. Legendre. The coefficient of restitution for air bubbles colliding against solid walls in viscous liquids. *Phys. Fluids*, 21:083306, 2009.

[319] C. Zhang, S. Eckert, and G. Gerbeth. Experimental study of single bubble motion in a liquid metal column exposed to a DC magnetic field. *Int. J. Multiphase Flow*, 106:824–842, 2005.

[320] C. Zhang, S. Eckert, and G. Gerbeth. The flow structure of a bubble-driven liquid metal jet in a horizontal magnetic field. *J. Fluid Mech.*, 557:57–82, 2007.

[321] Z.C. Zheng and N. Zhang. Frequency effects on lift and drag for flow past an oscillating cylinder. *J. Fluids Struct.*, 24(3):382–399, 2008.

Appendix

A Discretized MHD equations

This appendix presents the discretized electrodynamic equations for the computations of the Lorentz force in Section 2.1. The indices i, j, k are cell centered and $i+\frac{1}{2}, j+\frac{1}{2}, k+\frac{1}{2}$ correspond to faces of the cuboidal control volume sketched in Figure 2.1. The finite volume discretization can also be interpreted in terms of finite differences. Central difference schemes are employed. First, the electric fields originating from the gradient of the electric potential, $\mathbf{e}_\Phi$, and the electric field based on the cross product of velocity and magnetic field, $\mathbf{e}_\mathbf{u}$, are computed. Constant indices are dropped on the rhs:

$$\begin{aligned} e_{\Phi,x}^{i+\frac{1}{2},j,k} &= -\frac{\Phi^{i+1}-\Phi^{i}}{x^{i+1}-x^{i}}, \\ e_{\Phi,y}^{i,j+\frac{1}{2},k} &= -\frac{\Phi^{j+1}-\Phi^{j}}{y^{j+1}-y^{j}}, \\ e_{\Phi,z}^{i,j,k+\frac{1}{2}} &= -\frac{\Phi^{k+1}-\Phi^{k}}{z^{k+1}-z^{k}}, \end{aligned} \tag{A.1}$$

$$\begin{aligned} e_u^{i+\frac{1}{2},j,k} &= v^{i+\frac{1}{2}}B_z^{i+\frac{1}{2}} - w^{i+\frac{1}{2}}B_y^{i+\frac{1}{2}}, \\ e_v^{i+\frac{1}{2},j,k} &= w^{j+\frac{1}{2}}B_x^{j+\frac{1}{2}} - u^{j+\frac{1}{2}}B_z^{j+\frac{1}{2}}, \\ e_w^{i,j,k+\frac{1}{2}} &= u^{k+\frac{1}{2}}B_y^{k+\frac{1}{2}} - v^{k+\frac{1}{2}}B_x^{j+\frac{1}{2}}. \end{aligned} \tag{A.2}$$

Herein, $\mathbf{u}$ and $\mathbf{B}$ have to be interpolated to the required face positions. The preliminary current density is then computed from

$$\begin{aligned} j_x^{*,i+\frac{1}{2},j,k} &= \sigma_e\left(e_{\Phi,x}^{i+\frac{1}{2}} + e_u^{i+\frac{1}{2}}\right), \\ j_y^{*,i+\frac{1}{2},j,k} &= \sigma_e\left(e_{\Phi,y}^{j+\frac{1}{2}} + e_v^{j+\frac{1}{2}}\right), \\ j_z^{*,i,j,k+\frac{1}{2}} &= \sigma_e\left(e_{\Phi,z}^{k+\frac{1}{2}} + e_w^{k+\frac{1}{2}}\right), \end{aligned} \tag{A.3}$$

where $\mathbf{j}^*$ does not necessarily fulfill charge conservation. A divergence free current field is obtained by a projection method solving a correction Poisson equation for $\delta\Phi$. This is done in a similar way as for the pressure. Indeed the same Poisson solver [134] is used and therefore not discussed in further detail here,

$$\nabla^2\delta\Phi = \frac{j_x^{*,i+\frac{1}{2}} - j_x^{*,i-\frac{1}{2}}}{x^{i+\frac{1}{2}} - x^{i-\frac{1}{2}}} + \frac{j_y^{*,j+\frac{1}{2}} - j_y^{*,j-\frac{1}{2}}}{y^{j+\frac{1}{2}} - y^{j-\frac{1}{2}}} + \frac{j_z^{*,k+\frac{1}{2}} - j_z^{*,k-\frac{1}{2}}}{z^{k+\frac{1}{2}} - z^{k-\frac{1}{2}}}, \tag{A.4}$$

$$\Phi^{i,j,k} = \Phi_0^{i,j,k} + \delta\Phi^{i,j,k}\,, \tag{A.5}$$

where Φ_0 is the previous value of the electric potential.
The divergence free current density follows from:

$$\begin{aligned} j_x^{i+\frac{1}{2},j,k} &= j_x^{*,i+\frac{1}{2}} - \frac{\delta\Phi^{i+1} - \delta\Phi^{i}}{x^{i+1} - x^{i}}\,, \\ j_y^{i+\frac{1}{2},j,k} &= j_y^{*,j+\frac{1}{2}} - \frac{\delta\Phi^{j+1} - \delta\Phi^{j}}{y^{j+1} - y^{j}}\,, \\ j_z^{i,j,k+\frac{1}{2}} &= j_z^{*,k+\frac{1}{2}} - \frac{\delta\Phi^{k+1} - \delta\Phi^{k}}{z^{k+1} - z^{k}}\,, \end{aligned} \tag{A.6}$$

Finally, the Lorentz force is calculated [94] by

$$\begin{aligned} f_{L,x}^{i+\frac{1}{2},j,k} &= j_y^{i+\frac{1}{2}} B_z^{i+\frac{1}{2}} - j_z^{i+\frac{1}{2}} B_y^{i+\frac{1}{2}}\,, \\ f_{L,y}^{i+\frac{1}{2},j,k} &= j_z^{j+\frac{1}{2}} B_x^{j+\frac{1}{2}} - j_x^{j+\frac{1}{2}} B_z^{j+\frac{1}{2}}\,, \\ f_{L,z}^{i,j,k+\frac{1}{2}} &= j_x^{k+\frac{1}{2}} B_y^{k+\frac{1}{2}} - j_y^{k+\frac{1}{2}} B_x^{j+\frac{1}{2}}\,, \end{aligned} \tag{A.7}$$

where $\mathbf{j}$ has to be interpolated to the required face positions.
The face-to-face interpolation is illustrated by an example for $v^{i+\frac{1}{2},j,k}$, the other quantities are interpolated in the same fashion. On a uniform grid, interpolation with a 4-point stencil yields

$$v^{i+\frac{1}{2},j,k} = \frac{1}{4}\left(v^{i,j+\frac{1}{2},k} + v^{i,j-\frac{1}{2},k} + v^{i+1,j+\frac{1}{2},k} + v^{i+1,j-\frac{1}{2},k}\right). \tag{A.8}$$

The face-to-face interpolation can be interpreted as a two-step procedure. This is now outlined by the interpolation with a 4-point stencil on a non-uniform grid. The face-to-face interpolation can be seen as two face-to-center interpolations where the results are then combined by a center-to-face interpolation. First, the velocity fluxes at the faces of the control volume are used to obtain the velocity at the center,

$$\begin{aligned} v^{i,j,k} &= \frac{1}{2}\left(v^{i,j+\frac{1}{2},k} + v^{i,j-\frac{1}{2},k}\right), \\ v^{i+1,j,k} &= \frac{1}{2}\left(v^{i+1,j+\frac{1}{2},k} + v^{i+1,j-\frac{1}{2},k}\right). \end{aligned} \tag{A.9}$$

Second, the velocity at the face position is calculated,

$$v^{i+\frac{1}{2},j,k} = v^{i,j,k} + \frac{\left(v^{i+1,j,k} - v^{i,j,k}\right)}{x^{i+1} - x^{i}}\left(x^{i+\frac{1}{2}} - x^{i}\right). \tag{A.10}$$

B Number of forcing points

The number of forcing points is provided for one Lagrangian forcing point controlling one Eulerian cell, i.e. $V_E = V_L$. A shell with a thickness of one mesh width h is formed by two tri-axial ellipsoids with the respective semi-axes $a \pm h/2$, $b \pm h/2$ and $c \pm h/2$. The shell volume is obtained from

$$V_{shell} = \frac{4}{3}\pi h\left(ab + ac + bc + \frac{1}{4}h^2\right). \tag{B.1}$$

Given an even distribution of N_L Lagrangian surface markers, a single forcing point volume shall be given by $V_L = V_{shell}/N_L$. For an equidistant Eulerian grid and thus $V_E = h^3 = V_L$, we get

$$N_{L,ellipsoid} = \frac{\pi}{3}\left(\frac{4(ab+ac+bc)}{h^2}+1\right). \tag{B.2}$$

Equation (B.2) yields for an oblate ellipsoid, $a = c > b$,

$$N_{L,oblate} = \frac{\pi}{3}\left(\frac{8ab+4a^2}{h^2}+1\right) = \frac{\pi}{3}\left(\frac{4r_{eq}^2\left(2X^{-1/3}+X^{2/3}\right)}{h^2}+1\right), \tag{B.3}$$

and for a prolate ellipsoid, $a > b = c$,

$$N_{L,prolate} = \frac{\pi}{3}\left(\frac{8ac+4c^2}{h^2}+1\right) = \frac{\pi}{3}\left(\frac{4r_{eq}^2\left(2X^{1/3}+X^{-2/3}\right)}{h^2}+1\right). \tag{B.4}$$

It simplifies for a sphere $a = b = c$ to the known form [282]

$$N_{L,sphere} = \frac{\pi}{3}\left(\frac{12a^2}{h^2}+1\right). \tag{B.5}$$

In the above equations, we employed the ellipsoid aspect ratio $X = a/b$ and the volume-equivalent radius obtained from $V = 4/3\,\pi abc = 4/3\,\pi r_{eq}^3$. This also yields:

$$\text{oblate ellipsoid:}\quad a = c = r_{eq}X^{1/3} \quad\text{and}\quad b = r_{eq}X^{-2/3},$$

$$\text{prolate ellipsoid:}\quad a = r_{eq}X^{2/3} \quad\text{and}\quad b = c = r_{eq}X^{-1/3}.$$

The provided number of forcing points, N_L, constitutes a lower limit for the present implementation in PRIME.

C Normal vector and curvature for triangulated surface

For particles of ellipsoidal shape or particles represented by spherical harmonics, the analytical solution can be used for the surface normal and the surface curvature. For general geometries, the triangulated surface is used to compute these quantities at the forcing point location. Two approaches were considered,

i) based on discrete surface triangles,

ii) based on a local Monge patch [302].

Figure 3.2 provides the nomenclature. For variant i), the unit surface normal of a triangle is obtained from

$$\mathbf{n}_{tri} = \frac{\mathbf{a}_\triangle \times \mathbf{b}_\triangle}{\|\mathbf{a}_\triangle \times \mathbf{b}_\triangle\|}. \tag{C.1}$$

It is further ensured that the normal vector points outwards. The surface normal vector at the forcing point is obtained from a weighted average of the triangles associated with this point,

$$\mathbf{n}_{fp} = \frac{1}{\sum_{n=1}^{N_{nt}} w^{(n)}} \sum_{n=1}^{N_{nt}} w^{(n)} \mathbf{n}_{tri}^{(n)} , \tag{C.2}$$

where $w^{(n)} = 1$ gives an arithmetic average, $w^{(n)} = 1/A_{tri}^{(n)}$ weights with the inverse of the triangle area. Further details are provided in [304].
The general definition of the mean curvature is

$$H = \frac{1}{2} \nabla \cdot \mathbf{n} , \tag{C.3}$$

where the evaluation is rather delicate on a unstructured triangular surface grid. For suitable approximations, see [304]. The computation of curvature for the surface tension force is also addressed in [275].
A simple measure of the curvature is obtained by fitting a sphere to two triangles sharing one edge and adjacent to the forcing point of interest. The procedure is repeated for all neighbor triangles and the mean curvature is approximated from $H_{fp} = 1/2 \left(1/\min(R_s^{(i)}) + 1/\max(R_s^{(i)})\right)$ where $R_s^{(i)}$ denotes the radii of the fitted spheres with $i = 1, \ldots, N_{nt}$.
With a more sophisticated approach, a differentiable representation is created based on the discrete triangulation which leads to variant ii), the Monge patch. A triangle and its neighbor triangles which share one edge are considered providing in total six corner points (see Figure 3.2). A local coordinate system, shifted and rotated to the triangle reference frame, is employed. We fit a quadratic surface to define the Monge patch $\mathbf{x} = (u,\, v\,,\, h(u,v))$ with

$$h(u,v) = c_0 + c_1 u + c_2 v + c_3 uv + c_4 u^2 + c_5 v^2 . \tag{C.4}$$

The surface normal in the projected center of mass of the triangle, $\tilde{\mathbf{S}}_{tri}$, is then computed from

$$\mathbf{n}_{tri}(\tilde{\mathbf{S}}_{tri}(u,v)) = \frac{\mathbf{x}_u \times \mathbf{x}_v}{\|\mathbf{x}_u \times \mathbf{x}_v\|} , \tag{C.5}$$

where the indices denote partial derivatives with respect to u and v. To obtain the surface normal at the forcing point, $\mathbf{n}_{fp}$, the result is rotated back to the particle reference frame, an outward orientation is ensured and again a weighted average for the triangles is employed. The mean curvature can now be expressed by

$$H(\tilde{\mathbf{S}}_{tri}(u,v)) = \frac{(1 + h_v^2)\, h_{uu} - 2 h_u h_v h_{uv} + (1 + h_u^2)\, h_{vv}}{2\,(1 + h_u^2 + h_v^2)^{3/2}} . \tag{C.6}$$

The curvature at the forcing point H_{fp} location is again obtained from the weighted average of H_{tri}. Equation (C.6) also yields the analytical solution for the mean curvature of an ellipsoid (4.10). Ellipsoids of various aspect ratios were used for validation purposes. The agreement with the analytical values is good for moderate aspect ratios and $N_L \geq 500$ [275]. However, larger errors in curvature arise for stronger deformations or locally poor resolution. This motivated the analytical representation of the entire particle by spherical harmonics.

D WALE Model

The Wall-Adapted Local Eddy viscosity model [53, 198] has the advantage of reproducing the scaling $\nu_t \sim y^{+3}$ near a solid wall and therefore no additional wall damping, like van Driest damping for the Standard Smagorinsky model, is needed. The turbulent eddy viscosity vanishes in case of pure shear. These properties should be advantageous in connection with the application of an immersed boundary method. Modeling of the eddy viscosity is realized by

$$\nu_t = C_\nu V_{st}^{2/3} \frac{(S_{ij}^d S_{ij}^d)^{3/2}}{(\overline{S}_{ij}\overline{S}_{ij})^{5/2} + (S_{ij}^d S_{ij}^d)^{5/4}} , \tag{D.1}$$

where $V_{st} = \Delta x\, \Delta y\, \Delta z$ is the cell volume of the staggered grid. A model constant of $C_\nu = 0.1$ was used as proposed in [72, 198].
The model is based on the traceless symmetric part of the square of the velocity gradient tensor

$$S_{ij}^d = \frac{1}{2}\left(\overline{g}_{ij}^2 + \overline{g}_{ji}^2\right) - \frac{1}{3}\delta_{ij}\overline{g}_{kk}^2 , \tag{D.2}$$

with the velocity gradients

$$\overline{g}_{ij} = \frac{\partial \overline{u_i}}{\partial x_j} . \tag{D.3}$$

The deformation tensor $\bar{S}_{ij}$ reads

$$\bar{S}_{ij} = \frac{1}{2}\left(\frac{\partial \bar{u}_i}{\partial x_j} + \frac{\partial \bar{u}_j}{\partial x_i}\right) . \tag{D.4}$$

An overbar denotes the resolved / low pass filtered scales.

E Conversion between Euler parameters and Euler angles

The conversion between a given set of Euler angles in zyx-convention and Euler parameters q is given according to [91] by:

$$\begin{aligned}
q_0 &= \cos\left(\frac{1}{2}\phi_1\right)\cos\left(\frac{1}{2}\phi_2\right)\cos\left(\frac{1}{2}\phi_3\right) + \sin\left(\frac{1}{2}\phi_1\right)\sin\left(\frac{1}{2}\phi_2\right)\sin\left(\frac{1}{2}\phi_3\right) , \\
q_1 &= \sin\left(\frac{1}{2}\phi_1\right)\cos\left(\frac{1}{2}\phi_2\right)\cos\left(\frac{1}{2}\phi_3\right) - \cos\left(\frac{1}{2}\phi_1\right)\sin\left(\frac{1}{2}\phi_2\right)\sin\left(\frac{1}{2}\phi_3\right) , \\
q_2 &= \cos\left(\frac{1}{2}\phi_1\right)\sin\left(\frac{1}{2}\phi_2\right)\cos\left(\frac{1}{2}\phi_3\right) + \sin\left(\frac{1}{2}\phi_1\right)\cos\left(\frac{1}{2}\phi_2\right)\sin\left(\frac{1}{2}\phi_3\right) , \\
q_3 &= \cos\left(\frac{1}{2}\phi_1\right)\cos\left(\frac{1}{2}\phi_2\right)\sin\left(\frac{1}{2}\phi_3\right) - \sin\left(\frac{1}{2}\phi_1\right)\sin\left(\frac{1}{2}\phi_2\right)\cos\left(\frac{1}{2}\phi_3\right) .
\end{aligned} \tag{E.1}$$

Inversely, the conversion from Euler parameters q to Euler angles in zyx-convention is obtained from:

$$\begin{aligned}
\phi_1 &= \operatorname{atan2}\left(2\left(q_0 q_1 + q_2 q_3\right), 1 - 2\left(q_1^2 + q_2^2\right)\right) , \\
\phi_2 &= \operatorname{asin}\left(2\left(q_0 q_2 - q_3 q_1\right)\right) , \\
\phi_3 &= \operatorname{atan2}\left(2\left(q_0 q_3 + q_1 q_2\right), 1 - 2\left(q_2^2 + q_3^2\right)\right) .
\end{aligned} \tag{E.2}$$

F Further time integration schemes for the virtual mass approach

Additional time integration schemes are listed for the virtual mass approach. The generic test case of a sphere rising under Stokes flow conditions is considered with the nomenclature introduced in Section 3.3.2.

Multistep methods

The Crank-Nicolson method (*CN-vm*) is a two-step, implicit scheme and reads for constant time step size

$$f_v^n = c_a \frac{3u^n - 4u^{n-1} + u^{n-2}}{2\Delta t}, \tag{F.1}$$

$$f^n = c_u u^n + c_g g + f_v^n, \tag{F.2}$$

$$u^{n+1} = u^n + \frac{\Delta t}{2}\left(f^n + f^{n+1}\right). \tag{F.3}$$

Second order convergence is obtained (Table 3.3). An iterative solution is necessary due to f^{n+1} on the rhs.

The Leap-Frog method (*LF-vm*) uses u^{n-1} as the basis for the advancement. It is explicit and for a constant time step it is given by

$$f_v^n = c_a \frac{3u^n - 4u^{n-1} + u^{n-2}}{2\Delta t}, \tag{F.4}$$

$$u^{n+1} = u^{n-1} + 2\Delta t \left(c_u u^n + c_g g + f_v^n\right). \tag{F.5}$$

It combines easy implementation and second order accuracy. However, it tends to oscillations and instability in more general applications.

Predictor-Corrector schemes

The scheme *RK3-1-AM3-vm* uses a 3-stage Runge-Kutta predictor and a three-step Adams-Moulton corrector. Within the *RK3*-predictor, f_v^n is used in all substeps:

$$f_v^n = c_a \frac{11u^n - 18u^{n-1} + 9u^{n-2} - 2u^{n-3}}{6\Delta t}, \tag{F.6}$$

$$k_1 = c_u u^n + c_g g + f_v^n, \tag{F.7}$$

$$k_2 = c_u \left(u^n + a_{21}\Delta t k_1\right) + c_g g + f_v^n, \tag{F.8}$$

$$k_3 = c_u \left(u^n + a_{31}\Delta t k_1 + a_{32}\Delta t k_2\right) + c_g g + f_v^n, \tag{F.9}$$

$$u_{pred}^{n+1} = u^n + \Delta t \left(b_1 k_1 + b_2 k_2 + b_3 k_3\right). \tag{F.10}$$

The Adams-Moulton corrector reads

$$f_{v,pred}^n = c_a \frac{11u_{pred}^{n+1} - 18u^n + 9u^{n-1} - 2u^{n-2}}{6\Delta t}, \tag{F.11}$$

$$f^n = c_u u^n + c_g g + f_{v,pred}^n, \tag{F.12}$$

$$u^{n+1} = u^n + \frac{\Delta t}{12}\left(5f^{n+1} + 8f^n - f^{n-1}\right) . \tag{F.13}$$

The combined method is still only first order accurate as the predicted u^{n+1}_{pred} is only of first order and so is $f_{v,pred}$. No iterative solution is necessary and a reduction of the absolute error is achieved compared to the pure predictor scheme in the considered test case (Table 3.3).

If a Leap-Frog-predictor is applied and the trapezoidal rule is used for the virtual force in the Runge-Kutta method, this yields the scheme

$$f_v^n = c_a \frac{3u^n - 4u^{n-1} + u^{n-2}}{2\Delta t} , \tag{F.14}$$

$$u^{n+1} = u^{n-1} + 2\Delta t\left(c_u u^n + c_g g + f_v^n\right) , \tag{F.15}$$

$$f^{n+1}_{v,pred} = c_a \frac{-u^{n-1} + 4u^n - 3u^{n+1}}{-2\Delta t} , \tag{F.16}$$

$$\tilde{f}_v^{n+\frac{1}{2}} = \frac{1}{2}\left(f_v^n + f^{n+1}_{v,pred}\right) , \tag{F.17}$$

$$k_1 = c_u u^n + c_g g + \tilde{f}_v^{n+\frac{1}{2}} , \tag{F.18}$$

$$k_2 = c_u\left(u^n + a_{21}\Delta t\, k_1\right) + c_g g + \tilde{f}_v^{n+\frac{1}{2}} , \tag{F.19}$$

$$k_3 = c_u\left(u^n + a_{31}\Delta t\, k_1 + a_{32}\Delta t\, k_2\right) + c_g g + \tilde{f}_v^{n+\frac{1}{2}} , \tag{F.20}$$

$$u^{n+1} = u^n + \Delta t\left(b_1 k_1 + b_2 k_2 + b_3 k_3\right) . \tag{F.21}$$

termed *LF-RK3-vm* here. Stability issues might arise from the predictor step. Lower absolute errors are obtained than for scheme LF alone. Overall second order convergence is found for the present test case (Table 3.3).

Employing a second order Lagrange approximation for the virtual force $f_v\left(t^n + c_i\Delta t\right)$ in the Runge-Kutta substeps yields the scheme

$$f_v\left(t^n + c_i\Delta t\right) = c_a P_2'\left(t^n + c_i\Delta t\right) , \tag{F.22}$$

$$k_1 = c_u u^n + c_g g + f_v^n , \tag{F.23}$$

$$k_2 = c_u\left(u^n + a_{21}\Delta t\, k_1\right) + c_g g + f_v\left(t^n + c_2\Delta t\right) , \tag{F.24}$$

$$k_3 = c_u\left(u^n + a_{31}\Delta t\, k_1 + a_{32}\Delta t\, k_2\right) + c_g g + f_v\left(t^n + c_3\Delta t\right) , \tag{F.25}$$

$$u^{n+1} = u^n + \Delta t\left(b_1 k_1 + b_2 k_2 + b_3 k_3\right) , \tag{F.26}$$

termed *Lag-RK3c-vm* here, which is thus very similar to the scheme *RK3-a-vm*. The advantages of this scheme are simple implementation for variable time steps, low absolute errors and second order convergence (Table 3.3).

G Derivation of Runge-Kutta coefficients

The derivation of the coefficients of a three stage Runge-Kutta method is adapted in the present nomenclature from [50] for the unmodified problem, $\dot{u} = f(u(t))$ and $u_0 = u(t=0)$, i.e. without virtual force.
The local discretization error is given by

$$\epsilon^{n+1} = u^{n+1} - u^n - \Delta t \left(b_1 k_1 + b_2 k_2 + b_3 k_3 \right) . \tag{G.1}$$

Taylor series expansion for u^{n+1} and k_i at t^n yields:

$$\begin{aligned} u^{n+1} &= u\left(t^n + \Delta t\right) , \\ &= u^n + \Delta t\, u' + \frac{1}{2}\Delta t^2 u'' + \frac{1}{6}\Delta t^3 u''' + \mathcal{O}\left(\Delta t^4\right) , \\ &= u^n + \Delta t\, f + \frac{1}{2}\Delta t^2 F + \frac{1}{6}\Delta t^3 \left(F\, f_u + G\right) + \mathcal{O}\left(\Delta t^4\right) , \end{aligned} \tag{G.2}$$

where

$$F = f_t + f_u\, f\,, \quad G = f_{tt} + 2 f_{tu} f + f_{uu} f^2 \,, \tag{G.3}$$

and

$$k_1 = f\left(t^n,\ u^n\right) = f\,, \tag{G.4}$$

$$\begin{aligned} k_2 =& f\left(t^n + c_2\Delta t,\ u^n + c_2 \Delta t\, k_1\right) , \\ =& f + c_2 \Delta t \left(f_t + f\, f_u\right) + \frac{1}{2} c_2^2 \Delta t^2 \left(f_{tt} + f\, f_{tu} + f^2 f_{uu}\right) + \mathcal{O}\left(\Delta t^3\right) , \\ =& f + c_2 \Delta t\, F + \frac{1}{2} c_2^2 \Delta t^2 G + \mathcal{O}\left(\Delta t^3\right) , \end{aligned} \tag{G.5}$$

$$\begin{aligned} k_3 =& f\left(t^n + c_3\Delta t,\ u^n + \Delta t \left(a_{31} k_1 + a_{32} k_2\right)\right) , \\ =& f + c_3 \Delta t\, f_t + \Delta t \left(a_{31} k_1 + a_{32} k_2\right) f_u + \frac{1}{2} c_3^2 \Delta t^2 f_{tt} \\ & + c_3 \Delta t^2 \left(a_{31} k_1 + a_{32} k_2\right) f_{tu} + \frac{1}{2} \Delta t^2 \left(a_{31} k_1 + a_{32} k_2\right)^2 f_{uu} + \mathcal{O}\left(\Delta t^3\right) , \\ =& f + c_3 \Delta t\, F + \Delta t^2 \left(c_2 a_{32} F\, f_u + \frac{1}{2} c_3^2 G \right) + \mathcal{O}\left(\Delta t^3\right) . \end{aligned} \tag{G.6}$$

The discretization error is thus expressed by:

$$\begin{aligned} \epsilon^{n+1} =& \Delta t\, f \left(1 - b_1 - b_2 - b_3\right) \\ & + \Delta t^2\, F \left(\frac{1}{2} - b_2 c_2 - b_3 c_3 \right) \\ & + \Delta t^3 \left(F\, f_u \left(\frac{1}{6} - b_3 c_2 a_{32} \right) + G \left(\frac{1}{6} - \frac{1}{2} b_2 c_2^2 - \frac{1}{2} b_3 c_3^2 \right) \right) \\ & + \mathcal{O}\left(\Delta t^4\right) . \end{aligned} \tag{G.7}$$

For the method to be consistent, the condition

$$\sum_{i=1}^{3} b_i k_i = b_1 + b_2 + b_3 = 1 \tag{G.8}$$

has to be fulfilled.
The following additional constraints result, if the higher order terms in (G.7) shall vanish:

$$\sum_{i=1}^{3} b_i c_i = b_2 c_2 + b_3 c_3 = \frac{1}{2}, \tag{G.9}$$

$$\sum_{i=1}^{3} b_i c_i^2 = b_2 c_2^2 + b_3 c_3^3 = \frac{1}{3}, \qquad \sum_{i,j=1}^{3} b_i a_{ij} c_j = b_3 a_{32} c_2 = \frac{1}{6}, \tag{G.10}$$

with $c_1 = 0$ and $a_{ij} = 0$ for $j \geq i$.

H Additional data with respect to the virtual mass concept

The effect of the choice of the density ratios $\pi_\rho = 0.5$ and $\pi_\rho = 0.1$, as well as the choice of the virtual mass coefficients $C_v = 0.5$ and $C_v = 1.0$ are studied. The parameters are chosen as in Section 3.3.2, i.e. the physical properties are $\mu_f = 0.001$, $\rho_f = 1000$, $g = 9.81$, $r_p = 0.0007$ and the discretization is conducted with $\Delta t = t_{end}/N_{\Delta t}$ where $t_{end} = 0.5$.
Additional to the maximum or L_∞-error (3.52) used above, also the L_1-error $\epsilon_{ave} = 1/N_{\Delta t} \sum |u(t) - u_{ref}(t)|$ is provided here, with u_{ref} obtained from (3.50).

Table H.1 Error ϵ_{ave} and ϵ_{max} in $u(t)$ for the density ratios $\pi_\rho = 0.5$ and $\pi_\rho = 0.1$, as well as the virtual mass coefficients $C_v = 0.5$ and $C_v = 1.0$. The scheme is *Lag-RK3tp-vm* throughout.

$\pi_\rho = 0.5$, $C_v = 0.5$	$N_{\Delta t} = 400$	800	1600	3200	Order q
ϵ_{ave}	2.81E-06	7.18E-06	1.81E-06	4.56E-07	1.99
ϵ_{max}	9.45E-05	2.42E-05	6.13E-06	1.54E-06	1.99
$\pi_\rho = 0.1$, $C_v = 0.5$					
ϵ_{ave}	1.10E-03	3.01E-04	8.00E-05	2.06E-05	1.96
ϵ_{max}	2.62E-02	5.60E-03	1.35E-03	3.48E-04	1.82
$\pi_\rho = 0.5$, $C_v = 1.0$					
ϵ_{ave}	5.70E-05	1.46E-05	3.68E-06	9.25E-07	1.99
ϵ_{max}	1.92E-04	4.91E-05	1.24E-05	3.13E-06	1.99
$\pi_\rho = 0.1$, $C_v = 1.0$					
ϵ_{ave}	2.30E-03	6.18E-04	1.61E-04	4.15E-05	1.96
ϵ_{max}	6.22E-02	1.38E-02	2.93E-03	7.03E-04	2.06

I IBM patch and stretched grid

The present IBM [136] requires an equidistant grid. This is mainly founded on the symmetry properties of the delta function involved in the forcing procedure of the boundary condition [225]. In the original implementation, a globally equidistant grid is assumed. Very high numerical efficiency can be achieved by the usage of such a structured grid. Further, many operations can be executed using the index space, e.g. the x-coordinate of the cell i is obtained from $x_i = i\,\Delta x_{i=1}$ employing the spacing of the first grid cell. Vice versa the index of a cell is easily found from a given coordinate by $i = \mathrm{nint}\left(x_i\,\Delta x_{i=1}^{-1}\right)$. In the original implementation of the IBM, these efficient formulation are incorporated in the interpolation of velocities and spreading of forces with the regularized delta function, the determination of bounding boxes around the particle for computation of cut cell volumes, the master-slave distribution in the parallelization etc. In general, a grid stretching outside the range of the delta function is permitted. For some simulations of fundamental character, it is beneficial to create a locally fine resolution, e.g. around a single fixed particle, and to enable large computational domains by stretching the grid towards the outer boundaries. An example for such a grid is provided in Figure I.2a) with the nomenclature used in the actual implementation in PRIME for the x-direction. A local equidistant patch is situated around the particle location. All of the operations connected to the IBM are conducted in a coordinate system translated from the laboratory frame to O_{eq} - the origin of the patch.

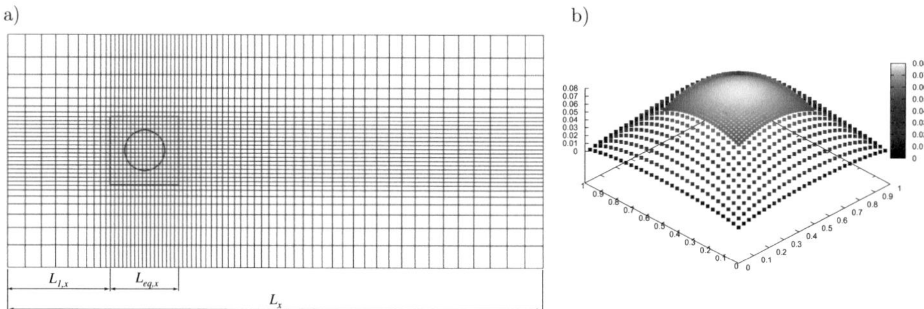

Figure I.2 IBM patch. a) Local patch with equidistant grid and grid stretching away from patch. b) Local patch refinement with hanging nodes for solution of Poisson equation $\nabla^2 p = -1$ with Dirichlet boundary conditions by M. Jurisch (student project).

With the stretched grid, a substantial reduction of grid cells can be achieved for specific setups. For example, the simulation of the flow around a fixes sphere at $Re = 100$ with a globally equidistant grid with $\mathbf{N} = (256, 128, 128)$ (4.2 million cells) for a domain of extent $\mathbf{L} = (12.8, 6.4, 6.4)\,d_p$ yields a drag coefficient of $C_D = 1.179$. The same result is obtained with the same local resolution around the sphere and a grid with $\mathbf{N} = (128, 72, 72)$ which is stretched in all three directions away from the patch and has only about 1/6 of the cells. However, a bit of caution has to be taken as the convergence speed of the solver, as implemented from the *hypre* library, changes compared to an equidistant grid and stability issues might arise for large aspect ratios of the cells and strong grid stretching.
Future developments could include patches moving with the particle since found knowledge on moving structured meshes is available within the group [106]. Adaptive mesh refinement with hanging nodes and hierarchical patches [225, 93] is currently progress in the making. Figure I.2a) illustrates the solution of the Poisson problem $\nabla^2 p = -1$ with Dirichlet boundary conditions on such a patch. The actual performance plus of such an approach has to be scrutinized in a highly parallel environment with many particles.

J Spherical Harmonics

In a local spherical coordinate system, the location of the interface is represented by

$$r(\theta,\phi) = \sum_{n=0}^{\infty}\sum_{m=-n}^{n} a_{nm} Y_n^m(\theta,\phi)\,, \tag{J.1}$$

with the spherical harmonic function [77]

$$Y_n^m(\theta,\phi) = \sqrt{\left(\frac{(2n+1)(n-m)!}{4\pi(n+m)!}\right)} P_n^m(\,\cos(\theta)\,)\, e^{im\phi}\,, \tag{J.2}$$

and P_n^m being associated Legendre functions of the argument $\cos(\theta)$ consisting of a set of orthogonal polynomials. The angles are spherical coordinates defined within $0 \leq \theta \leq \pi$ and $0 \leq \phi \leq 2\pi$. In the actual implementation in PRIME, the spherical coordinate θ is mapped to the standard interval $[-1,\ 1]$ as also indicated by the argument of the Legendre polynomials $P_n^m(\,\cos(\theta)\,)$.

For a given surface the coefficients of the series are found from

$$a_{nm} = \int_0^{2\pi}\int_0^{\pi} d\phi\, d\theta\ \sin(\theta)\, r(\theta,\phi)\, Y_n^{m*}\,, \tag{J.3}$$

with the star denoting the conjugate complex.

The partial derivatives of r with respect to ϕ and θ are

$$\begin{aligned}
r_\phi &= \sum_{n=0}^{\infty}\sum_{m=-n}^{n} (im)\, a_{nm} Y_n^m(\theta,\phi)\,, \\
r_{\phi\phi} &= \sum_{n=0}^{\infty}\sum_{m=-n}^{n} -m^2\, a_{nm} Y_n^m(\theta,\phi)\,, \\
r_\theta &= \sum_{n=0}^{\infty}\sum_{m=-n}^{n} \frac{-a_{nm} f_{nm}}{\sin\theta}\ \left[(n+1)\cos\theta P_n^m - (n-m+1)P_{n+1}^m\right]\, e^{im\phi}\,, \\
r_{\theta\theta} &= \sum_{n=0}^{\infty}\sum_{m=-n}^{n} \frac{a_{nm} f_{nm}}{\sin^2\theta} \\
&\quad [(n+1+(n+1)^2\cos^2\theta)P_n^m - 2\cos\theta(n-m+1)(n+2)P_{n+1}^m \\
&\quad +(n-m+1)(n-m+2)P_{n+2}^m]\, e^{im\phi}\,, \\
r_{\phi\theta} &= \sum_{n=0}^{\infty}\sum_{m=-n}^{n} \frac{-im\, a_{nm}\, f_{nm}}{\sin\theta}\ \left[(n+1)\cos\theta P_n^m - (n-m+1)P_{n+1}^m\right]\, e^{im\phi}\,, \\
&= r_{\theta\phi}\,,
\end{aligned} \tag{J.4}$$

with the auxiliary function

$$f_{nm} = \sqrt{\frac{(2n+1)(n-m)!}{4\pi(n+m)!}}\,. \tag{J.5}$$

The surface vector reads as

$$\begin{aligned} x_S &= r(\theta,\phi)\sin(\theta)\cos(\phi)\,, \\ y_S &= r(\theta,\phi)\sin(\theta)\sin(\phi)\,, \\ z_S &= r(\theta,\phi)\cos(\theta)\,, \\ \mathbf{x} &= \mathbf{x}_S\,. \end{aligned} \tag{J.6}$$

The index S for surface is dropped from now on as well as the arguments of r.

Differential surface element:

$$dA = Sd\theta d\phi\,, \tag{J.7}$$

$$S = \|\mathbf{x}_\theta \times\ \mathbf{x}_\phi\|\,, = r[r_\phi^2 + r_\theta^2\sin^2(\theta) + r^2\sin^2(\theta)]^{1/2}\,. \tag{J.8}$$

Derivatives of the surface vector:

$$\mathbf{x}_\phi = \begin{pmatrix} r_\phi\sin\theta\cos\phi - r\sin\theta\sin\phi \\ r_\phi\sin\theta\sin\phi + r\sin\theta\cos\phi \\ r_\phi\cos\theta \end{pmatrix}\,, \tag{J.9}$$

$$\mathbf{x}_\theta = \begin{pmatrix} r\cos\theta\cos\phi + r_\theta\sin\theta\cos\phi \\ r\cos\theta\sin\phi + r_\theta\sin\theta\sin\phi \\ r_\theta\cos\theta - r\sin\theta \end{pmatrix}\,. \tag{J.10}$$

Surface normal vector:

$$\mathbf{n} = \frac{\mathbf{x}_\theta \times \mathbf{x}_\phi}{\|\mathbf{x}_\theta \times \mathbf{x}_\phi\|}\,. \tag{J.11}$$

Derivatives of the surface normal vector (component i):

$$n_{i,\phi} = S^{-1}\left[a_i - b_i\left(\frac{r_\phi}{r} + \frac{c_i}{S^2}\right)\right]\,, \tag{J.12}$$

where

$$\begin{aligned} a_1 &= r_\phi^2\sin\phi + rr_{\phi\phi}\sin\phi + rr_\phi\cos\phi - r_\phi r_\theta\sin\theta\cos\theta\cos\phi - rr_{\theta\phi}\sin\theta\cos\theta\cos\phi + \\ &\quad rr_\theta\sin\theta\cos\theta\sin\phi + 2rr_\phi\sin^2\theta\cos\phi - r^2\sin^2\theta\sin\phi \\ b_1 &= rr_\phi\sin\phi - rr_\theta\sin\theta\cos\theta\cos\phi + r^2\sin^2\theta\cos\phi \\ c_1 &= r^2(r_\phi r_{\phi\phi} + r_\theta r_{\theta\phi}\sin^2\theta + rr_\phi\sin^2\theta \\ a_2 &= -r_\phi^2\cos\phi - rr_{\phi\phi}\cos\phi + rr_\phi\sin\phi - r_\phi r_\theta\sin\theta\cos\theta\sin\phi - rr_{\theta\phi}\sin\theta\cos\theta\sin\phi - \\ &\quad rr_\theta\sin\theta\cos\theta\cos\phi + 2rr_\phi\sin^2\theta\sin\phi + r^2\sin^2\theta\cos\phi \\ b_2 &= -rr_\phi\cos\phi - rr_\theta\sin\theta\cos\theta\sin\phi + r^2\sin^2\theta\sin\phi \\ c_2 &= c_1 \\ a_3 &= r_\phi r_\theta\sin^2\theta + rr_{\theta\phi}\sin^2\theta + 2rr_\phi\sin\theta\cos\theta \\ b_3 &= rr_\theta\sin^2\theta + r^2\sin\theta\cos\theta \\ c_3 &= c_1 \end{aligned} \tag{J.13}$$

$$n_{i,\theta} = S^{-1}\left[a_i - b_i\left(\frac{r_\theta}{r} + \frac{c_i}{S^2}\right)\right]\,, \tag{J.14}$$

where

$$
\begin{aligned}
a_1 &= r_\theta r_\phi \sin\phi + r r_{\theta\phi} \sin\phi - r_\theta^2 \sin\theta\cos\theta\cos\phi - r r_{\theta\theta}\sin\theta\cos\theta\cos\phi \\
&\quad - r r_\theta \cos^2\theta\cos\phi + r r_\theta \sin^2\theta\cos\phi + 2 r r_\theta \sin^2\theta\cos\phi + 2r^2\sin\theta\cos\theta\cos\phi \\
b_1 &= r r_\phi \sin\phi - r r_\theta \sin\theta\cos\theta\cos\phi + r^2\sin^2\theta\cos\phi \\
c_1 &= r_\phi r_{\theta\phi} + r_\theta r_{\theta\theta}\sin^2\theta + r r_\theta \sin^2\theta + r_\theta^2 \sin\theta\cos\theta + r^2\sin\theta\cos\theta \\
a_2 &= -r_\phi r_\theta \cos\phi - r r_{\theta\phi}\cos\phi - r_\theta^2\sin\theta\cos\theta\sin\phi - r r_{\theta\theta}\sin\theta\cos\theta\sin\phi \\
&\quad - r r_\theta\cos^2\theta\sin\phi + r r_\theta \sin^2\theta\sin\phi + 2 r r_\theta\sin^2\theta\sin\phi + 2r^2\sin\theta\cos\theta\sin\phi \\
b_2 &= -r r_\phi\cos\phi - r r_\theta\sin\theta\cos\theta\sin\phi + r^2\sin^2\theta\sin\phi \\
c_2 &= c_1 \\
a_3 &= r_\theta^2\sin^2\theta + r r_{\theta\theta}\sin^2\theta + 2 r r_\theta \sin\theta\cos\theta \\
&\quad + 2 r r_\theta\sin\theta\cos\theta + r^2\cos^2\theta - r^2\sin^2\theta \\
b_3 &= r r_\theta\sin^2\theta + r^2\sin\theta\cos\theta \\
c_3 &= c_1
\end{aligned}
\tag{J.15}
$$

Elements of first and second fundamental tensor:

$$
\begin{aligned}
E &= \mathbf{x}_\theta \cdot \mathbf{x}_\theta\,, \\
F &= \mathbf{x}_\theta \cdot \mathbf{x}_\phi\,, \\
G &= \mathbf{x}_\phi \cdot \mathbf{x}_\phi\,, \\
L &= -\mathbf{x}_\theta \cdot \mathbf{n}_\theta\,, \\
M &= \frac{1}{2}\left(\mathbf{x}_\theta \cdot \mathbf{n}_\phi + \mathbf{x}_\phi \cdot \mathbf{n}_\theta\right)\,, \\
N &= -\mathbf{x}_\phi \cdot \mathbf{n}_\phi\,.
\end{aligned}
\tag{J.16}
$$

Mean curvature:

$$H(\theta,\phi) = \frac{EN + GL - 2FM}{2(EG - F^2)} = \frac{1}{2}\,\kappa\,. \tag{J.17}$$

Gaussian curvature:

$$K(\theta,\phi) = \frac{LN - M^2}{EG - F^2}\,. \tag{J.18}$$

Integral quantities:

Surface:

$$A = \int_0^{2\pi}\int_0^{\pi} r[r_\phi^2 + r_\theta^2\sin^2(\theta) + r^2\sin^2(\theta)]^{1/2} d\theta\, d\phi\,. \tag{J.19}$$

Volume:

$$V = \frac{1}{3}\int_0^{2\pi}\int_0^{\pi} r^3(\theta,\phi)\sin(\theta) d\theta\, d\phi\,. \tag{J.20}$$

Volume correction:
To fulfill the constraint of constant bubble volume, $V_b = const.$, the obtained shape coefficients are linearly scaled, $a_{nm}^V = C_V\, a_{nm}$, to correct the current volume, V, to the desired volume, V_b. From (J.1), one obtains $r_V(\theta,\phi) = C_V\, r(\theta,\phi)$ which can be employed in (J.20)

to determine the correction constant C_V:

$$\begin{aligned} V_b &= \frac{1}{3}\int_0^{2\pi}\int_0^{\pi} r_V^3(\theta,\phi)\,\sin(\theta)\,\mathrm{d}\theta\,\mathrm{d}\phi\,,\\ &= \frac{1}{3}\int_0^{2\pi}\int_0^{\pi} (C_V\,r(\theta,\phi))^3\,\sin(\theta)\,\mathrm{d}\theta\,\mathrm{d}\phi\,, \qquad (\mathrm{J.21})\\ &= C_V^3\,V\,, \end{aligned}$$

$$\Rightarrow \qquad C_V = \left(\frac{V_b}{V}\right)^{\frac{1}{3}}. \qquad (\mathrm{J.22})$$

Inertial tensor:

$$\begin{aligned} I_{11} &= \frac{1}{5}\rho_p\int_0^{2\pi}\int_0^{\pi} r^5(\theta,\phi)\sin(\theta)[1-\sin^2(\theta)\cos^2(\phi)]\;d\theta\;d\phi\,,\\ I_{22} &= \frac{1}{5}\rho_p\int_0^{2\pi}\int_0^{\pi} r^5(\theta,\phi)\sin(\theta)[1-\sin^2(\theta)\sin^2(\phi)]\;\;d\phi\;d\theta\,,\\ I_{33} &= \frac{1}{5}\rho_p\int_0^{2\pi}\int_0^{\pi} r^5(\theta,\phi)\sin^3(\theta)d\theta\;d\phi\,,\\ I_{12} &= -\frac{1}{5}\rho_p\int_0^{2\pi}\int_0^{\pi} r^5(\theta,\phi)\sin^3(\theta)\cos(\phi)\sin(\phi)\;d\theta\;d\phi\,,\\ I_{13} &= -\frac{1}{5}\rho_p\int_0^{2\pi}\int_0^{\pi} r^5(\theta,\phi)\cos(\theta)\sin^2(\theta)\cos(\phi)\;d\theta\;d\phi\,,\\ I_{23} &= -\frac{1}{5}\rho_p\int_0^{2\pi}\int_0^{\pi} r^5(\theta,\phi)\cos(\theta)\sin^2(\theta)\sin(\phi)\;d\theta\;d\phi\,. \qquad (\mathrm{J.23}) \end{aligned}$$

Center of mass:

$$x_{i,cm} = \frac{1}{V}\int_0^{2\pi}\int_0^{\pi}\int_0^{r} x_i\;r^2(\theta,\phi)\sin(\theta)\;dr\;d\theta\;d\phi\,. \qquad (\mathrm{J.24})$$

K Moving-Least-Squares method

The Moving-Least-Squares method [151, 158, 193] provides a continuously differentiable, global approximation $f_g(\mathbf{x})$ of data f_i scattered at $\mathbf{x}_i$ locations. The global function consists of a set of local approximations, $f_g(\mathbf{x}) = f(\mathbf{x})$, obtained from a weighted least squares fit around a specific point. This point is then *moved* over the entire domain. The local approximating function is obtained here from $f \in P_M^D$ being polynomials of degree M in D dimensions and it reads as

$$f(\mathbf{x}) = \mathbf{a}\cdot\mathbf{b}(\mathbf{x})\,, \qquad (\mathrm{K.1})$$

where $\mathbf{a} = (a_1,\,\ldots,\,a_k)^T$ contains the coefficients to be determined locally and $\mathbf{b} = (b_1,\,\ldots,\,b_k)^T$ holds the polynomial basis, e.g. $\mathbf{b} = (1,\,x,\,y,\,z)^T$ is a linear basis in three dimensions, i.e. $M = 1$, $D = 3$.

The coefficients are obtained minimizing the error functional J in the weighted least squares formulation with

$$J(\mathbf{a}) = \sum_i^N w(\|\mathbf{x}_i - \mathbf{x}\|)\;\|\mathbf{a}\cdot\mathbf{b}(\mathbf{x}_i) - f_i\|^2 + \sum_j^{N_\mathcal{C}} w_\mathcal{C}\left[\mathcal{C}(\mathbf{a},\mathbf{b}(\mathbf{x}_j))\right]^2. \qquad (\mathrm{K.2})$$

Herein, the first term relates to the approximation of the N data points with the weighting function $w(\|\mathbf{x}_i - \mathbf{x}\|)$, $w \geq 0$. The weighting function needs to be continuously differentiable and decreasing away from $\mathbf{x}_i$. Suitable choices are, e.g. cubic splines [12] or the Gaussian

$$w\left(\|\mathbf{x}_i - \mathbf{x}\|\right) = \exp\left(-\frac{\|\mathbf{x}_i - \mathbf{x}\|^2}{r_G^2}\right), \tag{K.3}$$

where r_G is a spacing parameter [193].
The second term in (K.2) relates to additional constraints $\mathcal{C}$, weighted with $w_{\mathcal{C}}$, to be fulfilled in a least squares sense at $N_{\mathcal{C}}$ locations $\mathbf{x}_j$. These constraints are expressed in terms of $\mathbf{a}$ and $\mathbf{b}(\mathbf{x}_j)$ (or its derivatives) and e.g. comprise a vanishing surface normal pressure derivative or a divergence free velocity field [12].
The second term is *not* considered from here on, i.e. $w_{\mathcal{C}} = 0$, as the further notation depends on the nature of $\mathcal{C}$.
The minimum in the error functional J is obtained from

$$\nabla\left(J(\mathbf{a})\right) = 0\,, \qquad \text{where} \qquad \nabla = \left(\frac{\partial}{\partial a_1}, \ldots, \frac{\partial}{\partial a_k}\right)^T, \tag{K.4}$$

yielding a linear system of equations to solve for the coefficients

$$\mathbf{a} = \left[\sum_i^N w_i\, \mathbf{b}(\mathbf{x}_i)\mathbf{b}(\mathbf{x}_i)^T\right]^{-1} \sum_i^N w_i\, \mathbf{b}(\mathbf{x}_i)\, f_i\,. \tag{K.5}$$

The involved matrix inversion is performed using LAPACK [5].

Nomenclature

Latin symbols	
A	surface area
$\mathbf{A}$	rotation matrix
a, b, c	semi-axes of ellipsoid
a_{nm}	coefficients in spherical harmonic expansion
a_{nm}^0, a_{nm}^e	shape coefficients at the beginning and end of coalescence modeling
$\mathbf{B}$	magnetic field
c	absolute value of velocity
C_{AM}	added mass coefficient
C_D, F_D	drag coefficient and force, respectively
C_L, F_L	lift coefficient and force, respectively
C_v	virtual mass coefficient
d_{12}, r_{12}	equivalent diameter and radius of two interacting bubbles
d_{eq}, r_{eq}	sphere volume-equivalent diameter and radius, respectively
d_p, r_p	particle diameter and radius, respectively
E	ellipsoid function, exponent in collision model
$\mathbf{e}$	unit vector
e_n	restitution coefficient
$\mathbf{e}_u$, $\mathbf{e}_\Phi$	electric field contributed to velocity and electric potential
E_{uu}, E_{vv}	energy spectra of u and v, respectively
E_{uu}^{2D}	two-dimensional energy spectrum of u
$\mathbf{f}$	volumetric force
$\mathbf{F}_{AM}$	added mass force
$\mathbf{F}_B$	buoyancy force
f_b	bubble detachment frequency
$\mathbf{F}_{col}$, $\mathbf{M}_{col}$	collision force and collision torque acting on the particle
$\mathbf{f}_L$	Lorentz force density
$\mathbf{F}_p$, $\mathbf{M}_p$	force and torque acting on the particle
f_{ref}	reference frequency
$\mathbf{f}_v$	specific virtual mass force
$\mathbf{F}_v$, $\mathbf{M}_v$	virtual mass force and torque
$\mathbf{g}$	gravitational acceleration
H	channel half-width, height of container, mean curvature
h	spacing of the equidistant Cartesian grid
$\mathbf{I}_p$	particle tensor of inertia
$\mathbf{j}$	electric current density
j_0	free stream current density

k_{col}, d_{col}	spring stiffness and damping coefficient in collision model
k_x, k_z	wave number in x and z, respectively
$\mathbf{L}$	domain size $\mathbf{L} = (L_x, L_y, L_z)$
L_{ref}	reference length
l_τ	viscous length scale in turbulent channel flow
m_p	particle mass
$\mathbf{N}$	Cartesian grid size $\mathbf{N} = (N_x, N_y, N_z)$
n_f	viscous forcing loops of IBM
N_L	number of Lagrangian forcing points
N_p	number of particles
N_{proc}	number of processes
N_{SH}	number of modes in axisymmetric spherical harmonic expansion
n_τ	number of pseudo-time loops in pseudo-compressibility approach
n_θ	number of quadrature points in θ-direction
N_{tri}	number of surface triangles
$\mathbf{n}_S$	normal vector of bubble surface
P_n	polynomial in series expansion (Lagrange in vm and Legendre in SH)
p	pressure
p_o, p_i	inner and outer pressure, respectively
q	quaternion
q_n	coefficients of series expansion
$\mathbf{r}$	point on particle surface with respect to particle center $\mathbf{x}_p$
r_s	saddle point radius
s	shear parameter
S	particle surface
T	period
t	time
t_{col}	collision time
t_{ref}	reference time
$\mathbf{u}$	fluid velocity, $\mathbf{u} = (u, v, w)$
$\mathbf{u}^d$, $\mathbf{u}^i$	desired and interpolated velocity at particle surface, respectively
u_0	centerline velocity
u_b	bulk velocity
u_g	gravitational velocity scale
$\mathbf{u}_p$	velocity of particle center, $\mathbf{u}_p = (u_p, v_p, w_p)$
u_{ref}	reference velocity
u_{rel}	relative velocity
$\mathbf{u}_S$	local velocity of particle surface
$\mathbf{u}_{sh}$	local velocity of particle surface originating from shape deformation
u_t	tangential velocity
u_τ	wall shear velocity in turbulent channel flow
V_p	particle volume
W	potential displacement energy
$\mathbf{x}$	Cartesian coordinates $\mathbf{x} = (x, y, z)$
$\mathbf{x}_p$	position of particle center, $\mathbf{x}_p = (x_p, y_p, z_p))$
$\mathbf{x}_{sub,A}$	sub-contact point on surface of particle A for distance calculation
Y_n^m	associated Legendre polynomial in spherical harmonic expansion

Greek symbols	
α	phase indicator, color function
α^k, γ^k, ζ^k	coefficients of low-storage Runge-Kutta scheme
β_T	constant for pseudo-compressibility concept
Γ	circulation
Γ_y	average contribution of ω_y to absolute value of vorticity
Δt	size of time step
ΔT_e	time interval of coalescence modeling
Δt_{sh}	size of time step for shape adaptation
ΔV_L, ΔV_E	Lagrangian (forcing point) volume, Eulerian (cell) volume
Δx, Δy, Δz	grid spacing
δ_h	regularized delta function
δ_p	correction of pseudo-pressure
δ_Φ	correction of electric potential
ε	local residuum, dissipation rate of turbulent kinetic energy
ε_e	ellipticity
ϵ	error
$\zeta_{A,B}$, $\zeta_{A,w}$	inter-particle distance, particle-wall distance
ζ_{col}	range of collision model
η, ξ	surface parameters for ellipsoid
θ, ϕ	angles of spherical coordinates
κ	twice the mean curvature
λ	curve parameter
λ_{12}	coalescence efficiency
λ_{por}	average porosity
μ	dynamic viscosity
ν	kinematic viscosity
ν_t	turbulent (eddy) viscosity
ρ_f	density of the continuous fluid phase
ρ_p	density of the dispersed particle phase
σ	surface tension
σ_e	electric conductivity
σ_{Re}	standard deviation in oscillation of $Re(t)$
$\boldsymbol{\tau}$	hydrodynamic stress tensor, including pressure, divided by ρ_f
τ_W	wall shear stress
Φ	electric potential
ϕ	signed-distance level set function
ϕ_{Roma}, ϕ_{Peskin}	continuous function for δ_h of Roma and Peskin
$\boldsymbol{\phi}_p$	particle orientation, list of Euler angles
Ψ, Ψ_n	sphericity, crosswise sphericity
$\boldsymbol{\psi}_p$	increment function
$\boldsymbol{\omega}_p$	particle angular velocity
$\boldsymbol{\omega}$	fluid vorticity

Dimensionless groups	
Ca	capillary number
CFL	Courant-Friedrichs-Lewy number
C_D	drag coefficient
C_L	lift coefficient
C_p	pressure coefficient
Eo	Eötvös number
G	Galilei number
Ha	Hartmann number
I^*	dimensionless moment of inertia, disks
M	Morton number
N	magnetic interaction parameter
Oh	Ohnesorg number
π_ρ	density ratio
R_m	magnetic Reynolds number
Re	Reynolds number
Sr	Strouhal number
St	Stokes number
We	Weber number
X	particle aspect ratio, shape parameter

Miscellaneous	
$\langle a \rangle$	average of a
a^+	inner scaling of a in turbulent channel flow
a'	fluctuation of a, representation of a in body-fixed system
$\dot{a}$	temporal derivative of a
$a_{i,j,k}$	discrete representation of a on Cartesian grid with indices i, j, k
a_{fp}	discrete representation of a at Lagrangian forcing points
a_{tri}	discrete representation of a for surface triangle

Abbreviations	
ACM	adaptive collision model
BC	boundary condition
DFT	discrete Fourier transform
DNS	direct numerical simulation
FFT	fast Fourier transform
HPLS	hybrid particle level set
IBM	immersed boundary method
LES	large eddy simulation
MHD	magnetohydrodynamic
MLS	moving least squares

PRIME	phase-resolving simulation environment
PIV	particle image velocimetry
RANS	Reynolds-averaged Navier-Stokes
RMS	root mean square
SEF	shape evolution function
SH	spherical harmonics
TKE	turbulent kinetic energy
UDV	ultrasound Doppler velocimetry
UTTT	ultrasound transit time technique
VOF	volume of fluid